Elasticsearch
实战与原理解析

牛冬 编著

电子工业出版社
Publishing House of Electronics Industry
北京·BEIJING

内 容 简 介

本书基于 Elasticsearch 7.X 版本编写，内容由浅入深，先教会初学者使用，再介绍背后的原理。本书共分为三大部分，分别是 Elasticsearch 前传、Elasticsearch 实战、Elasticsearch 生态。Elasticsearch 前传部分主要介绍搜索技术发展史和基本知识，并介绍搜索引擎技术原理，为读者构建搜索引擎全景。Elasticsearch 实战部分主要介绍 Elasticsearch 的核心概念和架构设计，并重点介绍客户端、文档、搜索和索引等实战内容，待读者能上手实战后，再介绍这些内容的背后实现原理和关联知识，为读者构建知识网络。Elasticsearch 生态部分主要介绍插件的使用和管理，以及 Elastic Stack 生态圈。

本书适合有一定基础知识的初、中级 Elasticsearch 学习者阅读。

未经许可，不得以任何方式复制或抄袭本书之部分或全部内容。
版权所有，侵权必究。

图书在版编目（CIP）数据

Elasticsearch 实战与原理解析 / 牛冬编著. —北京：电子工业出版社，2020.3
ISBN 978-7-121-38380-9

Ⅰ. ①E… Ⅱ. ①牛… Ⅲ. ①搜索引擎－程序设计 Ⅳ. ①TP391.3

中国版本图书馆 CIP 数据核字(2020)第 021836 号

责任编辑：安　娜
印　　刷：三河市君旺印务有限公司
装　　订：三河市君旺印务有限公司
出版发行：电子工业出版社
　　　　　北京市海淀区万寿路 173 信箱　邮编：100036
开　　本：787×980　1/16　印张：26.5　字数：554 千字
版　　次：2020 年 3 月第 1 版
印　　次：2020 年 8 月第 3 次印刷
定　　价：109.00 元

凡所购买电子工业出版社图书有缺损问题，请向购买书店调换。若书店售缺，请与本社发行部联系，联系及邮购电话：（010）88254888，88258888。
质量投诉请发邮件至 zlts@phei.com.cn，盗版侵权举报请发邮件至 dbqq@phei.com.cn。
本书咨询联系方式：010-51260888-819，faq@phei.com.cn。

序

在信息大爆炸的当下，信息过载已成为越来越多的人的负担。

随着 5G 时代的到来，物联网和智慧城市将会随处可见，随之而来的是信息会更加复杂和庞大。如何挣脱信息的束缚，高效地找到自己需要的信息呢？答案就是搜索引擎，即借助搜索引擎来寻找我们想要的信息！

本书介绍的搜索引擎是 Elasticsearch——一个开源的搜索引擎。

目前，Elasticsearch 的功能已不局限于搜索，它还在不断地丰富和完善自己的生态。在 API 接口层面，除基本的数据索引和数据搜索外，Elasticsearch 还提供了 Elasticsearch 服务监控接口、推荐相关接口，以及机器学习相关接口。

本书目的

与追求知识点全部覆盖但都泛泛而谈的书不同，本书聚焦初学者的学习和实战需要，将初学者接触 Elasticsearch 从 0 到 1 过程中的必备知识点讲透。只有学透了基础知识，再学习更多的有关 Elasticsearch 的知识才成为可能。

这一点笔者在培训 Elasticsearch 初学者时深有体会。因此，本书重点结合笔者在 Elasticsearch 上的沉淀、实战、培训和 Elasticsearch 最新版本内容，帮助 Elasticsearch 初学者点破这层窗户纸！

正如王阳明在《传习录》中谈为学之道时所言："殊不知私欲日生，如地上尘，一日不扫便又有一层。着实用功，便见道无终穷，愈探愈深，必使精白无一毫不彻方可。"

对于知识与近代和现代高速发展的经济之间的关系，管理学大师德鲁克有一段精辟论述。他认为二者的关系可以分为三个发展阶段，即工业革命、生产力革命、管理革命。所谓工业革命，指的是知识应用于生产工具、生产流程和产品创新；所谓生产力革命，指的是知识以及被赋予的含义开始被应用于工作中；所谓管理革命，指的是知识正被用于知识本身。而管理革命的核心在于连接。在知识领域，连接意味着知识点关联。

很多人无法有效地将相似或关联的知识点进行关联，所以更谈不上构建网状知识体系。

因此，在本书行文过程中，笔者会基于自己构建的知识体系向读者进行必要的体系输出，力求帮助读者在快速上手的同时，构建搜索引擎全景，洞悉 Elasticsearch 生态，建立关联知识网络。

本书基于 Elasticsearch 7.X 系列版本编写，内容由浅入深，先让初学者会用、能用，再介绍背后的原理。这种方式在笔者主导过的 Elasticsearch 技术培训中效果较好。

本书结构

本书分为三大部分，分别是 Elasticsearch 前传、Elasticsearch 实战和 Elasticsearch 生态。

Elasticsearch 前传部分主要介绍搜索技术发展史和基本知识，并介绍搜索引擎技术原理，为读者构建搜索引擎全景。在技术发展史上，我们能看见多久的历史，就能看见多远的未来！

Elasticsearch 实战部分主要介绍 Elasticsearch 的核心概念和架构设计，并重点介绍客户端、文档、搜索、索引等实战内容，待读者能上手实战后，再介绍这些内容的背后实现原理和关联知识，为读者构建知识网络。

Elasticsearch 生态部分主要介绍插件的使用和管理，以及 Elastic Stack 生态圈。

本书特色

特色 1：基于 Elasticsearch 7.X 系列版本编写。

特色 2：聚焦初学者学习和实战需要，不求知识点全部覆盖，但求必备知识透彻易懂。

特色 3：让初学者快速上手的同时，帮助他们构建搜索引擎全景、洞悉 Elasticsearch 生态、建立关联知识网络。

特色 4：由浅入深，先让初学者会用，再介绍背后的原理。

在本书编写过程中，Elasticsearch 仍在升级版本，因此书中难免有理解和实践不足之处。"卑辞俚语，不揣谫陋"，欢迎读者和笔者交流学习，共同进步。

牛冬

2019 年 12 月

目 录

第一部分　Elasticsearch 前传

第 1 章　搜索技术发展史 2
    1.1　正说搜索技术发展史 2
    1.2　Elasticsearch 简介 5
    1.3　Lucene 简介 5
    1.4　知识点关联 7
    1.5　小结 15

第 2 章　搜索技术基本知识 16
    2.1　数据搜索方式 16
    2.2　搜索引擎工作原理 17
    2.3　网络爬虫工作原理 18
    2.4　网页分析 20
    2.5　倒排索引 23
    2.6　结果排序 26
    2.7　中文分词实战 27
        2.7.1　Ansj 中文分词 27
        2.7.2　Jcseg 轻量级 Java 中文分词器 30
    2.8　知识点关联 38
    2.9　小结 39

第二部分 Elasticsearch 实战

第 3 章 初识 Elasticsearch 42
3.1 Elasticsearch 简介 42
3.2 Elasticsearch 的安装与配置 43
3.2.1 安装 Java 环境 43
3.2.2 Elasticsearch 的安装 47
3.2.3 Elasticsearch 的配置 52
3.3 Elasticsearch 的核心概念 60
3.4 Elasticsearch 的架构设计 62
3.4.1 Elasticsearch 的节点自动发现机制 64
3.4.2 节点类型 66
3.4.3 分片和路由 66
3.4.4 数据写入过程 67
3.5 知识点关联 70
3.6 小结 75

第 4 章 初级客户端实战 76
4.1 初级客户端初始化 76
4.2 提交请求 83
4.3 对请求结果的解析 89
4.4 常见通用设置 91
4.5 高级客户端初始化 95
4.6 创建请求对象模式 98
4.7 知识点关联 98
4.8 小结 100

第 5 章 高级客户端文档实战一 101
5.1 文档 102
5.2 文档索引 103
5.3 文档索引查询 114
5.4 文档存在性校验 118
5.5 删除文档索引 121

5.6 更新文档索引 ... 125
5.7 获取文档索引的词向量 ... 131
5.8 文档处理过程解析 ... 138
　5.8.1 文档的索引过程 ... 138
　5.8.2 文档在文件系统中的处理过程 ... 140
5.9 知识点关联 ... 145
5.10 小结 ... 146

第6章 高级客户端文档实战二 ... 147
6.1 批量请求 ... 148
6.2 批量处理器 ... 154
6.3 MultiGet 批量处理实战 ... 158
6.4 文档 ReIndex 实战 ... 164
6.5 文档查询时更新实战 ... 171
6.6 文档查询时删除实战 ... 176
6.7 获取文档索引的多词向量 ... 180
6.8 文档处理过程解析 ... 185
　6.8.1 Elasticsearch 文档分片存储 ... 185
　6.8.2 Elasticsearch 的数据分区 ... 187
6.9 知识点关联 ... 188
6.10 小结 ... 189

第7章 搜索实战 ... 190
7.1 搜索 API ... 191
7.2 滚动搜索 ... 208
7.3 批量搜索 ... 220
7.4 跨索引字段搜索 ... 228
7.5 搜索结果的排序评估 ... 235
7.6 搜索结果解释 ... 243
7.7 统计 ... 251
7.8 搜索过程解析 ... 258
　7.8.1 对已知文档的搜索 ... 258

7.8.2　对未知文档的搜索 259
　　7.8.3　对词条的搜索 260
7.9　知识点关联 262
7.10　小结 262

第 8 章　索引实战 263

8.1　字段索引分析 264
8.2　创建索引 271
8.3　获取索引 277
8.4　删除索引 282
8.5　索引存在验证 285
8.6　打开索引 289
8.7　关闭索引 292
8.8　缩小索引 296
8.9　拆分索引 299
8.10　刷新索引 303
8.11　Flush 刷新 306
8.12　同步 Flush 刷新 310
8.13　清除索引缓存 314
8.14　强制合并索引 317
8.15　滚动索引 322
8.16　索引别名 326
8.17　索引别名存在校验 330
8.18　获取索引别名 333
8.19　索引原理解析 337
　　8.19.1　近实时搜索的实现 337
　　8.19.2　倒排索引的压缩 337
8.20　知识点关联 338
8.21　小结 339

第三部分　Elasticsearch 生态

第 9 章　Elasticsearch 插件 342

9.1　插件简介 342
9.2　插件管理 343
9.3　分析插件 346
9.3.1　分析插件简介 346
9.3.2　Elasticsearch 中的分析插件 347
9.3.3　ICU 分析插件 349
9.3.4　智能中文分析插件 360
9.4　API 扩展插件 367
9.5　监控插件 368
9.6　数据提取插件 368
9.7　常用插件实战 369
9.7.1　Head 插件 369
9.7.2　Cerebro 插件 385
9.8　知识点关联 393
9.9　小结 394

第 10 章　Elasticsearch 生态圈 395

10.1　ELK 395
10.1.1　Elastic Stack 395
10.1.2　Elastic Stack 版本的由来 396
10.1.3　ELK 实战的背景 397
10.1.4　ELK 的部署架构变迁 397
10.2　Logstash 400
10.2.1　Logstash 简介 400
10.2.2　Logstash 的输入模块 402
10.2.3　Logstash 过滤器 403
10.2.4　Logstash 的输出模块 404
10.3　Kibana 405
10.3.1　Kibana 简介 405
10.3.2　连接 Elasticsearch 406

10.4 Beats 410
   10.4.1 Beats 简介 410
   10.4.2 Beats 轻量级设计的实现 412
   10.4.3 Beats 的架构 412
10.5 知识点关联 413
10.6 小结 414

第一部分
Elasticsearch前传

第 1 章 搜索技术发展史

人事有代谢
往来成古今

1.1 正说搜索技术发展史

"我们面前无所不有，我们面前一无所有。"

正如查尔斯·狄更斯在《双城记》中所述。在信息大爆炸的当下，"我们面前无所不有"；而个人信息过载已成为越来越多的人的负担，"我们面前一无所有"。

如何挣脱过载的信息的束缚，高效地找到自己需要的信息呢？——答案是搜索引擎，借助搜索引擎来实现！

本书介绍的搜索引擎是 Elasticsearch——一个开源的搜索引擎（简称 ES）。

我们每天都在某种场景下使用搜索引擎，在电脑上、手机上，都可以找到自己惯用的搜索引擎，比如百度搜索、搜狗搜索、神马搜索、谷歌搜索、360 搜索、头条搜索，等等。

那么，搜索引擎是什么呢，它是如何发展到今天的样子呢？本章就介绍搜索技术发展，让我们沿着技术发展的脉络更深刻地认识搜索技术。

宏观而言，搜索引擎的发展经历了五个阶段和两大分类。五个阶段分别是 FTP 文件检索阶段、分类目录导航阶段、文本相关性检索阶段、网页链接分析阶段和用户意图识别阶段。具体情况汇总如下。

FTP 文件检索阶段

该阶段的搜索引擎只检索多个 FTP 服务器上存储的文件，代表作是 Archie。用户搜索文件时需输入精确的文件名来搜索查找，搜索引擎会告诉用户从哪一个 FTP 地址可以下载被搜索的文件。

分类目录导航阶段

该阶段的搜索引擎就是一个导航网站，网站中都是网址的分类陈列，用户在互联网上常用的网址在这里一应俱全。

在使用该类搜索引擎时，用户需要从各个分类目录里找到自己想要的网址，单击其网站链接后进入相应的网站。

直到今天，这类搜索引擎依然不过时，我们常用的网站如好 123、搜狗浏览器主页、UC 导航等均是这类导航页面。

文本相关性检索阶段

随着互联网内容的不断丰富，网页的内容和形态也越来越多样化，页面中开始出现内容可能与网页地址和网页标题大相径庭的情况。

为了解决这个问题，搜索引擎引入全文搜索技术，来保证搜索引擎检索到的网页标题与网页全文内容强一致，摒弃了单纯依靠网页标题和网页地址来判断网页内容的方法。

在使用这类搜索引擎查询信息时，用户将输入的查询信息提交给搜索引擎后台服务器，搜索引擎服务器通过查阅已经索引好的网页全文信息，返回一些相关程度高的页面信息。

计算输入的查询信息与网页内容相关性判断的模型主要有布尔模型、概率模型、向量空间模型等。

这个阶段的搜索引擎的主要代表作是 Alta Vista、Excite 等。

网页链接分析阶段

这个阶段的搜索引擎所使用的网站链接形式与当前基本相同。在该阶段，外部链接表示推荐。因此，通过计算每个网站的推荐链接的数量，就可以判断一个网站的流行性和重要性。

于是，搜索引擎通过结合网页内容的重要性和相似程度来改善搜索的信息质量。在这一阶段，搜索引擎的代表作是谷歌搜索。

这种模式是谷歌首创的，并且大获成功，随之引起了学术界和其他商业搜索引擎的极度关注和效仿。目前，网页链接分析算法及其改进优化的版本在主流搜索引擎中大行其道。

用户意图识别阶段

这个阶段的搜索引擎以用户为中心作为设计的初心，搜索引擎力求理解每一位用户的真正搜索诉求，力求做到千人千面，追求个性化识别和反馈。

在使用这类搜索引擎时，即便是同一个查询的请求关键词，不同的用户可能也会得到不同的查询结果。比如输入的是"小米"，那么一个想要购买小米电子设备的用户和一个想要购买小米食用的用户，他们的搜索意图显然天壤之别，因而得到不同的搜索结果是顺理成章的事情。不光是不同用户之间，同一个用户搜索同样的关键词也会因时因地的不同而有所差异。比如当用户在搜索引擎上首次输入"TAL"时，可能是想查找 TAL 股票代码对应的好未来公司的网站；当用户在好未来的办公区内搜索"TAL"时，有可能是想查看 TAL 股票代码的实时股价。

其实在这两个案例背后，搜索引擎都在致力于解决同一个问题，即怎样才能通过输入的简短的关键词来判断用户的真正查询诉求。这也是我们将其归类为用户意图识别的原因。这一阶段的搜索引擎典型代表就是百度。

在搜索引擎技术不断演进的过程中，为了更好地识别及满足用户的搜索需求，更多的新技术也在不断引入，如 AI 技术、地理位置信息、用户画像等。

两大分类是指站内搜索和站外搜索。

站外搜索就是全网搜索，现在主流的搜索引擎基本都是全网搜索，如谷歌、百度。随着技术的发展，搜索领域的生态圈搜索形态不断扩大。以谷歌为代表的搜索引擎推出了整合搜索、个人化搜索、实时搜索、地图服务、线上文件编辑、网站统计、浏览器、网管工具、超大容量电子邮件、即时通信等。百度上线了百度百科、百度知道、百度贴吧等服务，这些服务中嵌入了文字搜索、语音搜索、图像搜索、地图搜索等搜索形态。

站内搜索近几年发展比较迅猛，各大网站平台纷纷上线了站内搜索，如 SNS 平台中的微博、人人网等，如电商平台中的京东、饿了么、淘宝、美团等。

另外，区块链内容搜索是近两年新的站内搜索形式，如比特币区块链的搜索内容在比特币公链上，但比特币公链的节点所在地域却是分布式的，和常见的站内搜索大相径庭，如图 1-1 所示。

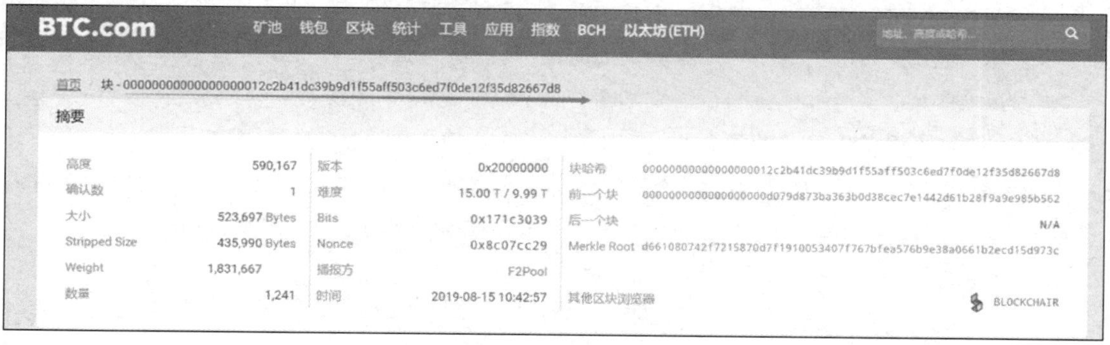

图 1-1

在未来，搜索引擎的发展会是什么样的呢？我们不妨畅想一下。随着 5G 时代的到来，物联网和智慧城市将会随处可见；AR/VR 技术会更加成熟，设备更加普及和便宜。与之对应的，除现在的文字搜索、语音搜索、图像搜索外，还会出现 AR/VR 搜索等搜索形态。

在 5G 的加持下，搜索引擎的搜索效率会更高；物联网和区块链中设备和信息搜索也会更加普遍，而搜索引擎的商业模式也可能随之升级，广告的效果可能会更好。

1.2 Elasticsearch 简介

Elasticsearch 是一个分布式、可扩展、近实时的高性能搜索与数据分析引擎。

Elasticsearch 提供了搜集、分析、存储数据三大功能，其主要特点有：分布式、零配置、易装易用、自动发现、索引自动分片、索引副本机制、RESTful 风格接口、多数据源和自动搜索负载等。

Elasticsearch 并非从零起步，而是站在巨人的肩膀上。Elasticsearch 基于 Java 编写，其内部使用 Lucene 做索引与搜索。通过进一步封装 Lucene，向开发人员屏蔽了 Lucene 的复杂性。开发人员无须深入了解检索的相关知识来理解它是如何工作的，只需使用一套简单一致的 RESTful API 即可，从此全文搜索变得简单。

除此之外，Elasticsearch 还解决了检索相关数据、返回统计结果、响应速度等相关的问题。因此，Elasticsearch 能做到分布式环境下的实时文档存储和实时分析搜索。实时存储的文档，每个字段都可以被索引与搜索。

最令人惊喜的是，Elasticsearch 能胜任上成百上千个服务节点的分布式扩展，支持 PB 级别的结构化或者非结构化海量数据的处理。

2019 年 4 月 10 日，Elasticsearch 发布了 7.0 版本。该版本的重要特性包含引入内存断路器、引入 Elasticsearch 的全新集群协调层——Zen2、支持更快的前 k 个查询、引入 Function score 2.0 等。

其中内存断路器可以更精准地检测出无法处理的请求，并防止它们使单个节点不稳定；Zen2 是 Elasticsearch 的全新集群协调层，提高了可靠性、性能和用户体验，使 Elasticsearch 变得更快、更安全，并更易于使用。

1.3 Lucene 简介

Lucene 是一个免费、开源、高性能、纯 Java 编写的全文检索引擎。

在业务开发场景中，Lucene 几乎适用任何需要全文检索的场景。因此，应普遍的搜索开发需求，各种编程语言的 Lucene 版本不断涌现。目前，Lucene 先后发展出了 C++、C#、Perl 和 Python 等语言的版本，Lucene 逐渐成为开源代码中最好的全文检索引擎工具包。

2005 年，Lucene 升级成为 Apache 顶级项目。

Lucene 包含大量相关项目，核心项目有 Lucene Core、Solr 和 PyLucene。

需要指出的是，Lucene 仅仅是一个工具包，它并非一个完整的全文检索引擎，这和 Lucene 的初衷相关。Lucene 主要为软件开发人员提供一个简单易用的工具包，主要提供倒排索引的查询结构，以方便软件开发人员在其业务系统中实现全文检索的功能。这也是我们常说全文检索引擎主要是 Solr 和 Elasticsearch 的原因，虽然二者均是以 Lucene 为基础建立的。

Lucene 作为一个全文检索引擎工具包，具有如下突出优点。

索引文件格式独立于应用平台

Lucene 定义了一套以 8 位字节为基础的索引文件格式，使得兼容系统或者不同平台的应用能够共享建立的索引文件。

索引速度快

在传统全文检索引擎的倒排索引的基础上，实现了分块索引，能够针对新的文件建立小文件索引，提升索引速度。然后通过与原有索引的合并，达到优化的目的。

简单易学

优秀的面向对象的系统架构，降低了 Lucene 扩展的学习难度，方便扩充新功能。

跨语言

设计了独立于语言和文件格式的文本分析接口，索引器通过接收 Token 流完成索引文件的创立，用户扩展新的语言和文件格式，只需实现文本分析的接口即可。

强大的查询引擎

Lucene 默认实现了一套强大的查询引擎，用户无须自己编写代码即可通过系统获得强大的查询能力。Lucene 默认实现了布尔操作、模糊查询、分组查询等。

Lucene 的主要模块有 Analysis 模块、Index 模块、Store 模块、QueryParser 模块、Search 模块和 Similarity 模块，各模块的功能分别汇总如下。

① Analysis 模块：主要负责词法分析及语言处理，也就是我们常说的分词，通过该模块可最终形成存储或者搜索的最小单元 Term。

② Index 模块：主要负责索引的创建工作。

③ Store 模块：主要负责索引的读和写，主要是对文件的一些操作，其主要目的是抽象出和平台文件系统无关的存储。

④ QueryParser 模块：主要负责语法分析，把查询语句生成 Lucene 底层可以识别的条件。

⑤ Search 模块：主要负责对索引的搜索工作。

⑥ Similarity 模块：主要负责相关性打分和排序的实现。

在 Lucene 中，还有一些核心术语，主要涉及 Term、词典（Term Dictionary，也叫作字典）、倒排表（Posting List）、正向信息和段（Segment），这些术语的含义汇总如下。

① Term：索引中最小的存储和查询单元。对于英文语境而言，一般是指一个单词；对于中文语境而言，一般是指一个分词后的词。

② 词典：是 Term 的集合。词典的数据结构有很多种，各有优缺点。如可以通过排序数组（通过二分查找来检索数据）、HashMap（哈希表，检索速度更快，属于空间换时间的模式）、FST（Finite-State Transducer，有很好的压缩率）等来实现。

③ 倒排表：一篇文章通常由多个词组成，倒排表记录的是某个词在哪些文章中出现过。

④ 正向信息：原始的文档信息，可以用来做排序、聚合、展示等。

⑤ 段：索引中最小的独立存储单元。一个索引文件由一个或者多个段组成。在 Lucene 中，段有不变性，段一旦生成，在段上只能读取、不可写入。

1.4 知识点关联

站在巨人的肩膀上

从上文中我们可以看到，Elasticsearch 并非从 0 起步，重复制造轮子，而是站在巨人 Lucene 的肩膀上构建的。其实很多新技术都是基于已有的技术发展演进而来的，如最近几年火热的区块链技术。

1. REST 协议

RESTful 编程风格是近年来前后端流行的交互模式，在工作和面试中，有人会将 REST 协议与 HTTP 混为一谈，因此这里介绍 REST 协议的由来。

下面从几个维度来展开叙述：

① Web 技术发展与 REST 的由来——讲历史。

② REST 架构风格的推导过程——讲过程。

③ REST 定义——讲定义。

④ REST 关键原则——讲原则。

⑤ 总结 REST 风格的架构特点——讲特点。
⑥ REST 架构风格的优点和缺点——讲辩证。

2. Web 技术发展与 REST 的由来

Web 技术基础有哪些？从技术架构层面上看，Web 技术架构包括四个基石：URI、HTTP、HyperText、MIME（Multipurpose Internet Mail Extensions，多用途互联网邮件扩展类型）。这四个基石相互支撑，相互助力，共同推动着 Web 这座宏伟的大厦以几何级数的速度迅猛发展起来。

在 Web 开发技术历程中，Web 开发技术经历了 5 个阶段，分别是静态内容阶段、CGI（Common Gateway Interface）程序阶段、脚本语言阶段、瘦客户端应用阶段、RIA（Rich Internet Application）应用阶段和移动 Web 应用阶段。其中：

在静态内容阶段，Web 由静态 HTML 文档组成，Web 服务器有点像支持超文本的共享文件服务器。

在 CGI 程序阶段，Web 服务器增加了编程 API，研发人员可以通过 CGI 协议完成程序开发，这类应用程序被称作 CGI 程序。

在脚本语言阶段，服务器端先后出现了 ASP、PHP、JSP、ColdFusion 等支持 session 的脚本语言技术，浏览器端则出现了 Java Applet、JavaScript 等技术。基于这些技术，网站可以提供更加丰富的动态内容。

在瘦客户端应用阶段，在服务器端出现了独立于 Web 服务器的应用服务器，出现了 Web MVC 开发模式。基于这些框架开发的 Web 应用，通常都是瘦客户端应用。

在 RIA 应用阶段，出现了多种 RIA 技术，大幅改善了 Web 应用的用户体验，如最为广泛的 RIA 技术 DHTML+Ajax，Ajax 技术支持前端在不刷新页面的情况下动态更新页面中的局部内容，显著提升了用户体验。这一时期也诞生了大量的 Web 前端 DHTML 开发库，如 Prototype、Dojo、ExtJS、jQuery/jQuery UI。其他的 RIA 技术还有 Adobe 公司的 Flex、微软公司的 Silverlight、Sun 公司的 JavaFX（现在为 Oracle 公司所有），等等。

在移动 Web 应用阶段，出现了大量面向移动设备的 Web 应用开发技术。除 Android、iOS、Windows Phone 等操作系统平台原生的开发技术外，基于 HTML5 的开发技术也变得非常流行。

从上述 Web 开发技术的发展过程可以得出结论，Web 从最初其设计者所构思的主要支持静态文档的阶段，逐渐变得越来越动态化；Web 应用的交互模式变得越来越复杂，从静态文档发展到以内容为主的门户网站、电子商务网站、搜索引擎、社交网站，再到以娱乐为主的大型多人在线游戏、手机游戏。

Web 开发技术 和 REST 之间有什么关系呢？上面提及 Web 技术架构包括四个基石，其中之一就是 HTTP，下面简要介绍 HTTP 的发展历程。

1995 年，CGI、ASP 等技术出现后，沿用多年、面向静态文档的 HTTP/1.0 已无法满足 Web 应用的开发需求，因此需要设计新版本的 HTTP。于是，HTTP/1.0 协议专家组成立。在这些专家之中，有一位名叫 Roy Fielding 的年轻人脱颖而出，显示出了不凡的洞察力，后来 Roy Fielding 成为 HTTP/1.1 协议专家组的负责人。而 Roy Fielding 还是 Apache HTTP 服务器的核心开发者，更是 Apache 软件基金会的合作创始人。

HTTP/1.1 的第一个草稿于 1996 年 1 月发布，后经三年多的反复修订和论证，于 1999 年 6 月成为 IETF 的正式规范。

2000 年，Roy Fielding 在其博士学位论文 *Architectural Styles and the Design of Network-based Software Architectures* 中，系统、严谨地阐述了这套理论框架，还使用这套理论框架推导出了一种新的架构风格，并且为这种架构风格取了一个令人轻松愉快的名字"REST"——Representational State Transfer（表述性状态转移）。

2007 年 1 月，支持 REST 开发的 Ruby on Rails 1.2 版发布。特别的是，Rails 将支持 REST 开发作为其未来发展的优先内容。Ruby on Rails 的创始人 DHH 做了一场名为"World of Resources"的精彩演讲。DHH 在 Web 开发技术社区中的强大影响力，使得 REST 一下子处在 Web 开发技术舞台的聚光灯之下。在 Rails 等新兴 Web 开发技术的推波助澜之下，Web 开发技术社区掀起了一场重归 Web 架构设计本源的运动。REST 架构风格得到了越来越多的关注，现在各种流行的 Web 开发框架几乎都支持 REST 开发。

而广大的开发人员也纷纷学习 REST 架构风格。当然大多数 Web 开发者都是通过阅读 REST 开发框架的文档，以及通过一些例子代码来学习 REST 开发的。不过，通过例子代码来学习 REST 有非常大的局限性，因为 REST 并不是一种具体的技术，也不是一种具体的规范，而是一种内涵非常丰富的架构风格。

虽然通过例子代码来学习 REST，能"短平快"地学习到一种有趣的 Web 开发技术，但并不能全面深入地理解 REST 究竟是什么，甚至还会误以为这些简单的例子代码就是 REST 本身，以为 REST 不过是一种简单的 Web 开发技术而已。这有点类似像盲人摸象，有的人摸到了象鼻子、有的人摸到了象耳朵、有的人摸到了象腿、有的人摸到了象尾巴。摸象的众人都坚信自认为感觉到的大象才是最真实的大象，而其他人的感觉都是错误的，但现实往往恰恰相反。

3. REST 架构风格的推导过程——讲过程

Roy Fielding 在批判性继承前人研究成果的基础上，建立起了一整套研究和评价软件架构的方法论。这套方法论的核心是"架构风格"这个概念。

架构风格是一种研究和评价软件架构设计的方法，它是比架构更加抽象的概念。一种架构风格是由一组相互协作的架构约束来定义的。架构约束是指软件的运行环境施加在架构设计之上的约束。

Roy Fielding 的 REST 架构风格的推导过程如图 1-2 所示。

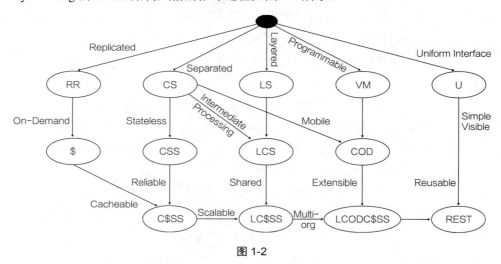

图 1-2

每一个椭圆形里面的缩写词代表了一种架构风格，而每一个箭头边的单词代表了一种架构约束。其中：

- RR（Replicated Repository）可译为复制仓库。RR 利用多个进程提供相同的服务，来改善数据的可访问和可伸缩。
- COD（Code-On-Demand），可译为按需代码。
- $表示 Cache，指的是复制个别请求的结果，以便可以被后面的请求重用。
- CSS（Client-Stateless-Server）可译为客户无状态服务器。
- LCS（Layered-Client-Server）可译为分层客户服务器。
- VM（Virtual Machine）可译为虚拟机风格。

通过推导，Roy Fielding 得出了 REST 架构风格中最重要的 6 个架构约束，即客户—服务器（Client-Server）、无状态（Stateless）、缓存（Cache）、统一接口（Uniform Interface）、分层系统（Layered System）和 按需代码（Code-on-Demand）。其中：

- 客户-服务器是指通信只能由客户端单方面发起，表现为请求-响应的形式。
- 无状态是指通信的会话状态（Session State）应该全部由客户端负责维护。
- 缓存是指响应内容可以在通信链的某处被缓存，以改善网络效率。

- 统一接口是指通信链的组件之间通过统一的接口相互通信,以提高交互的可见性。
- 分层系统是指通过限制组件的行为,将架构分解为若干等级的层。
- 按需代码是指支持通过下载并执行一些代码,对客户端的功能进行扩展。

HTTP/1.1 作为一种 REST 架构风格的架构实例,其架构演变经历了图 1-3 和图 1-4 所示的过程。

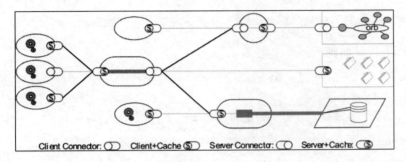

图 1-3

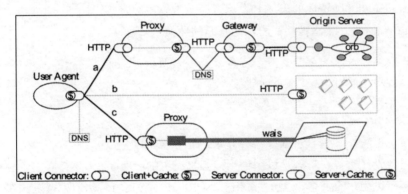

图 1-4

用户代理处在三个并行交互(a、b 和 c)的中间,用户代理的客户端连接器缓存无法满足请求,因此它根据每个资源标识符的属性和客户端连接器的配置,将每个请求路由到资源的来源。

请求 a 被发送到一个本地代理,代理随后访问一个通过 DNS 查找发现的缓存网关,该网关将这个请求转发到一个能够满足该请求的来源服务器,来源服务器的内部资源由一个封装过的对象请求代理(Object Request Broker)架构来定义。

请求 b 直接发送到一个来源服务器,它能够通过自己的缓存满足这个请求。

请求 c 被发送到一个代理,它能够直接访问 WAIS(一种与 Web 架构分离的信息服务),并将 WAIS 的响应翻译为一种通用的连接器接口能够识别的格式。

在图 1-4 中，每个组件只知道与它们自己的客户端或服务器连接器的交互。

4．REST 定义——讲定义

REST（REpresentational State Transfer，表述性状态转移）是所有 Web 应用都应该遵守的架构设计指导原则。

当然，REST 并不是法律法规。如果在软件开发实践中违反了 REST 的指导原则，虽然能够实现应用的功能，但是会付出很多代价，特别是对于大流量的网站。

5．REST 关键原则——讲原则

REST 有 5 个关键原则，分别是 Resource、Hypertext Driven、Uniform Interface、Representation 和 State Transfer。

- Resource（资源）表示为所有事物定义 ID。
- Hypertext Driven（超文本驱动）表示将所有事物链接在一起。
- Uniform Interface（统一接口）表示使用标准方法。
- Representation（资源的表述）表示资源多重表述的方式。
- State Transfer（状态转移）表示无状态的通信。

> 注：在翻译外文文献时，可参考严复在《天演论》中的"译例言"讲到的"译事三难：信、达、雅"。这三个原则的核心在于意译。

资源（Resource）

资源其实可以看作一种看待服务器的方式。

与面向对象设计类似，资源是以名词为核心来组织的。

一个资源可以由一个或多个 URI 来标识。URI 既是资源的名称，也是资源在 Web 上的地址。对某个资源感兴趣的客户端应用，可以通过资源的 URI 与其进行交互访问并获取。

将服务器看作由很多离散的资源组成，每个资源是服务器上一个可命名的抽象概念。资源可以是服务器文件系统中的一个文件，也可以是数据库中的一张表。

上文提到资源意味着为所有事物定义 ID，也就是说，每个事物都是可被标志的，都会拥有一个 ID 标识符。在 Web 开发中，代表 ID 的统一概念就是 URI。URI 构成了一个全局命名空间，使用 URI 标识你的关键资源意味着它们获得了一个唯一、全局的 ID。

这里要注意区分 URI、URL 和 URN。

- URI，即 Uniform Resource Identifier：统一资源标识符，包括 URL 和 URN。

- URL，即 Uniform Resource Locator：统一资源定位符，常见的有 Web URL 和 FTP URL。
- URN，即 Uniform Resource Name：统一资源名称，例如 tel:+1-888-888-8888。

设计 URI 的 4 个原则是：

- 它们是名词。
- 区分单复数。
- URI 有长度限制，建议小于 1KB。
- 在 URI 中不要放未经加密的敏感信息。

在 URI 中，有几个符号需要特别注意：

① / 表示层次关系，例如 https://www.×××/cn/products/elasticsearch。

②;，来表示并列关系，例如 https://www.×××.com/axis;x=0,y=9 sip:user@domain.com;foo=bar;x=y。

③ - 来提高可读性，最好全用小写，例如 https://www.×××-fan.com/。

④ 用参数或者 HTTP Range Header 来限定范围，例如 https://www.×××.com/sortbyAsc=name&fileds=email,title&limit=10&start=20。

超文本驱动

超文本驱动意味着将所有事物链接在一起。超媒体被当作应用状态引擎（Hypermedia As The Engine of Application State，HATEOAS），其核心是超媒体概念，换句话说是链接的思想。

引入链接后，我们可以将 Web 应用看作一个由很多状态组成的有限状态机。资源之间通过超链接相互关联，超链接既代表资源之间的关系，也代表可执行的状态迁移。

使用 URI 表示链接时，链接可以指向由不同应用、不同服务器甚至位于不同地域、不同洲际的不同公司提供的资源——因为 URI 命名规范是全球标准，所以构成 Web 的所有资源都可以互联互通。

超媒体原则还意味着应用"状态"，服务器端为客户端提供一组链接，使客户端能够通过链接将应用从一个状态改变为另一个状态。而链接恰恰是构成动态应用的非常有效的方式。

因此链接成就了现在的 Web。

统一接口

统一接口意味着使用标准方法。

按 REST 要求，必须通过统一接口对资源执行各种操作。对于每个资源只能执行一组有限的操作。

在 RESTful HTTP 方式中，一般开发人员通过组成 HTTP 应用协议的通用接口访问服务程序。在 HTTP/1.1 中，已经定义了一个操作资源的统一接口，主要包括以下内容。

- 7 个 HTTP 方法：POST、GET、PUT、DELETE、PATCH、HEAD 和 OPTIONS。
- HTTP 头信息（可自定义）。
- HTTP 响应状态代码（可自定义）。
- 一套标准的内容协商机制。
- 一套标准的缓存机制。
- 一套标准的客户端身份认证机制。

对开发人员而言，程序代码将从图 1-5 中转变。

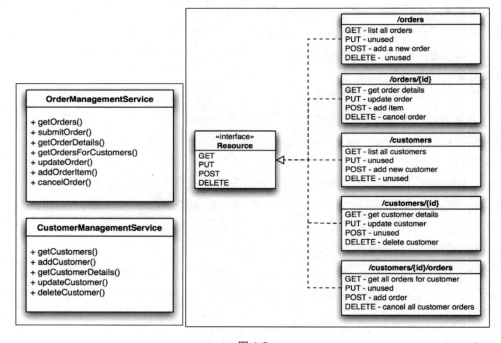

图 1-5

资源的表述

资源的表述是一段对资源在某个特定时刻的状态的描述。资源可以在客户端—服务器端之间转移传递。

资源的表述有多种格式，如 HTML、XML、JSON、文本、图片、视频、音频等。资源的表述格式可以通过协商机制来确定。请求—响应的表述通常使用不同的格式。

状态转移

状态转移意味着无状态通信。这里的状态转移是指在客户端和服务器端之间转移。通过转移和操作资源的表述，间接实现操作资源的目的。

无状态意味着服务器的变化对客户端是不可见的。这主要是为了保证架构设计的可伸缩性和可扩展性。试想一下，如果服务器需要存储每个客户端状态，那么大量的客户端交互会严重影响服务器的内存可用空间。

总结 REST 风格的架构特点

判断 REST 架构设计是否优秀的标准有 6 个：分别是面向资源、可寻址、连通性、无状态、统一接口和超文本驱动。

REST 架构风格的优点&缺点——讲辩证

REST 架构设计的优点是简单、可伸缩、松耦合。需要指出的是，REST 架构设计不是"银弹"。在实时性要求很高的应用中，REST 的表现不如 RPC。因此，在做架构设计和技术选型时，需要根据具体的运行环境对具体问题进行具体分析。

1.5 小结

本章主要介绍了搜索引擎技术发展的历史，帮助读者沿着技术发展的脉络更深刻地认识搜索技术，并畅想了未来搜索引擎技术的发展情况。

在生活中，搜索无处不在，搜索随时可在——在查字典时、在图书馆找书时。其实人类社会的发展史也是一部信息高效整理、高效查阅的历史，而信息的高效搜索与高效查阅，恰恰是搜索引擎的本质。

第 2 章
搜索技术基本知识

> 骐骥一跃，不能十步
> 驽马十驾，功在不舍

在正式介绍 Elasticsearch 的相关内容之前，先介绍搜索引擎的相关工作原理。当了解了通用搜索引擎工作原理之后，再学习 Elasticsearch 时会轻松得多，正所谓"磨刀不误砍柴工"！

2.1　数据搜索方式

在第 1 章中，我们已经了解到：搜索引擎主要是对数据进行检索。而在研发过程中不难发现，数据有两种类型，即结构化数据和非结构化数据。

对于软件研发人员来说，在做数据持久化时，对数据的结构化感知会特别强烈。如结构化数据一般我们会放入关系数据库（如 MySQL、Oracle 等），这是因为结构化数据有固定的数据格式和有限个数的字段，因此可以通过二维化的表结构来承载。

而非结构化数据一般会放入 MongoDB 中，这是因为非结构化的数据长度不定且无固定数据格式，显然在关系数据库中存储这类数据较为困难。

与数据形态相对应的，数据的搜索分为两种，即结构化数据搜索和非结构化数据搜索。

因为结构化数据可以基于关系数据库来存储，而关系数据库往往支持索引，因此结构化数据可以通过关系数据库来完成搜索和查找。常用的方式有顺序扫描、关键词精确匹配、关键词部分匹配等，对于较为复杂的关键词部分匹配，通常需要借助 like 关键字来实现，如左匹配关键词"TAL"时需要使用 like "TAL%"，右匹配关键词"TAL"时需要使用 like "%TAL"，完全模糊匹配关键词"TAL"时需要使用 like "%TAL%"。

对于非结构化数据，数据的搜索主要有顺序扫描和全文检索两种方法。显然，对于非结构化数据而言，顺序扫描是效率很低的方法，因此全文检索技术应运而生，而全文搜索就是本书所说的搜索引擎要做的事情。

在实现全文检索的过程中，一般都需要提取非结构化数据中的有效信息，重新组织数据的承载结构形式。而搜索数据时，需要基于新结构化的数据展开，从而达到提高搜索速度的目的。显而易见，全文检索是一种空间换时间的做法——前期进行数据索引的创建，需要花费一定的时间和空间，但能显著提高后期搜索的效率。

下面介绍搜索引擎的工作原理。

2.2 搜索引擎工作原理

通用搜索引擎的工作原理如图 2-1 所示。

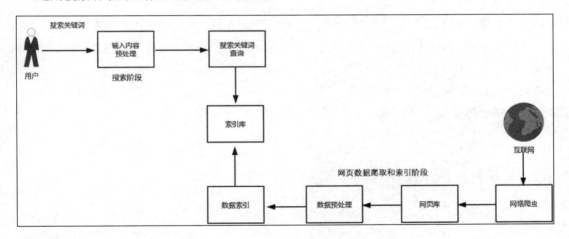

图 2-1

搜索引擎的工作原理分为两个阶段，即网页数据爬取和索引阶段、搜索阶段。其中网页数据爬取和索引阶段包含网络爬虫、数据预处理、数据索引三个主要动作，搜索阶段包含搜索关键词、输入内容预处理、搜索关键词查询三个主要动作。

其中，网络爬虫用于爬取互联网上的网页，爬取到一个新网页后还要继续通过该页面中的链接来爬取其他网页。因此网络爬虫是一个不停歇的工作，一般需要自动化手段来实施。网络爬虫的主要工作就是尽可能快、尽可能全地发现和抓取互联网上的各种网页。

网页被网络爬虫爬取后，会被存入网页库，以备下一阶段进行数据的预处理。需要指出的是，网页库里存储的网页信息与我们在浏览器看到的页面内容相同。此外，由于互联网上的网页有一定的重复性，因此在把新网页真正插入网页库之前，需要进行查重检查。

网页数据预处理程序不断地从网页库中取出网页进行必要的预处理。常见的预处理动作有去除噪声内容、关键词处理（如中文分词、去除停止词）、网页间链接关系计算等。其中，去除噪声内容包括版权声明文字、导航条、广告等。网页经过预处理后，会被浓缩成以关键词为核心的内容。

此外，互联网上的内容除常规的页面外，还有各类文件文档（如 PDF、Word、WPS、XLS、PPT、TXT 等）、多媒体文件（如图片、视频）等，这些内容均需进行相应的数据预处理动作。

数据预处理后，要进行数据索引过程。索引过程先后经历正向索引和倒排索引阶段，最终建立索引库。随着新的网页等内容不断地被加入网页库，索引库的更新和维护往往也是增量进行的。

以上是网页数据爬取和索引阶段的核心工作，下面介绍搜索阶段的核心工作。

用户输入的关键词同样会经过预处理，如删除不必要的标点符号、停用词、空格、字符拼写错误识别等，随后进行相关的分词，分词后搜索引擎系统将向索引库发出搜索请求。索引库会将包含搜索关键词的相关网页从索引库中找出来，搜索引擎根据索引库返回的内容进行排序处理，最终返回给用户。

2.3 网络爬虫工作原理

网络爬虫有多个不同的称谓，如网络探测器、Crawler 爬行器、Spider 蜘蛛、Robot 机器人等，其中网络爬虫或网络蜘蛛的叫法会更形象更生动一些，取意为网页爬取程序像虫子和蜘蛛一样在网络间爬来爬去，从一个网页链接爬到另一个网页链接。世界上第一个网络爬虫是 MIT Matthew Gray 的 World Wide Web Wanderer，Wanderer 主要用于追踪互联网发展规模。

网络蜘蛛在工作时，通过种子爬取网页的链接地址来寻找目标网页。随后从网站的 1 个页面，如首页，开始读取网页的内容和网页中其他网页的链接地址，然后通过这些链接地址继续寻找下一个网页。如此循环，直到所有内容都被抓取完成。

在网络爬虫爬取过程中，为了提高爬取效率，一般采用并行爬取的方式。多个网络爬虫在并行爬取过程中，不重复爬取同一个网页尤为重要，这将极大地提高爬取效率。

常规的做法如图 2-2 所示。

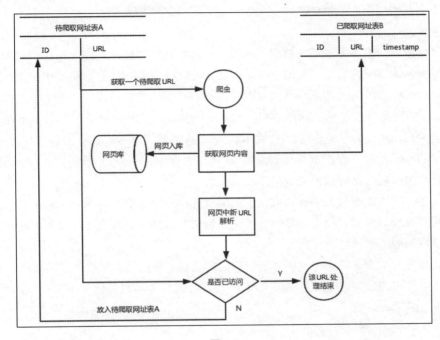

图 2-2

网络爬虫在爬取网页时，搜索引擎会建立两张不同的表，一张表记录已经访问过的网址，另一张表记录没有访问过的网址。当网络爬虫爬取某个外部链接页面 URL 时，需把该网站的 URL 下载回来分析，当网络爬虫处理好这个 URL 后，将该 URL 存入已经访问过的表中。当另一个网络爬虫从其他网站或页面中又发现了这个 URL 时，它会在已访问列表中有对比查看有没有该 URL 的访问记录，如果有，则网络爬虫会自动丢弃该 URL，不再访问。

网络爬虫在按照链接爬取网页的过程中，网页之间的关系有点类似于有向图。在有向图的节点遍历过程中，我们可以按照"先深度后广度"的方式遍历，也可以按照"先广度后深度"的方式遍历。同样，在网络爬虫爬取网页的过程中，网络爬虫需要根据一定的策略来爬取网页，一般采用"先深度后广度"的方式。

此外，在网络爬虫爬取网页的过程中，还需要注意网页的收录模式，一般有两种，即增量收集和全量搜集。全量搜集，顾名思义，每次爬取网页都更新全部数据内容。这种模式一般定期展开，因为全量搜集模式的资源开销大、付出成本高、内容更新的时效性不高、网络宽带消耗高，而且更新全量数据所需时间也比较长。

而增量收集可以避免全量收集模式的弊端，这种模式主要用于搜集新网页、搜集更新的网页、删除不存在的页面。当然，相较于全量收集，网络爬虫的系统设计也会复杂一些，但时效性好。

对于网站而言，被各家主流搜索引擎收录是共同的夙愿。因此，网站往往采取一些技术手段告知搜索引擎来抓取内容。一般网站可以通过 SiteMap 告知搜索引擎网站中可提供抓取的网址，SiteMap 的核心作用就是向网络抓取工具提供一些提示信息，以便它们更有效地抓取网站。SiteMap 的最简单实现形式就是 XML 文件。当然，各家搜索引擎定义的 SiteMap 不尽相同，如百度 SiteMap 分为三种格式：txt 文本格式、XML 格式和 SiteMap 索引格式。

对网站的维护人员而言，除了 SiteMap，还可以结合 SEO（Search Engine Optimization，搜索引擎优化）来改善网站的被抓取效果。

综上所述，网络爬虫的工作核心就在于在网页搜集效率、质量和对目标网站的友好程度上。网络爬虫要用最少的资源、最少的时间，搜集尽可能多的高质量网页；同时对目标网站的内容抓取不影响网站的正常运转和使用。

对软件开发人员来说，我们可以基于现有的爬虫框架来实现对网络数据的爬取。Java 语言栈的读者可以使用 WebMagic、Gecco；Python 语言栈的读者可以使用 Scrapy；Go 语言栈的读者可以使用 YiSpider。

2.4 网页分析

网络爬虫将网页数据爬取后，会将网页内容存储到网页库。随后，网页分析程序将自动对网页进行分析。主要的分析动作有网页内容摘要、链接分析、网页重要程度计算、关键词提取/分词、去除噪声等。经过网页分析后，网页数据将变成网页中关键词组、链接与关键词的相关度、网页重要程度等信息。

网页内容中的去除噪声主要是去除如广告、无关的导航条、版权信息、调查问卷等和文章主体内容无关的内容。这些内容如果纳入网页分析中，往往会让网页的主题发生偏移；同时更多的无关内容的索引，还会使得索引结构规模变大，拖累搜索的准确性，降低搜索速度，而这一点非常重要，因为搜索的准确性和检索速度是衡量搜索引擎的主要标准之一。在实践中，去除噪声可以基于 Doc View 模型实现。

网页内容摘要一般由网页正文生成，摘要一般会显示在搜索结果的展示区，如图 2-3 中框形区域所示。

图 2-3

一般来说，摘要的生成方式有静态生成方式和动态生成方式两种。其中静态生成方式比较简单，在网页分析阶段即可从网页内容中提取。虽然这种方式"短、平、快"，但缺点也很明显，即当呈现搜索结果时，展示的摘要可能与搜索的关键词无关。

现在的搜索引擎往往采用动态生成模方式生成，即根据查询关键词在文档中的位置，提取其周边的文字，并高亮显示。

网页重要程度计算用于衡量网站的权威性。一般越权威的网站，越容易被其他网站主动链接；换言之，网站被引用的次数越多，说明该网站越重要。对搜索引擎而言，在返回相关性强的内容时，应该尽量先返回权威网站的内容；对搜索引擎的用户而言，这样往往更能匹配他们的需要。因此这也是评价搜索引擎体验好坏的核心指标之一。

网站之间链接的关系，其实可以从算法的维度来解析。链接其实是一种投票、一种信任。网站被主动链接的次数越多，说明互联网环境下其他网站对该网站的投票越多、信任越多，该网站在互联网中越流行。本质上这就是一种分布式系统下的共识投票。如图 2-4 所示，网页 A 在多个网页间被主动链接、被主动投票的次数最多，因此我们认为网页 A 更权威。

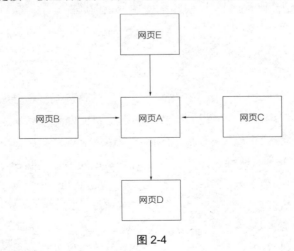

图 2-4

如果将网页间的链接关系视作有向图，则网页的链接关系就会变成入度和出度。这里的入度指的是网页能通过其他网页的链接来访问；出度指的是网页中链接了其他网页。因此，入度大网页，说明其被多个网页引用，这也意味着该网页比较权威、比较流行和热门，如图 2-4 中的网页 A。

在各家搜索引擎的实践中，谷歌提出了 PageRank（佩奇等级）算法，该算法也是谷歌搜索引擎的重要法宝。

在关键词提取/分词环节，基础技术是分词，因此分词是各家搜索引擎中非常关键的技术。不论中英文网站，在创建索引之前都需要对内容进行分词。分词不仅是关键词提取的前提，也是后续文本挖掘的基础。

在分词方面，中英文分词有天然的差异性。相较而言，英文分词会简单一些，因为英文有天然的空格作为单词的分隔符。中文不仅没有天然空格来分隔汉字，而且汉字的词组大部分由两个及以上的汉字组成，汉语语句也习惯连续性书写，因此增加了中文分词的难度。

此外，在中英文中，都有一类词称之为 "Stop Word"，即 "停用词"。停用词一般是指无内容指示意义的词。

在中文分词时，常用的算法可以分为两大类，一类是基于字典的机械式分词，另一类是基于统计的分词。

基于字典的分词方法，一般会按照一定的策略将待分析的汉字串与一个充分大的词典的词条进行匹配，若在词典中找到某个字符串词条，则匹配成功。因此基于字典的分词方法的核心字符串的匹配。

在匹配字符串时分为正向匹配和逆向匹配两种。正向匹配指的是在匹配字符串时从左向右匹配。如 "中国""中国人" 两个词条字符串在匹配时，从左向右可以依次匹配成功 "中" 和 "国"。而逆向匹配与之相反，一般是从右向左匹配。同样，当 "中国""中国人" 两个词条字符串从右向左匹配时，"中国" 的 "国" 会与 "中国人" 的 "人" 去匹配，显而易见，匹配失败；此时需要匹配的字就是 "中国人" 中 "人" 字左侧的 "国" 字，这时，两个 "国" 字是能匹配成功的，依此类推，两个字符串词条中的 "中" 字也能匹配成功。

由于中文的特性，多个词条往往有相关的前缀或后缀，如 "中国""中国人""中国话" 三个词条都有 "中国" 这个公共前缀。因此在某方向，如正向或逆向匹配过程中，按匹配长度的不同，还可以细分为最大/最长匹配和最小/最短匹配。我们仍然以 "中国""中国人""中国话" 三个词条举例，正向最短匹配的结果是 "中"，而正向最长匹配的结果是 "中国"。

因此，基于字典的分词算法一般常用正向最大匹配、逆向最大匹配，或者是组合模式。不过基于字典的分词算法对分词词典的更新有较强依赖，特别是需要对新词敏感的场景。

第二类分词算法是基于统计的分词算法。该算法无须词典，一般会根据汉字与汉字相邻出现的概率来进行分词。因此基于统计的分词算法往往需要构建一个语料库，并不断更新。在分词前，算法需要进行预处理，即对语料库中相邻出现的各个字的组合进行统计，计算两个汉字间的组合概率。

其实不论是哪类分词算法，都会对新词不敏感，都会分出一些经常出现但并非有效词组的内容，如"有点""有的""好像""非常"等。因此有必要对这类词语进行过滤，毕竟分词算法最主要的指标就是分词的准确率。

在开发实践中，各个语言栈常用的分词中间件汇总如下。Python 语言中的中文分词组件有 jieba 中文分词，Java 语言中常用 Jcseg、Ansj 和庖丁分词，Go 语言中常用 sego。

2.5 倒排索引

在中文信息检索领域，索引的发展经历了基于字的索引和基于词的索引两种。不论是基于字做索引，还是基于词做索引，在建立索引过程中，均涉及正排索引和倒排索引两个数据结构。

正排索引的数据结构如图 2-5 所示。

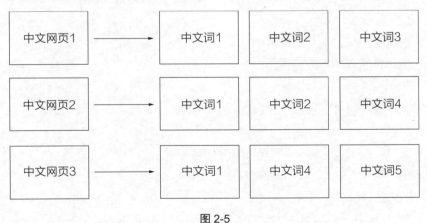

图 2-5

如图 2-5 所示，在正排索引中，以网页或文章映射关系为 Key、以分词的列表为 Value。而在实际搜索网页或文章时，恰恰与此结构相反，即在搜索时是以查询语句的分词列表为 Key 来进行搜索的。因此，为了提高搜索效率，我们需要对正排索引进行转化，将其转化为以分词为 Key、以网页或文章列表为 Value 的结构，而这个结构就是倒排索引，如图 2-6 所示。

图 2-6

在介绍完正排索引和倒排索引的结构后，下面我们以三句话为例，展示基于字的索引和基于词的索引的区别。

内容 1：英国是近代的世界强国。

内容 2：美国是当代的世界强国。

内容 3：中国是未来的世界强国。

当基于字做索引时，对上述三句话进行字的分解，分解结果如下所示。

内容 1：英 国 是 近 代 的 世 界 强 国。

内容 2：美 国 是 当 代 的 世 界 强 国。

内容 3：中 国 是 未 来 的 世 界 强 国。

倒排索引结构如表 2-1 所示。

表 2-1

字	内容列表		
英	内容 1		
国	内容 1	内容 2	内容 3
是	内容 1	内容 2	内容 3
近	内容 1		

续表

字	内容列表		
代	内容1	内容2	内容3
的	内容1	内容2	内容3
世	内容1	内容2	内容3
界	内容1	内容2	内容3
强	内容1	内容2	内容3
国	内容1	内容2	内容3
美		内容2	
中			内容3
当		内容2	
未			内容3
来			内容3

当基于词做索引时，对这三句话进行分词，分词结果如下所示。

内容1：英国 是 近代 的 世界 强国。

内容2：美国 是 当代 的 世界 强国。

内容3：中国 是 未来 的 世界 强国。

倒排索引结构如表2-2所示。

表2-2

词	内容列表		
英国	内容1		
是	内容1	内容2	内容3
近代	内容1		
的	内容1	内容2	内容3
世界	内容1	内容2	内容3
强国	内容1	内容2	内容3
美国		内容2	
当代		内容2	
中国			内容3
未来			内容3

通过对比不难发现，基于词做索引的索引内容明显比基于字做索引的索引内容要少，因而

查询时会更加高效。

此外，在倒排索引中一般还会记录更多的信息，如词汇在网页或文章中出现的位置、频率和权重等，这些信息在查询阶段会用到。

在倒排索引中，有词条（Term）、词典（Term Dictionary）、倒排表（Post List）三个名词，下面我们一一介绍。

词条是索引里面最小的存储和查询单元。一般来说，在英文语境中词条是一个单词，在中文语境中词条指的是分词后的一个词组。

词典又称字典，是词条的集合。单词词典一般是由网页或文章集合中出现过的所有词构成的字符串集合。

倒排表如图 2-2 所示，倒排表记录的是词出现在哪些文档里、出现的位置和频率等。

在倒排表中，每条记录被称为一个倒排项。

前文中提及，Elasticsearch 是基于 Lucene 实现的。在 Lucene 中，词典和倒排表是实现快速检索的重要基石。另外，词典和倒排表是分两部分存储的，词典存储在内存中，倒排表存储在磁盘上。

2.6 结果排序

搜索引擎除对网络爬虫爬取的网页进行处理，将它们结构化成倒排索引外，还有一项重要工作就是响应用户的查询需求。

搜索引擎系统接收用户提交的查询字符串后，对字符串进行分词，去除不必要的停用词等无意义词汇后，进行倒排索引的查询。多个关键词的倒排索引查询结果的交集即为搜索的结果。

而搜索的结果往往是需要进一步处理的，如一般都会进行排序。搜索结果排序是搜索引擎查询服务的核心所在。排序结果决定了搜索引擎体验的好与坏、用户的使用满意度和搜索引擎的口碑。

搜索结果的排序算法也是不断迭代发展的，早期主要基于查询词出现的频率来排序，随后出现了 PageRank 和相关性等算法排序。

一般而言，相关性算法主要考虑的因素有关键词的使用频率、关键词在网页中的词频、关键词出现在所在网页的位置、关键词间的距离、网页链接及重要性。其中，关键词的使用频率指的是日常生活用词的频率，如"有的""有点""可能""非常"这些词经常出现在日常交流过程中，但在搜索引擎看来，这些词汇的意义并不大。

关键词在网页中的词频越高，意味着出现次数越多，说明页面与搜索词的关系越密切。

关键词出现在所在网页的位置是指关键词是否出现在了比较重要的位置，如标题。关键词出现在所在网页的位置越重要，说明页面与关键词越相关。一般而言，倒排索引库在建立时，关键词出现在所在网页的位置是会被记录在其中的。

关键词间的距离指的是多个关键词在页面上出现的位置的接近程度，关键词间越接近，说明在该网页与搜索词字符串的相关度越高。

网页链接及重要性指的是页面有越多以搜索词为关键词的导入链接，说明页面的相关性越强；链接分析还包括了链接源页面本身的主题、目标文字周围的文字等。

当然，在目前的搜索引擎中，还都不约而同地引入了用户行为分析、数据挖掘等技术，来提升搜索结果的质量。

2.7 中文分词实战

前文提及了 Ansj 和 Jcseg 两种轻量级 Java 中文分词器，本节将介绍这两种分词组件的使用方法。

2.7.1 Ansj 中文分词

Ansj 中文分词基于 n-Gram+CRF+HMM 算法，用 Java 实现。Ansj 中文分词工具的分词速度可达到大约 200 万字/s，准确率达 96%以上。

目前 Ansj 中文分词工具已经实现了中文分词、中文姓名识别、用户自定义词典、关键字提取、自动摘要、关键字标记等功能，可以应用到自然语言处理等方面，适用于对分词效果要求较高的各种项目。

Ansj 中文分词工具支持的分词方式有 ToAnalysis（精准分词）、DicAnalysis（用户自定义词典优先策略的分词）、NlpAnalysis（带有新词发现功能的分词）、IndexAnalysis（面向索引的分词）、BaseAnalysis（最小颗粒度的分词）。

其中，ToAnalysis 是 Ansj 分词方式的首选项。该分词方式在易用性、稳定性、准确性及分词效率上，都取得了一个不错的平衡。如果初次尝试使用 Ansj，想"开箱即用"，那么建议使用 ToAnalysis。

在 DicAnalysis 中，用户可以自定义词典优先策略。如果自定义词典足够好，或者用户对自定义词典的要求比较高，那么强烈建议使用 DicAnalysis。在此种情景下，很多方面它会优于 ToAnalysis 的分词结果。

NlpAnalysis 可以识别出未登录词。它的缺点是速度比较慢、稳定性差一些。如果需要进行未登录词识别、对文本进行分析，则首推该分词方法。

IndexAnalysis 适合在 Lucene 等文本检索中使用，其召回率较高，准确率也较高。

BaseAnalysis 则保证了最基本的分词，词语颗粒度非常小，所涉及的词大约是 10 万（个）左右。这种分词方式的分词速度非常快，可达到 300 万字/s，同时准确率也很高。不过，对于新词，这种分词方式的效果不太理想。

下面展示上述分词方式的分词 API 的使用和分词效果。分词 API 的使用如下所示：

```
package com.niudong.esdemo.util;
import org.ansj.splitWord.analysis.BaseAnalysis;
import org.ansj.splitWord.analysis.DicAnalysis;
import org.ansj.splitWord.analysis.IndexAnalysis;
import org.ansj.splitWord.analysis.NlpAnalysis;
import org.ansj.splitWord.analysis.ToAnalysis;
public class AnsjSegUtil {
  public static void processString(String content) {
    // 最小颗粒度的分词
    System.out.println(BaseAnalysis.parse(content));
    // 精准分词
    System.out.println(ToAnalysis.parse(content));
    // 用户自定义词典优先策略的分词
    System.out.println(DicAnalysis.parse(content));
    // 面向索引的分词
    System.out.println(IndexAnalysis.parse(content));
    // 带有新词发现功能的分词
    System.out.println(NlpAnalysis.parse(content));
  }
  public static void main(String[] args) {
    String content ="15 年来首批由深海探险家组成的国际团队五次潜入大西洋海底 3800 米深处，对泰坦尼克号沉船残骸进行调查。探险队发现，虽然沉船的部分残骸状况良好，也有一些部分已消失在大海中。强大的洋流、盐蚀和细菌正不断侵蚀着这艘沉船。英媒曝光的高清图片可见，沉船的部分残骸遭腐蚀情况严重。";
    processString(content);
  }
}
```

在当前类中执行 main 方法，各个分词方式的运行结果如下所示。

最小颗粒度的分词

15/m,年来/t,首批/n,由/p,深海/n,探险家/n,组成/v,的/u,国际/n,团队/n,五/m,次/q,潜入/v,大西洋

/ns,海底/n,3800/m,米/q,深处/s,,/w,对/p,泰坦尼克号/nz,沉船/n,残骸/n,进行/v,调查/vn,。/w,探险队/n,发现/v,,/w,虽然/c,沉船/n,的/u,部分/n,残骸/n,状况/n,良好/a,,/w,也/d,有/v,一些/m,部分/n,已/d,消失/v,在/p,大海/n,中/f,。/w,强大/a,的/u,洋流/n,、/w,盐/n,蚀/vg,和/c,细菌/n,正/d,不断/d,侵蚀/vn,着/u,这/r,艘/q,沉船/n,。/w,英/j,媒/ng,曝光/v,的/u,高清/n,图片/n,可见/v,,/w,沉船/n,的/u,部分/n,残骸/n,遭/v,腐蚀/v,情况/n,严重/a,。/w

该方式累计有效分词 71 个，其中单字分词结果 24 个。

精准分词

15/m,年来/t,首批/n,由/p,深海/n,探险家/n,组成/v,的/u,国际/n,团队/n,五次/mq,潜入/v,大西洋/ns,海底/n,3800 米/mq,深处/s,,/w,对/p,泰坦尼克号/nz,沉船/n,残骸/n,进行/v,调查/vn,。/w,探险队/n,发现/v,,/w,虽然/c,沉船/n,的/u,部分/n,残骸/n,状况/n,良好/a,,/w,也/d,有/v,一些/m,部分/n,已/d,消失/v,在/p,大海/n,中/f,。/w,强大/a,的/u,洋流/n,、/w,盐/n,蚀/vg,和/c,细菌/n,正/d,不断/d,侵蚀/vn,着/u,这/r,艘/q,沉船/n,。/w,英/j,媒/ng,曝光/v,的/u,高清/n,图片/n,可见/v,,/w,沉船/n,的/u,部分/n,残骸/n,遭/v,腐蚀/v,情况/n,严重/a,。/w

该方式累计有效分词 68 个，其中单字分词结果 21 个。

用户自定义词典优先策略的分词

15/m,年来/t,首批/n,由/p,深海/n,探险家/n,组成/v,的/u,国际/n,团队/n,五次/mq,潜入/v,大西洋/ns,海底/n,3800 米/mq,深处/s,,/w,对/p,泰坦尼克号/nz,沉船/n,残骸/n,进行/v,调查/vn,。/w,探险队/n,发现/v,,/w,虽然/c,沉船/n,的/u,部分/n,残骸/n,状况/n,良好/a,,/w,也/d,有/v,一些/m,部分/n,已/d,消失/v,在/p,大/a,海中/s,。/w,强大/a,的/u,洋流/n,、/w,盐/n,蚀/vg,和/c,细菌/n,正/d,不断/d,侵蚀/vn,着/u,这/r,艘/q,沉船/n,。/w,英/j,媒/ng,曝光/v,的/u,高清/n,图片/n,可见/v,,/w,沉船/n,的/u,部分/n,残骸/n,遭/v,腐蚀/v,情况/n,严重/a,。/w

该方式累计有效分词 68 个，其中单字分词结果 21 个。

面向索引的分词

15/m,年来/t,首批/n,由/p,深海/n,探险家/n,组成/v,的/u,国际/n,团队/n,五次/mq,潜入/v,大西洋/ns,海底/n,3800 米/mq,深处/s,,/w,对/p,泰坦尼克号/nz,沉船/n,残骸/n,进行/v,调查/vn,。/w,探险队/n,发现/v,,/w,虽然/c,沉船/n,的/u,部分/n,残骸/n,状况/n,良好/a,,/w,也/d,有/v,一些/m,部分/n,已/d,消失/v,在/p,大海/n,中/f,。/w,强大/a,的/u,洋流/n,、/w,盐/n,蚀/vg,和/c,细菌/n,正/d,不断/d,侵蚀/vn,着/u,这/r,艘/q,沉船/n,。/w,英/j,媒/ng,曝光/v,的/u,高清/n,图片/n,可见/v,,/w,沉船/n,的/u,部分/n,残骸/n,遭/v,腐蚀/v,情况/n,严重/a,。/w

该方式累计有效分词 68 个，其中单字分词结果 21 个。

带有新词发现功能的分词

15 年/t,来/v,首/m,批/q,由/p,深海/n,探险家/n,组成/v,的/u,国际/n,团队/n,五次/mq,潜入/v,大西洋/ns,海底/n,3800 米/mq,深处/s,,,/w,对/p,泰坦尼克号/nz,沉船/n,残骸/n,进行/v,调查/vn,。/w,探险队/n,发现/v,,,/w,虽然/c,沉船/n,的/u,部分/n,残骸/n,状况/n,良好/a,,,/w,也/d,有/v,一些/m,部分/n,已/d,消失/v,在/p,大海/n,中/f,。/w,强大/a,的/u,洋流/n,、/w,盐蚀/nw,和/c,细菌/n,正/d,不断/d,侵蚀/vn,着/u,这/r,艘/q,沉船/n,。/w,英媒/nw,曝光/v,的/u,高清/n,图片/n,可见/v,,/w,沉船/n,的/u,部分/n,残骸/n,遭/v,腐蚀/v,情况/n,严重/a,。/w

该方式累计有效分词 67 个,其中单字分词结果 20 个。

通过上述分词结果我们可以看到,精准分词在整体的分词效果上确实较好,在新词识别上,如上面文字中的"盐蚀""英媒"的识别,是带有新词发现功能的分词,即 NlpAnalysis 的效果更好。

2.7.2 Jcseg 轻量级 Java 中文分词器

Jcseg 是基于 MMSEG 算法的一个轻量级中文分词器,集成了关键字提取、关键短语提取、关键句子提取和文章自动摘要等功能,并且提供了一个基于 Jetty 的 Web 服务器,方便各大语言直接 HTTP 调用,同时提供了最新版本的 Lucene、Solr 和 Elasticsearch 的分词接口。

此外,Jcseg 还自带了一个 jcseg.properties 文件,用于快速配置,从而得到适合不同场合的分词应用,如最大匹配词长、是否开启中文人名识别、是否追加拼音、是否追加同义词等。

Jcseg 支持的中文分词模式有六种,即简易模式、复杂模式、检测模式、检索模式、分隔符模式、NLP 模式,各自特点如下。

(1)简易模式:基于 FMM 算法实现,适合分词速度要求较高的场合。

(2)复杂模式:基于 MMSEG 四种过滤算法实现,可去除较高的歧义,分词准确率达 98.41%。

(3)检测模式:只返回词库中已有的词条,很适合某些应用场合。

(4)检索模式:转为细粒度切分,专为检索而生。除中文处理外(不具备中文的人名、数字识别等智能功能),其他与复杂模式一致(英文,组合词等)。

(5)分隔符模式:按照给定的字符切分词条,默认是空格,特定场合的应用。

(6)NLP 模式:继承自复杂模式,更改了数字、单位等词条的组合方式,增加了电子邮件、手机号码、网址、人名、地名、货币等无限种自定义实体的识别与返回。

Jcseg 的核心功能有中文分词、关键字提取、关键短语提取、关键句子提取、文章自动摘要、自动词性标注和命名实体标注,并提供了 RESTful API。各个功能的介绍如下。

① 中文分词:基于 MMSEG 算法和 Jcseg 独创的优化算法实现,有四种切分模式。

② 关键字提取：基于 textRank 算法实现。

③ 关键短语提取：基于 textRank 算法实现。

④ 关键句子提取：基于 textRank 算法实现。

⑤ 文章自动摘要：基于 BM25+textRank 算法实现。

⑥ 自动词性标注：基于词库实现。目前效果不是很理想，对词性标注结果要求较高的应用不建议使用。

⑦ 命名实体标注：基于词库实现，可以识别电子邮件、网址、手机号码、地名、人名、货币、datetime 时间、长度、面积、距离单位等。

⑧ RESTful API：嵌入 Jetty 并提供了一个绝对高性能的 Server 模块，包含全部功能的 HTTP 接口，标准化 JSON 输出格式，方便各种语言客户端直接调用。

Jcseg 支持自定义词库。在 jcseg\vendors\lexicon 文件夹下，可以随便添加、删除、更改词库和词库内容，并且对词库进行了分类。此外，Jcseg 还支持词库多目录加载，在配置 lexicon.path 时使用';'隔开多个词库目录即可。

词库中的汉字类型可以分为简体、繁体、简繁体混合类型，Jcseg 可以专门适用于简体切分、繁体切分、简繁体混合切分，并且可以利用同义词实现简繁体的相互检索。Jcseg 同时提供了两个简单的词库管理工具来进行简繁体的转换和词库的合并。

Jcseg 还支持中英文同义词追加、同义词匹配、中文词条拼音追加。Jcseg 的词库中整合了《现代汉语词典》和 cc-cedict 辞典中的词条，并且依据 cc-cedict 词典为词条标上了拼音。使用时，可以更改 jcseg.properties 配置文档，在分词的时候，把拼音和同义词加入分词结果中。

Jcseg 还支持中文数字和中文分数识别，如"一百五十个人都来了，四十分之一的人。"中的"一百五十"和"四十分之一"均可有效识别。在输出结果时，Jcseg 会自动将其转换为阿拉伯数字加入分词结果中，如 150、1/40。

Jcseg 支持中英混合词和英中混合词的识别，如 B 超、X 射线、卡拉 OK、奇都 KTV、哆啦 A 梦等。此外，Jcseg 还支持阿拉伯数字/小数/中文数字基本单字单位的识别，如 2012 年、1.75 米、38.6℃、五折（Jcseg 会将其转换为"5 折"加入分词结果中）。

Jcseg 对智能圆角半角、英文大小写转换、特殊字母识别（如Ⅰ、Ⅱ、①、⑩）等均能有效支持。

Jcseg 可以配置 Elasticsearch 使用，具体步骤如下。

（1）拉取 Jcseg 工程代码并进行编译打包。其中，拉取代码的命令为 git clone https://gitee.com/lionsoul/jcseg.git，如图 2-7 所示。

```
牛冬@LAPTOP-1S8BALK3 MINGW64 ~/Desktop/ES/code
$ git clone https://gitee.com/lionsoul/jcseg.git
Cloning into 'jcseg'...
remote: Enumerating objects: 6378, done.
remote: Counting objects: 100% (6378/6378), done.
remote: Compressing objects: 100% (2317/2317), done.
remote: Total 6378 (delta 2868), reused 6056 (delta 2575)
Receiving objects: 100% (6378/6378), 8.91 MiB | 901.00 KiB/s, done.
Resolving deltas: 100% (2868/2868), done.
```

图 2-7

将 Jcseg 工程代码拉取到本地后切换到 Jcseg 目录，即在 DOS 窗口中执行命令 cd jcseg，随后在该目录下执行编译打包命令 mvn package，编译执行的输出结果如图 2-8 所示。

```
牛冬@LAPTOP-1S8BALK3 MINGW64 ~/Desktop/ES/code/jcseg (master)
$ mvn package
[INFO] Scanning for projects...
[INFO]
[INFO] ------------------------------------------------------------
[INFO] Reactor Build Order:
[INFO]
[INFO] jcseg
[INFO] jcseg-core
[INFO] jcseg-analyzer
[INFO] jcseg-elasticsearch
[INFO] jcseg-server
[INFO]
[INFO] ------------------------------------------------------------
[INFO] Building jcseg 2.4.1
[INFO] ------------------------------------------------------------
[INFO]
[INFO] ------------------------------------------------------------
[INFO] Building jcseg-core 2.4.1
[INFO] ------------------------------------------------------------
```

图 2-8

编译打包成功后，输出的内容如图 2-9 所示。

```
[INFO] ------------------------------------------------------------
[INFO] Reactor Summary:
[INFO]
[INFO] jcseg .............................................. SUCCESS [  0.279 s]
[INFO] jcseg-core ......................................... SUCCESS [ 26.282 s]
[INFO] jcseg-analyzer ..................................... SUCCESS [  4.896 s]
[INFO] jcseg-elasticsearch ................................ SUCCESS [  6.274 s]
[INFO] ------------------------------------------------------------
[INFO] BUILD SUCCESS
[INFO] ------------------------------------------------------------
[INFO] Total time: 37.968 s
[INFO] Finished at: 2019-08-23T22:49:04+08:00
[INFO] Final Memory: 44M/618M
[INFO] ------------------------------------------------------------
Picked up JAVA_TOOL_OPTIONS: -Dfile.encoding=UTF-8
```

图 2-9

（2）在 Elasticsearch 工程中配置 Jcseg 分词，具体方法如下。

先在\elasticsearch-7.2.0\plugins 目录下新建文件夹 jcseg，如图 2-10 所示。

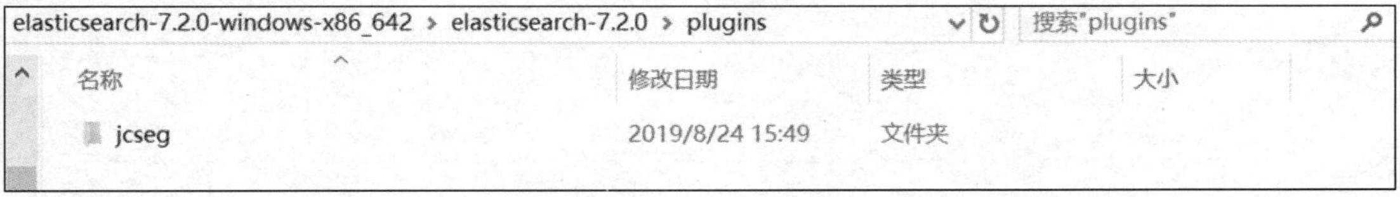

图 2-10

将（1）中打包的 3 个 jar 文件 jcseg-analyzer-2.4.1.jar、jcseg-core-2.4.1.jar 和 jcseg-elasticsearch-2.4.1.jar 复制到{ES_HOME}/plugins/jcseg 目录下，如图 2-11 所示。

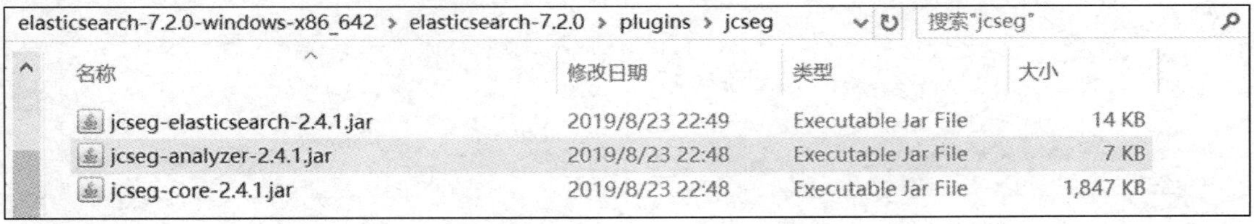

图 2-11

将 Jcseg 工程中的配置文件\jcseg-core\jcseg.properties 复制到{ES_HOME}/plugins/jcseg 目录下，如图 2-12 所示。

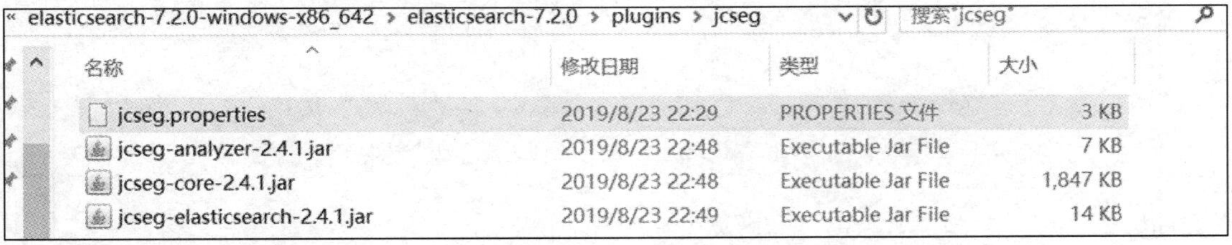

图 2-12

将 Jcseg 工程中的配置文件 jcseg-elasticsearch/plugin/plugin-descriptor.properties 复制到 {ES_HOME}/plugins/jcseg 目录下，如图 2-13 所示。

图 2-13

将 Jcseg 工程中的 lexicon 文件夹复制到{ES_HOME}/plugins/jcseg 目录下，如图 2-14 所示。

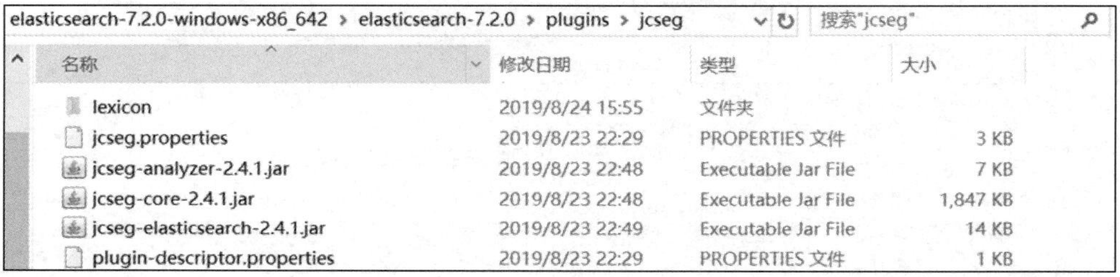

图 2-14

随后配置 jcseg.properties，主要是配置 lexicon.path，以便指向正确的词库，笔者本机的配置如图 2-15 所示。

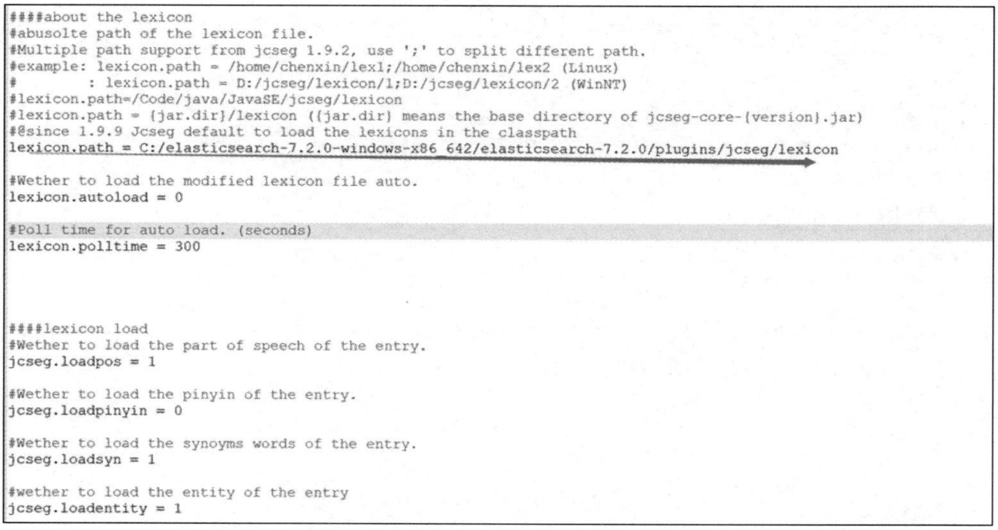

```
####about the lexicon
#abusolte path of the lexicon file.
#Multiple path support from jcseg 1.9.2, use ';' to split different path.
#example: lexicon.path = /home/chenxin/lex1;/home/chenxin/lex2 (Linux)
#        : lexicon.path = D:/jcseg/lexicon/1;D:/jcseg/lexicon/2 (WinNT)
#lexicon.path=/Code/java/JavaSE/jcseg/lexicon
#lexicon.path = {jar.dir}/lexicon ({jar.dir} means the base directory of jcseg-core-{version}.jar)
#@since 1.9.9 Jcseg default to load the lexicons in the classpath
lexicon.path = C:/elasticsearch-7.2.0-windows-x86_642/elasticsearch-7.2.0/plugins/jcseg/lexicon

#Wether to load the modified lexicon file auto.
lexicon.autoload = 0

#Poll time for auto load. (seconds)
lexicon.polltime = 300




####lexicon load
#Wether to load the part of speech of the entry.
jcseg.loadpos = 1

#Wether to load the pinyin of the entry.
jcseg.loadpinyin = 0

#Wether to load the synoyms words of the entry.
jcseg.loadsyn = 1

#wether to load the entity of the entry
jcseg.loadentity = 1
```

图 2-15

启动 Elasticsearch，在启动过程中会加载 Jcseg，如图 2-16 所示。

```
[2019-08-24T17:58:53,712][INFO ][o.e.p.PluginsService     ] [LAPTOP-1S8BALK3] loaded plugin [jcseg]
[2019-08-24T17:59:02,210][INFO ][o.e.x.s.a.s.FileRolesStore] [LAPTOP-1S8BALK3] parsed [0] roles from file [C:\elasticsea
rch-7.2.0-windows-x86_642\elasticsearch-7.2.0\config\roles.yml]
[2019-08-24T17:59:04,341][INFO ][o.e.x.m.p.l.CppLogMessageHandler] [LAPTOP-1S8BALK3] [controller/28728] [Main.cc@110] co
ntroller (64 bit): Version 7.2.0 (Build 65aefcbfce449b) Copyright (c) 2019 Elasticsearch BV
[2019-08-24T17:59:05,737][DEBUG][o.e.a.ActionModule       ] [LAPTOP-1S8BALK3] Using REST wrapper from plugin org.elastic
search.xpack.security.Security
[2019-08-24T17:59:06,471][INFO ][o.e.d.DiscoveryModule    ] [LAPTOP-1S8BALK3] using discovery type [zen] and seed hosts
providers [settings]
[2019-08-24T17:59:08,465][INFO ][o.e.n.Node               ] [LAPTOP-1S8BALK3] initialized
[2019-08-24T17:59:08,466][INFO ][o.e.n.Node               ] [LAPTOP-1S8BALK3] starting ...
[2019-08-24T17:59:10,563][INFO ][o.e.t.TransportService   ] [LAPTOP-1S8BALK3] publish_address {127.0.0.1:9300}, bound_ad
dresses {127.0.0.1:9300}, {[::1]:9300}
[2019-08-24T17:59:10,641][WARN ][o.e.b.BootstrapChecks    ] [LAPTOP-1S8BALK3] the default discovery settings are unsuita
ble for production use; at least one of [discovery.seed_hosts, discovery.seed_providers, cluster.initial_master_nodes] m
ust be configured
[2019-08-24T17:59:10,657][INFO ][o.e.c.c.Coordinator      ] [LAPTOP-1S8BALK3] cluster UUID [6UkGmpPrSwyXlzPHgSaivg]
[2019-08-24T17:59:10,693][INFO ][o.e.c.c.ClusterBootstrapService] [LAPTOP-1S8BALK3] no discovery configuration found, wi
ll perform best-effort cluster bootstrapping after [3s] unless existing master is discovered
[2019-08-24T17:59:11,135][INFO ][o.e.c.s.MasterService    ] [LAPTOP-1S8BALK3] elected-as-master ([1] nodes joined)[{LAPT
OP-1S8BALK3}{3gNRbEL_TJSocpgRnc2wjA}{6Pqmhu5_QbC-AFSkD6kc2w}{127.0.0.1}{127.0.0.1:9300}{ml.machine_memory=8467296256, xp
ack.installed=true, ml.max_open_jobs=20} elect leader, _BECOME_MASTER_TASK_, _FINISH_ELECTION_], term: 7, version: 63, r
eason: master node changed {previous [], current [{LAPTOP-1S8BALK3}{3gNRbEL_TJSocpgRnc2wjA}{6Pqmhu5_QbC-AFSkD6kc2w}{127.
0.0.1}{127.0.0.1:9300}{ml.machine_memory=8467296256, xpack.installed=true, ml.max_open_jobs=20}]}
[2019-08-24T17:59:11,295][INFO ][o.e.c.s.ClusterApplierService] [LAPTOP-1S8BALK3] master node changed {previous [], curr
ent [{LAPTOP-1S8BALK3}{3gNRbEL_TJSocpgRnc2wjA}{6Pqmhu5_QbC-AFSkD6kc2w}{127.0.0.1}{127.0.0.1:9300}{ml.machine_memory=8467
296256, xpack.installed=true, ml.max_open_jobs=20}]}, term: 7, version: 63, reason: Publication{term=7, version=63}
[2019-08-24T17:59:12,298][INFO ][o.e.l.LicenseService     ] [LAPTOP-1S8BALK3] license [bbcef3ff-378f-4c13-bb43-962a9dff2
841] mode [basic] - valid
[2019-08-24T17:59:12,345][INFO ][o.e.g.GatewayService     ] [LAPTOP-1S8BALK3] recovered [2] indices into cluster_state
[2019-08-24T17:59:15,005][INFO ][o.e.h.AbstractHttpServerTransport] [LAPTOP-1S8BALK3] publish_address {127.0.0.1:9200},
bound_addresses {127.0.0.1:9200}, {[::1]:9200}
[2019-08-24T17:59:15,006][INFO ][o.e.n.Node               ] [LAPTOP-1S8BALK3] started
[2019-08-24T17:59:15,790][INFO ][o.e.c.r.a.AllocationService] [LAPTOP-1S8BALK3] Cluster health status changed from [RED]
 to [YELLOW] (reason: [shards started [[ultraman1][0], [ultraman][0]] ...]).
```

图 2-16

在 DOS 窗口运行以下命令，测试分词效果。

```
curl 'http://localhost:9200/_analyze?pretty=true' -H 'Content-Type: application/
json' -d'
{
    "analyzer": "jcseg_search",
    "text": "一百美元等于多少人民币"
}'
```

分词结果打印如下所示。

```
{
  "tokens" : [
    {
      "token" : "一",
      "start_offset" : 0,
      "end_offset" : 1,
      "type" : "word",
      "position" : 0
    },
    {
      "token" : "一百",
      "start_offset" : 0,
```

```
      "end_offset" : 2,
      "type" : "word",
      "position" : 1
    },
    {
      "token" : "百",
      "start_offset" : 1,
      "end_offset" : 2,
      "type" : "word",
      "position" : 2
    },
    {
      "token" : "美",
      "start_offset" : 2,
      "end_offset" : 3,
      "type" : "word",
      "position" : 3
    },
    {
      "token" : "美元",
      "start_offset" : 2,
      "end_offset" : 4,
      "type" : "word",
      "position" : 4
    },
    {
      "token" : "元",
      "start_offset" : 3,
      "end_offset" : 4,
      "type" : "word",
      "position" : 5
    },
    {
      "token" : "等",
      "start_offset" : 4,
      "end_offset" : 5,
      "type" : "word",
      "position" : 6
    },
    {
      "token" : "等于",
      "start_offset" : 4,
      "end_offset" : 6,
      "type" : "word",
```

```
    "position" : 7
  },
  {
    "token" : "于",
    "start_offset" : 5,
    "end_offset" : 6,
    "type" : "word",
    "position" : 8
  },
  {
    "token" : "多",
    "start_offset" : 6,
    "end_offset" : 7,
    "type" : "word",
    "position" : 9
  },
  {
    "token" : "多少",
    "start_offset" : 6,
    "end_offset" : 8,
    "type" : "word",
    "position" : 10
  },
  {
    "token" : "少",
    "start_offset" : 7,
    "end_offset" : 8,
    "type" : "word",
    "position" : 11
  },
  {
    "token" : "人",
    "start_offset" : 8,
    "end_offset" : 9,
    "type" : "word",
    "position" : 12
  },
  {
    "token" : "人民",
    "start_offset" : 8,
    "end_offset" : 10,
    "type" : "word",
    "position" : 13
  },
```

```
    {
      "token" : "人民币",
      "start_offset" : 8,
      "end_offset" : 11,
      "type" : "word",
      "position" : 14
    },
    {
      "token" : "民",
      "start_offset" : 9,
      "end_offset" : 10,
      "type" : "word",
      "position" : 15
    },
    {
      "token" : "币",
      "start_offset" : 10,
      "end_offset" : 11,
      "type" : "word",
      "position" : 16
    }
  ]
}
```

2.8 知识点关联

倒排索引的思想在日常生活中十分常见。比如在某个公交车站或者地铁站，我们会以站点为起始点做路线查询与规划，而非按公交线路或地铁线路——显然站点就是倒排索引中的词组，而文章就是倒排索引中的公交线路或地铁线路。

同样，我们在做读书笔记的过程中，通常是将书中的内容拆解为知识点，按知识点复习和思考——显然知识点就是倒排索引中的词组，而文章就是倒排索引中以书名为单位的书。

在本章开篇，我们提及了结构化数据和非结构化数据。结构化不仅是一种数据表现形式，更是一种思维方式。

结构化思维的核心是逻辑性。一般常用的逻辑有因果关系、时空关系、优先级关系等。而对这些逻辑性内容，我们其实并不陌生，并且从小就接触。比如在小学的作文练习中，"总分总"结构是必练的思维之一；在解答应用题时，老师教的"套路"也是结构化思维的一种外化形式；而在工作中，常用的思维导图工具更是结构化思维的产物。对程序员而言，架构设计一般采用自上而下的模式也是结构化思维的一种体现。

结构化思维不仅体现在日常的沟通表达上、在敲代码的过程中，更渗透在我们分析问题、解决问题的点滴过程中。

2.9 小结

本章以数据的检索为切入点，主要介绍了搜索引擎的工作原理，对搜索引擎的核心模块如网络爬虫、网页分析、倒排索引、结果排序和中文分词等进行了详细说明，并介绍了近几年开源分词组件中 Java 语言栈的优秀代表 Ansj 和 Jcseg。

第二部分

Elasticsearch实战

第 3 章
初识 Elasticsearch

> 与君初相识
> 犹如故人归

本章主要介绍 Elasticsearch 的基础知识，如 Elasticsearch 的安装、配置，另外，还会介绍 Elasticsearch 的相关术语及架构设计，以方便读者学习后续章节。

3.1 Elasticsearch 简介

Elasticsearch 是一个分布式、可扩展、近实时的高性能搜索与数据分析引擎。Elasticsearch 基于 Apache Lucene 构建，采用 Java 编写，并使用 Lucene 构建索引、提供搜索功能。Elasticsearch 的目标是让全文搜索功能的落地变得简单。

Elasticsearch 的特点和优势如下：

① 分布式实时文件存储。Elasticsearch 可将被索引文档中的每一个字段存入索引，以便字段可以被检索到。

② 实时分析的分布式搜索引擎。Elasticsearch 的索引分拆成多个分片，每个分片可以有零个或多个副本。集群中的每个数据节点都可承载一个或多个分片，并且协调和处理各种操作；负载再平衡和路由会自动完成。

③ 高可拓展性。大规模应用方面，Elasticsearch 可以扩展到上百台服务器，处理 PB 级别的结构化或非结构化数据。当然，Elasticsearch 也可以运行在单台 PC 上。

④ 可插拔插件支持。Elasticsearch 支持多种插件，如分词插件、同步插件、Hadoop 插件、可视化插件等。

根据最新的数据库引擎排名显示，Elasticsearch、Splunk 和 Solr 分别占据了数据库搜索引擎的前三位，如图 3-1 所示。

Rank Aug 2019	Rank Jul 2019	Rank Aug 2018	DBMS	Database Model	Score Aug 2019	Score Jul 2019	Score Aug 2018
1.	1.	1.	Oracle	Relational, Multi-model	1339.48	+18.22	+27.45
2.	2.	2.	MySQL	Relational, Multi-model	1253.68	+24.16	+46.87
3.	3.	3.	Microsoft SQL Server	Relational, Multi-model	1093.18	+2.35	+20.53
4.	4.	4.	PostgreSQL	Relational, Multi-model	481.33	-1.94	+63.83
5.	5.	5.	MongoDB	Document	404.57	-5.36	+53.59
6.	6.	6.	IBM Db2	Relational, Multi-model	172.95	-1.19	-8.89
7.	7.	↑8.	Elasticsearch	Search engine, Multi-model	149.08	+0.27	+10.97
8.	8.	↓7.	Redis	Key-value, Multi-model	144.08	-0.18	+5.51
9.	9.	9.	Microsoft Access	Relational	135.33	-1.98	+6.24
10.	10.	10.	Cassandra	Wide column	125.21	-1.80	+5.63
11.	11.	11.	SQLite	Relational	122.72	-1.91	+8.99
12.	12.	↑13.	Splunk	Search engine	85.88	+0.39	+15.39
13.	13.	↑14.	MariaDB	Relational, Multi-model	84.95	+0.52	+16.66
14.	14.	↑18.	Hive	Relational	81.80	+0.93	+23.86
15.	15.	↓12.	Teradata	Relational, Multi-model	76.64	-1.18	-0.77
16.	16.	↓15.	Solr	Search engine	59.12	-0.52	-2.78
17.	17.	↑19.	FileMaker	Relational	58.02	+0.12	+1.96
18.	↑20.	↑21.	Amazon DynamoDB	Multi-model	56.57	+0.15	+4.91
19.	↓18.	↓17.	HBase	Wide column	56.54	-1.00	-2.27
20.	↓19.	↓16.	SAP Adaptive Server	Relational	55.86	-0.79	-4.57
21.	21.	↓20.	SAP HANA	Relational, Multi-model	55.43	-0.11	+3.50
22.	22.	22.	Neo4j	Graph	48.39	-0.59	+7.47
23.	23.	23.	Couchbase	Document, Multi-model	33.83	+0.12	+0.88
24.	24.	↑29.	Microsoft Azure Cosmos DB	Multi-model	29.94	+0.85	+10.41

图 3-1

3.2 Elasticsearch 的安装与配置

常言道：工欲善其事，必先利其器。因此在使用 Elasticsearch 之前，我们需要安装 Elasticsearch。下面介绍 Elasticsearch 在 Windows 环境下和在 Linux 环境下的安装方法。由于 Elasticsearch 依赖 Java 环境，因此首先介绍 Java 环境的安装方法。

3.2.1 安装 Java 环境

首先下载并安装 JDK（JAVA Development Kit）。JDK 是整个 Java 开发的核心，它包含了

Java 的运行环境、Java 工具和 Java 基础类库。

当前最新版本是 JDK 12.0。

1．在 Windows 环境下安装

下面以 jdk-12.0.2_windows-x64_bin.zip 为例，展示 JDK 的安装过程。

首先，将 jdk-12.0.2_windows-x64_bin.zip 下载并解压缩后放到某一目录下，如在 C:\Program Files\下新建 Java 文件夹，并将压缩包放到该目录下，即 C:\Program Files\Java 文件夹下。

然后，进入 C:\Program Files\Java\jdk-12.0.2_windows-x64_bin\jdk-12.0.2，即可看到解压缩后的文件目录，如图 3-2 所示。

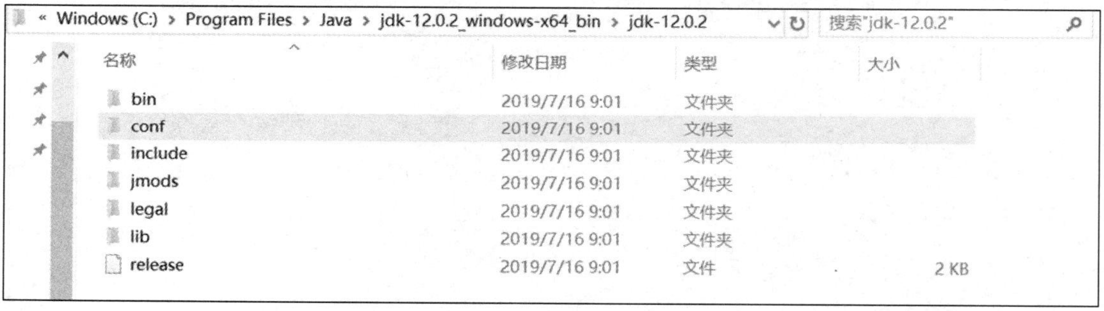

图 3-2

jdk-12.0.2 文件夹下包含 bin、conf、include、jmods、legal、lib 文件夹和 release 文件。

- bin 文件夹下存放的是可执行程序——Java 运行时环境（JRE）的实现。JRE 包括 Java 虚拟机（JVM）、类库和支持 Java 编程语言编写程序执行的其他文件。该目录还包括工具和实用程序，这些工具和实用程序将帮助软件开发人员开发、执行、调试和记录用 Java 编写的程序。
- conf 文件夹下存放的是配置文件，包含可配置选项的文件。读者可以编辑此目录中的文件以更改 JDK 的访问权限，还可以配置安全算法，设置可用于限制 JDK 密码强度的 Java 加密扩展策略文件。
- include 文件夹下存放的是 C 语言编写的头文件——基于 Java 本地接口和 Java 虚拟机（JVM）调试器接口，以支持本地代码编程的 C 语言头文件。
- jmods 文件夹下存放的是编译好的 Java 模块，读者可以基于 J-link 创建自定义运行时的编译模块。

- legal 文件夹下存放的是版权和许可文件，其中包含 JDK 每个模块的许可证和版权文件，以及作为.md 文件的第三方使用须知。
- lib 文件夹下存放的是其他类库，主要是 JDK 所需的附加类库和支持文件，这些文件不供外部使用。
- release 文件中包含的是 JDK 版本相关信息。

随后，开始配置 JDK 所需的环境变量。笔者所用的电脑的环境为 Windows 10，相关环境变量配置过程如下。

首先，打开一个文件夹，在左侧导航窗口选中"此电脑"，单击鼠标右键，在弹出的快捷菜单中单击"属性"选项，弹出"控制面板"窗口。

在"控制面板"窗口中，单击"高级系统设置"→"高级"→"环境变量(N)…"，弹出"环境变量配置"对话框，如图 3-3 所示。

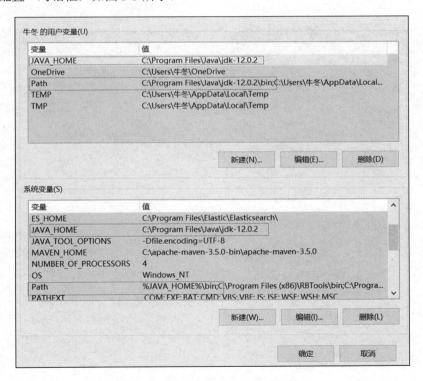

图 3-3

单击"新建(W)…"按钮，即可打开"新建用户变量"对话框，如图 3-4 所示。

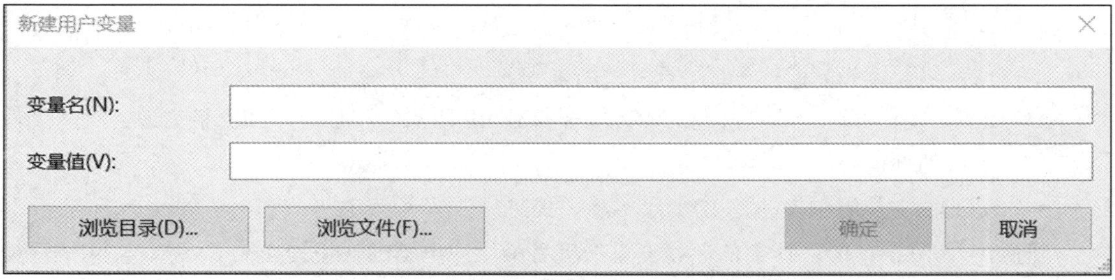

图 3-4

在该对话框中，我们分别配置 JAVA_HOME 和 Path，具体参数详见图 3-3。配置后，打开 Windows 10 自带的 PowerShell 窗口，输入命令"java -version"，并按"Enter"键，即可验证 JDK 是否安装成功。如果安装成功，则执行"java -version"命令，窗口中会显示如图 3-5 所示内容。

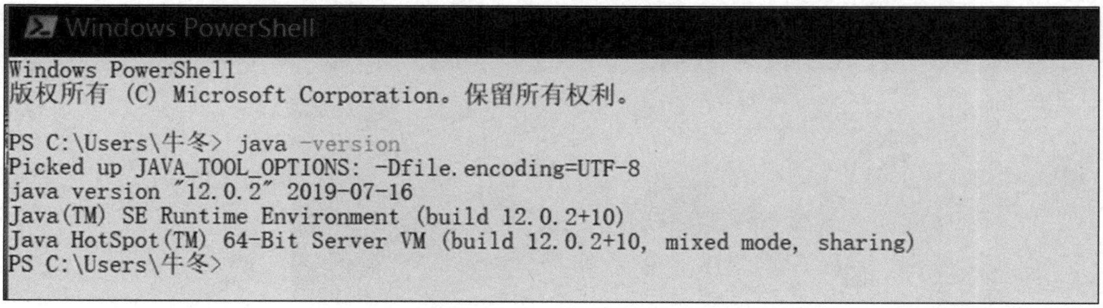

图 3-5

2. 在 Linux 环境下安装

首先将 jdk-12.0.2_linux-x64_bin.tar.gz 下载到本地。在本地打开 SecureCRT 软件，连接 Linux 服务器。切换到 java 目录下之后，用 rz 命令将 jdk-12.0.2_linux-x64_bin.tar.gz 上传到该目录。

随后使用命令 tar -zxvf jdk-12.0.1_linux-x64_bin.tar.gz 将压缩包解压缩到当前目录下。解压缩成功之后，使用 ll 命令查看文件列表，此时压缩包解压缩成了 jdk-12.0.1 文件夹。jdk-12.0.1 文件夹中的文件和 Windows 环境下的相同。

随后配置环境变量，具体方法为使用 vim 命令修改/etc/profile 文件。在 profile 文件中添加如下环境变量：

```
JAVA_HOME=/usr/java/jdk-12.0.1
```

```
CLASSPATH=$JAVA_HOME/lib/
PATH=$PATH:$JAVA_HOME/bin
export PATH JAVA_HOME CLASSPATH
```

保存修改后，需重新加载/etc/profile 配置文件，具体命令为 source /etc/profile。命令执行成功后，检查 JDK 是否安装成功。检查命令的方法与在 Windows 环境下的相同，不再赘述。

3.2.2 Elasticsearch 的安装

Elasticsearch 支持多平台，我们可以在 Elasticsearch 官网找到 Elasticsearch 官方支持的操作系统和 JVM 的矩阵，如图 3-6 所示。

	CentOS/RHEL 6.x/7.x	Oracle Enterprise Linux 6/7 with RHEL Kernel only	Ubuntu 14.04	Ubuntu 16.04	Ubuntu 18.04	SLES 11 SP4**	SLES 12	openSUSE Leap 42	Windows Server 2012/R2	Windows Server 2016
Elasticsearch 5.0.x	✓	✓	✓	✓	✗	✓	✓	✓	✓	✗
Elasticsearch 5.1.x	✓	✓	✓	✓	✗	✓	✓	✓	✓	✗
Elasticsearch 5.2.x	✓	✓	✓	✓	✗	✓	✓	✓	✓	✗
Elasticsearch 5.3.x	✓	✓	✓	✓	✗	✓	✓	✓	✓	✗
Elasticsearch 5.4.x	✓	✓	✓	✓	✗	✓	✓	✓	✓	✗
Elasticsearch 5.5.x	✓	✓	✓	✓	✗	✓	✓	✓	✓	✓
Elasticsearch 5.6.x	✓	✓	✓	✓	✗	✓	✓	✓	✓	✓
Elasticsearch 6.0.x	✓	✓	✓	✓	✗	✗	✓	✓	✓	✓
Elasticsearch 6.1.x	✓	✓	✓	✓	✗	✗	✓	✓	✓	✓
Elasticsearch 6.2.x	✓	✓	✓	✓	✗	✗	✓	✓	✓	✓
Elasticsearch 6.3.x	✓	✓	✓	✓	✗	✗	✓	✓	✓	✓
Elasticsearch 6.4.x	✓	✓	✓	✓	✗	✗	✓	✓	✓	✓
Elasticsearch 6.5.x	✓	✓	✓	✓	✓	✗	✓	✓	✓	✓
Elasticsearch 6.6.x	✓	✓	✓	✓	✓	✗	✓	✓	✓	✓
Elasticsearch 6.7.x	✓	✓	✓	✓	✓	✗	✓	✓	✓	✓
Elasticsearch 6.8.x	✓	✓	✓	✓	✓	✗	✓	✓	✓	✓
Elasticsearch 7.0.x	✓	✓	✗	✓	✓	✗	✓	✓	✓	✓
Elasticsearch 7.1.x	✓	✓	✗	✓	✗	✓	✓	✓	✓	✓

图 3-6

软件开发人员既可以选择自行安装 Elasticsearch 来构建搜索服务，也可以选择使用云端托管的 Elasticsearch 服务。Elasticsearch 服务可以在 AWS 和 GCP 上使用。

下面介绍如何安装 Elasticsearch，本书主要介绍在 Windows 系统和常用的 Linux 系统下的安装方法。

1．在 Windows 系统下安装 Elasticsearch

在 Windows 系统中，我们可以基于 Windows 下的 zip 安装包来构建 Elasticsearch 服务。该 zip 安装包附带了一个 elasticsearch-service.bat 命令文件，执行该命令文件，即可将 Elasticsearch 作为服务运行。

下载 Elasticsearch V7.2.0 的 .zip 安装包，

解压缩后，在同级目录下将创建一个名为 elasticsearch-7.2.0 的文件夹，我们称之为%ES_HOME%，如图 3-7 所示。

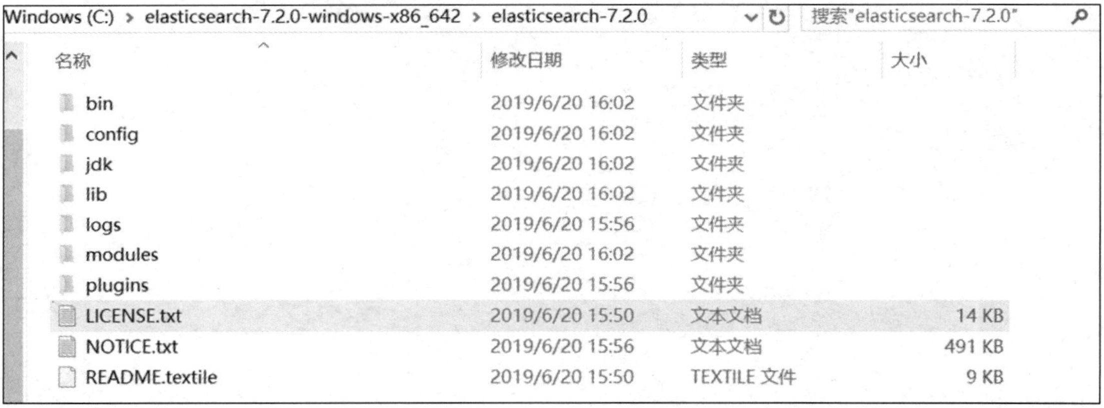

图 3-7

在 elasticsearch-7.2.0 文件夹中有 bin、config、jdk、lib、logs、modules、plugins 和文件夹。

- bin 文件夹下存放的是二进制脚本，包括启动 Elasticsearch 节点和安装的 Elasticsearch 插件。
- config 文件夹下存放的是包含 elasticsearch.yml 在内的配置文件。
- jdk 文件夹下存放的是 Java 运行环境。
- lib 文件夹下存放的是 Elasticsearch 自身所需的 jar 文件。
- logs 文件夹下存放的是日志文件。
- modules 文件夹下存放的是 Elasticsearch 的各个模块。

- plugins 文件夹下存放的是配置插件，每个插件都包含在一个子目录中。

启动 Elasticsearch 服务。

首先切换到终端窗口，如 PowerShell 窗口，在命令行窗口下执行 cd 命令 cd c:\elasticsearch-7.2.0，以便切换到%ES_HOME%目录。

然后从命令行启动 Elasticsearch，启动命令如下所示。

```
PS C:\elasticsearch-7.2.0-windows-x86_642\elasticsearch-7.2.0> .\bin\elasticsearch.bat
```

命令执行后，我们可以在窗口中看到 Elasticsearch 的启动过程，如下所示。当看到节点 started 的输出后，说明 Elasticsearch 服务已经启动。

```
[2019-07-29T19:16:12,579][INFO ][o.e.c.m.MetaDataIndexTemplateService] [LAPTOP-1S8BALK3] adding template [.watch-history-9] for index patterns [.watcher-history-9*]
[2019-07-29T19:16:12,646][INFO ][o.e.c.m.MetaDataIndexTemplateService] [LAPTOP-1S8BALK3] adding template [.triggered_watches] for index patterns [.triggered_watches*]
[2019-07-29T19:16:12,720][INFO ][o.e.c.m.MetaDataIndexTemplateService] [LAPTOP-1S8BALK3] adding template [.monitoring-logstash] for index patterns [.monitoring-logstash-7-*]
[2019-07-29T19:16:12,797][INFO ][o.e.c.m.MetaDataIndexTemplateService] [LAPTOP-1S8BALK3] adding template [.monitoring-es-7] for index patterns [.monitoring-es-7-*]
[2019-07-29T19:16:12,875][INFO ][o.e.c.m.MetaDataIndexTemplateService] [LAPTOP-1S8BALK3] adding template [.monitoring-beats] for index patterns [.monitoring-beats-7-*]
[2019-07-29T19:16:12,934][INFO ][o.e.c.m.MetaDataIndexTemplateService] [LAPTOP-1S8BALK3] adding template [.monitoring-alerts-7] for index patterns [.monitoring-alerts-7]
[2019-07-29T19:16:13,003][INFO ][o.e.c.m.MetaDataIndexTemplateService] [LAPTOP-1S8BALK3] adding template [.monitoring-kibana] for index patterns [.monitoring-kibana-7-*]
[2019-07-29T19:16:13,071][INFO ][o.e.x.i.a.TransportPutLifecycleAction] [LAPTOP-1S8BALK3] adding index lifecycle policy [watch-history-ilm-policy]
[2019-07-29T19:16:13,286][INFO ][o.e.l.LicenseService     ] [LAPTOP-1S8BALK3] license [bbcef3ff-378f-4c13-bb43-962a9dff2841] mode [basic] - valid
[2019-07-29T19:16:13,420][INFO ][o.e.h.AbstractHttpServerTransport] [LAPTOP-1S8BALK3] publish_address {127.0.0.1:9200}, bound_addresses {127.0.0.1:9200}, {[::1]:9200}
[2019-07-29T19:16:13,421][INFO ][o.e.n.Node               ] [LAPTOP-1S8BALK3] started
```

我们可以看到名为"LAPTOP_1S8BALK3"的节点（不同的电脑显示不同）已经启动，并且选举它自己作为单个集群中的 Master 主节点。

Elasticsearch 启动后，在默认情况下，Elasticsearch 将在前台运行，并将其日志打印到标准输出（stdout）。可以按 Ctrl+C 组合键停止运行 Elasticsearch。

在 Elasticsearch 运行过程中，如果需要将 Elasticsearch 作为守护进程运行，则需要在命令行上指定命令参数"-d"，并使用"-p"选项将 Elasticsearch 的进程 ID 记录在文件中。此时的启动命令如下：

```
./bin/elasticsearch -d -p pid
```

此时 Elasticsearch 的日志消息可以在$ES_HOME/logs/目录中找到。

在启动 Elasticsearch 的过程中，我们可以通过命令行对 Elasticsearch 进行配置。一般来说，在默认情况下，Elasticsearch 会从$ES_HOME/config/elasticsearch.yml 文件加载其配置内容。我们还可以在命令行上指定配置，此时需要使用"-e"语法。在命令行配置 Elasticsearch 参数

时,启动命令如下:

```
./bin/elasticsearch -d -Ecluster.name=my_cluster -Enode.name=node_1
```

在 Elasticsearch 启动后,我们可以在浏览器的地址栏输入 http://localhost:9200/ 来验证 Elasticsearch 的启动情况。按"Enter"键后,浏览器的页面会显示如下内容:

```
{
  "name" : "LAPTOP-1S8BALK3",
  "cluster_name" : "elasticsearch",
  "cluster_uuid" : "6UkGmpPrSwyXlzPHgSaivg",
  "version" : {
    "number" : "7.2.0",
    "build_flavor" : "default",
    "build_type" : "zip",
    "build_hash" : "508c38a",
    "build_date" : "2019-06-20T15:54:18.811730Z",
    "build_snapshot" : false,
    "lucene_version" : "8.0.0",
    "minimum_wire_compatibility_version" : "6.8.0",
    "minimum_index_compatibility_version" : "6.0.0-beta1"
  },
  "tagline" : "You Know, for Search"
}
```

此外,我们还可以设置 Elasticsearch 是否自动创建 x-pack 索引。x-pack 将尝试在 Elasticsearch 中自动创建多个索引。在默认情况下,Elasticsearch 是允许自动创建索引的,且不需要其他步骤。

如果需要在 Elasticsearch 中禁用自动创建索引,则必须在 Elasticsearch.yml 中配置 action.auto_create_index,以允许 x-pack 创建以下索引:

```
action.auto_create_index: .monitoring*,.watches,.triggered_watches,.watcher-history*,.ml*
```

2. 在 Linux 系统下安装 Elasticsearch

有 tar.gz 文件格式的安装包可以安装在任何 Linux 发行版和 macOS 上。

在 Linux 系统下,获取 Elasticsearch V7.2.0 版本安装包的命令如下所示。

方法 1:

```
wget https://artifacts.elastic.co/downloads/elasticsearch/elasticsearch-7.2.0-linux-x86_64.tar.gz
```

第 3 章 初识 Elasticsearch

方法 2：

```
wget https://artifacts.elastic.co/downloads/elasticsearch/elasticsearch-7.2.0-linux-x86_64.tar.gz.sha512
shasum -a 512 -c elasticsearch-7.2.0-linux-x86_64.tar.gz.sha512
```

其中，方法 2 中的 shasum 命令用于比较已下载的 tar.gz 安装包的 sha 和已发布的校验和，如果 shasum 命令能够执行并返回 OK，则证明所下载的安装包是正确的。

获取安装包后，可以使用如下命令解压缩安装包：

```
tar -xzf elasticsearch-7.2.0-linux-x86_64.tar.gz
```

该命令执行完毕后，会在当前目录显示 elasticsearch-7.2.0 文件夹，随后执行如下命令切换到 elasticsearch-7.2.0 目录下：

```
cd elasticsearch-7.2.0/
```

elasticsearch-7.2.0 目录被称为$ES_HOME，该目录下的文件结构与 Windows 下的目录结构相同。

安装完 elasticsearch-7.2.0 后，在 elasticsearch-7.2.0 目录中，我们可以执行如下命令启动 Elasticsearch：

```
./bin/elasticsearch
```

在默认情况下，Elasticsearch 在前台运行，并将其日志打印到标准输出（stdout）。在 Elasticsearch 运行过程中，如需停止服务，则可以通过按组合键 Ctrl+C 停止服务。

在 Elasticsearch 启动后，需要检查 Elasticsearch 是否能够运行。我们可以通过向本地主机上的端口 9200 发送 HTTP 请求来测试本地 Elasticsearch 节点是否正在运行，发送请求如下：

```
curl http://localhost:9200/
```

我们会看到如下输出结果：

```
{
  "name" : "Cp8oag6",
  "cluster_name" : "elasticsearch",
  "cluster_uuid" : "AT69_T_DTp-1qgIJlatQqA",
  "version" : {
    "number" : "7.2.0",
    "build_flavor" : "default",
    "build_type" : "tar",
    "build_hash" : "f27399d",
    "build_date" : "2016-03-30T09:51:41.449Z",
```

```
    "build_snapshot" : false,
    "lucene_version" : "8.0.0",
    "minimum_wire_compatibility_version" : "1.2.3",
    "minimum_index_compatibility_version" : "1.2.3"
  },
  "tagline" : "You Know, for Search"
}
```

在 Elasticsearch 运行过程中，如果需要将 Elasticsearch 作为守护进程运行，则需要在命令行上指定命令参数"-d"，并使用 "-p"选项将 Elasticsearch 的进程 ID 记录在文件中，启动命令如下：

```
./bin/elasticsearch -d -p pid
```

此时 Elasticsearch 的日志消息可以在$es_home/logs/ 目录中找到。

在关闭 Elasticsearch 时，可以根据 PID 文件中记录的进程 ID 执行 pkill 命令，具体命令如下：

```
pkill -F pid
```

在启动 Elasticsearch 过程中，我们还可以通过命令行对 Elasticsearch 进行配置。一般来说，在默认情况下，Elasticsearch 会从$es_home/config/elasticsearch.yml 文件加载其配置内容。我们还可以在命令行上指定配置，此时需要使用"-e"语法。当命令行配置 Elasticsearch 参数时，启动命令如下：

```
./bin/elasticsearch -d -Ecluster.name=my_cluster -Enode.name=node_1
```

3.2.3　Elasticsearch 的配置

与近年来很多流行的框架和中间件一样，Elasticsearch 的配置同样遵循"约定大于配置"的设计原则。Elasticsearch 具有极好的默认值设置，用户仅需要很少的配置即可使用 Elasticsearch。用户既可以使用群集更新设置 API 在正在运行的群集上更改大多数设置，也可以通过配置文件对 Elasticsearch 进行配置。

一般来说，配置文件应包含特定节点的设置，如 node.name 和 paths 路径等信息，还会包含节点为了能够加入 Elasticsearch 群集而需要做出的设置，如 cluster.name 和 network.host 等。

1．配置文件位置信息

在 Elasticsearch 中有三个配置文件，分别是 elasticsearch.yml、jvm.options 和 log4j2.properties，这些文件位于 config 目录下，如图 3-8 所示。

图 3-8

其中，elasticsearch.yml 用于配置 Elasticsearch，jvm.options 用于配置 Elasticsearch 依赖的 JVM 信息，log4j2.properties 用于配置 Elasticsearch 日志记录中的各个属性。

注：上述文件位于 config 目录下，这是默认位置。默认位置取决于我们安装 Elasticsearch 时是否基于下载的 tar.gz 包或 zip 包，如果是，则配置目录默认位置为 $es_home/config。如果用户想自定义配置目录的位置，则可以通过 es_path_conf 环境变量进行更改，如下所示：

```
ES_PATH_CONF=/path/to/my/config ./bin/elasticsearch
```

或者通过命令行或 shell 概要文件导出 es-path-conf 环境变量进行更改。

2. 配置文件的格式

Elasticsearch 的配置文件格式为 yaml。下面展示一些更改数据和日志目录路径的示例：

```
path:
    data: /var/lib/elasticsearch
    logs: /var/log/elasticsearch
```

除上述层级方式配制外，也可将层级路径参数整合为一条参数路径配置，如下所示：

```
path.data: /var/lib/elasticsearch
path.logs: /var/log/elasticsearch
```

如果需要在配置文件中引用环境变量的值，则可以在配置文件中使用 ${...} 符号。引用的环境变量会替换环境变量原有的值，如下所示：

```
node.name:      ${HOSTNAME}
network.host:   ${ES_NETWORK_HOST}
```

3. 设置 JVM 选项

在 Elasticsearch 中，用户很少需要更改 Java 虚拟机（JVM）选项。一般来说，最可能的更改是设置堆大小。在默认情况下，Elasticsearch 设置 JVM 使用最小堆空间和最大堆空间的大小均为 1GB。

设置 JVM 选项（包括系统属性和 jvm 标志）的首选方法是通过 jvm.options 配置文件设置。此文件的默认位置为 config/jvm.options。

在 Elasticsearch 中，我们通过 xms（最小堆大小）和 xmx（最大堆大小）这两个参数设置 jvm.options 配置文件指定的整个堆大小，一般应将这两个参数设置为相等。

在 jvm.options 配置文件中，包含了以下特殊语法行来分隔 JVM 参数列表。

（1）忽略由空白组成的行。

（2）以"#"开头的行被视为注释并被忽略，如下所示：

```
# this is a comment
```

（3）以"-"开头的行被视为独立于本机 JVM 版本号的 JVM 选项，如下所示：

```
-Xmx2g
```

（4）以数字开头，且后面为"："的行被视为一个 JVM 选项，该选项仅在本机 JVM 的版本号相互匹配时适用，如下所示：

```
8:-Xmx2g
```

（5）以数字开头，且后面为"-"的行被视为一个 JVM 选项，仅当本机 JVM 的版本号大于或等于该数字版本号时才适用，如下所示：

```
8-:-Xmx2g
```

（6）以数字开头，且后面为"-"，再后面为数字的行被视为一个 JVM 选项，仅当本机 JVM 的版本号在这两个数字版本号的范围内时才适用，如下所示：

```
8-9:-Xmx2g
```

（7）所有其他行都被拒绝解析。

此外，用户还可以通过 ES_JAVA_OPTS 环境变量来设置 Java 虚拟机选项，如下所示：

```
export ES_JAVA_OPTS="$ES_JAVA_OPTS -Djava.io.tmpdir=/path/to/temp/dir"
./bin/elasticsearch
```

4. 安全设置

在 Elasticsearch 中，有些设置信息是敏感且需要保密的，此时单纯依赖文件系统权限来保护这些信息是不够的，因此需要配置安全维度的信息。Elasticsearch 提供了一个密钥库和相应的密钥库工具来管理密钥库中的设置。这里的所有命令都适用于 Elasticsearch 用户。

需要指出的是，对密钥库所做的所有修改，都必须在重新启动 Elasticsearch 之后才会生效。此外，在当前 Elasticsearch 密钥库中只提供模糊处理，以后会增加密码保护。

安全设置就像 elasticsearch.yml 配置文件中的常规设置一样，需要在集群中的每个节点上指定。当前，所有安全设置都是特定于节点的设置，每个节点上必须有相同的值。

安全设置的常规操作有创建密钥库、查看密钥库中的设置列表、添加字符串设置、添加文件设置、删除密钥设置和可重新加载的安全设置等，下面一一介绍。

创建密钥库

想要创建 elasticsearch.keystore，需要使用 create 命令，如下所示：

```
bin/elasticsearch-keystore create
```

命令执行后，将创建 2 个文件，文件名分别为 elasticsearch.keystore 和 elasticsearch.yml。

查看密钥库中的设置列表

使用 list 命令可以查看密钥库中的设置列表，如下所示：

```
bin/elasticsearch-keystore list
```

添加字符串设置

如果需要设置敏感的字符串，如云插件的身份验证凭据，则可以使用 add 命令添加，如下所示：

```
bin/elasticsearch-keystore add the.setting.name.to.set
```

命令执行后将提示输入设置值。

用户可以使用--stdin 标志在窗口 stdin 中输出待设置的目标值，如下所示：

```
cat /file/containing/setting/value | bin/elasticsearch-keystore add --stdin the.setting.name.to.set
```

添加文件设置

用户可以使用添加文件命令添加敏感信息文件，如云插件的身份验证密钥文件。配置时需确保将文件路径作为参数包含在设置名称之后，如下所示：

```
bin/elasticsearch-keystore add-file the.setting.name.to.set /path/example-
```

```
file.json
```

删除密钥设置

如果需要从密钥库中删除设置，则使用 remove 命令，如下所示：

```
bin/elasticsearch-keystore remove the.setting.name.to.remove
```

可重新加载的安全设置

就像 elasticsearch.yml 中的设置值一样，对密钥库内容的更改不会自动应用于正在运行的 Elasticsearch 节点，因此需要重新启动节点才能重新读取设置。

对于某些安全设置，我们可以标记为可重新加载，这样设置后，就可以在正在运行的节点上重新读取和应用了。

需要指出的是，所有安全设置的值（不论是否可重新加载），在所有群集节点上必须相同。更改所需的安全设置后，使用 bin/elasticsearch keystore add 命令，调用：

```
POST _nodes/reload_secure_settings
```

该 API 接口将解密并重新读取每个集群节点上的整个密钥库，但只限于可重载的安全设置，对其他设置的更改将在下次重新启动之后生效。

该 API 接口调用返回后，重新加载就完成了，这意味着依赖于这些设置的所有内部数据结构都已更改，一切设置信息看起来好像从一开始就有了新的值。

当更改多个可重新加载的安全设置时，用户需要在每个群集节点上都修改所有设置，然后发出重新加载安全设置调用，而不是在每次修改后就重新加载。

5. 日志记录配置

在 Elasticsearch 中，使用 log4j2 来记录日志。用户可以使用 log4j2.properties 文件配置 log4j2。

Elasticsearch 公开了三个属性信息，分别是$sys:es.logs.base_path、$sys:es.logs.cluster_name 和$sys:es.logs.node_name，用户可以在配置文件中引用这些属性来确定日志文件的位置。

属性$sys:es.logs.base_path 将解析为日志文件目录地址，$sys:es.logs.cluster_name 将解析为群集名称（在默认配置中，用作日志文件名的前缀），$sys:es.logs.node_name_将解析为节点名称（如果显式地设置了节点名称）。

例如，假设用户的日志目录（path.logs）是/var/log/elasticsearch，集群命名为 production，那么$sys:es.logs.base_path_将解析为/var/log/elasticsearch，$sys:es.logs.base_path/sys:file. Separator/$sys:es.logs.cluster_name.log 将解析为/var/log/elasticsearch/production.log。

下面我们结合 log4j2.properties 文件的主要配置信息来介绍各个属性的含义。log4j2.properties 文件的配置信息如下所示：

```
######## Server JSON ############################
appender.rolling.type = RollingFile       编号1
appender.rolling.name = rolling
appender.rolling.fileName = ${sys:es.logs.base_path}${sys:file.separator}${sys:es.logs.cluster_name}_server.json 编号2
appender.rolling.layout.type = ESJsonLayout 编号3
appender.rolling.layout.type_name = server  编号4
appender.rolling.filePattern = ${sys:es.logs.base_path}${sys:file.separator}${sys:es.logs.cluster_name}-%d{yyyy-MM-dd}-%i.json.gz 编号5
appender.rolling.policies.type = Policies
appender.rolling.policies.time.type = TimeBasedTriggeringPolicy 编号6
appender.rolling.policies.time.interval = 1 编号7
appender.rolling.policies.time.modulate = true  编号8
appender.rolling.policies.size.type = SizeBasedTriggeringPolicy 编号9
appender.rolling.policies.size.size = 256MB 编号10
appender.rolling.strategy.type = DefaultRolloverStrategy
appender.rolling.strategy.fileIndex = nomax
appender.rolling.strategy.action.type = Delete 编号11
appender.rolling.strategy.action.basepath = ${sys:es.logs.base_path}
appender.rolling.strategy.action.condition.type = IfFileName 编号12
appender.rolling.strategy.action.condition.glob = ${sys:es.logs.cluster_name}-* 编号13
appender.rolling.strategy.action.condition.nested_condition.type = IfAccumulatedFileSize 编号14
appender.rolling.strategy.action.condition.nested_condition.exceeds = 2GB 编号15
####################################################
```

其中，上述被编号的配置属性含义如下所示。

编号1：配置 RollingFile 的 appender 属性。

编号2：日志信息将输出到 /var/log/elasticsearch/production.json 中。

编号3：使用 JSON 格式输出。

编号4：type_name 是填充 ESJsonLayout 的类型字段的标志，该字段可以让我们在解析不同类型的日志时更加简单。

编号5：将日志滚动输出到/var/log/elasticsearch/production-yyyy-MM-dd-i.json 文件。日志文件会被压缩处理，i 呈递增状态。

编号6：使用基于时间戳的新增日志滚动策略。

编号 7：按天滚动新增日志。

编号 8：在日期时间上对齐标准，而不是按每 24 小时来新增一次滚动日志文件。

编号 9：按日志文件大小的策略来滚动新增日志文件。

编号 10：每生成 256MB 的日志文件，就滚动新增日志一次。

编号 11：每次新增滚动日志时执行删除日志文件动作。

编号 12：仅当文件匹配时才删除日志文件。

编号 13：该配置仅用于删除日志文件。

编号 14：只有当日志目录下积累了较多日志时才删除。

编号 15：压缩日志的条件是日志文件大小达到 2 GB。

在 log4j2.properties 文件中，我们还可以配置日志记录级别。配置日志记录级别有四种方法，每种方法都有适合使用的场景。这四种配置方法分别是通过命令行配置、通过 elasticsearch.yml 文件配置、通过集群配置和通过 log4j2.properties 配置。

（1）通过命令行配置。

```
-e<name of logging hierarchy>=<level> （如 -e logger.org.elasticsearch.
transport=trace）。
```

适用场景：

当在单个节点上临时调试一个问题（如在后启动时或在开发过程中）时，这是最适合的方法。

（2）通过 elasticsearch.yml 文件配置。

所需要的配置属性如下所示：

```
<name of logging hierarchy>: <level>
```

如 logger.org.elasticsearch.transport:trace。

适用场景：

当临时调试一个问题，但没有通过命令行启动 Elasticsearch；或者希望在更持久的基础上调整日志级别时，这是最适合的方法。

（3）通过群集配置。

在集群中设置日志级别的方法如下所示：

```
PUT /_cluster/settings
{
  "transient": {
    "<name of logging hierarchy>": "<level>"
```

    }
}
```

示例如下所示：

```
PUT /_cluster/settings
{
  "transient": {
    "logger.org.elasticsearch.transport": "trace"
  }
}
```

适用场景：

当需要动态调整活动运行的集群上的日志级别时，这是最适合的方法。

（4）通过 log4j2.properties 配置。

在 log4j2.properties 中需要配置的属性如下所示：

```
logger.<unique_identifier>.name = <name of logging hierarchy>
logger.<unique_identifier>.level = <level>
```

示例如下所示：

```
logger.transport.name = org.elasticsearch.transport
logger.transport.level = trace
```

适用场景：

当需要对日志程序进行细粒度的控制时（如将日志程序发送到另一个文件，或者以不同的方式管理日志程序），这是最适合的方法。

deprecation 日志

除常规日志记录外，Elasticsearch 还允许用户启用不推荐操作的日志记录。如果用户需要迁移某些功能，则可以提前确定这部分属性的配置。

在默认情况下，启动警告级别日志后，所有禁用日志均可输出到控制台和日志文件中。具体配置如下所示：

```
logger.deprecation.level = warn
```

该配置生效后，将在日志目录中创建每日滚动 deprecation 日志文件。用户需要定期检查此文件，尤其是准备升级到新的主要版本时。
```

默认日志记录配置已将取消 deprecation 日志的滚动策略设置为在 1GB 后滚动和压缩，并最多保留五个日志文件（四个滚动日志和一个活动日志）。

用户可以在 config/log4j2.properties 文件中通过将取消 deprecation 日志级别设置为 error 来禁用它。

6. JSON 日志格式

为了便于分析 Elasticsearch 的日志，日志默认以 JSON 格式打印。这是由 log4j 布局属性 appender.rolling.layout.type=esjsonlayout 配置的。此布局需要设置一个 type_name 属性，用于在分析时区分日志流，具体配置如下所示：

```
appender.rolling.layout.type = ESJsonLayout
appender.rolling.layout.type_name = server
```

在配置生效后，日志的每一行就是一个 JSON 格式的字符串。

如果使用自定义布局，则需要用其他布局替换 appender.rolling.layout.type 行的配置，示例如下：

```
appender.rolling.type = RollingFile
appender.rolling.name = rolling
appender.rolling.fileName = ${sys:es.logs.base_path}${sys:file.separator}${sys:es.logs.cluster_name}_server.log
appender.rolling.layout.type = PatternLayout
appender.rolling.layout.pattern = [%d{ISO8601}][%-5p][%-25c{1.}][%node_name]%marker %.-10000m%n
appender.rolling.filePattern = ${sys:es.logs.base_path}${sys:file.separator}${sys:es.logs.cluster_name}-%d{yyyy-MM-dd}-%i.log.gz
```

3.3 Elasticsearch 的核心概念

想要学好、用好 Elasticsearch，首先要了解其核心概念、名词和属性。这就好比想要看懂地图，首先要知道地图里常用的标记符号一样。

Elasticsearch 的核心概念有 Node、Cluster、Shards、Replicas、Index、Type、Document、Settings、Mapping 和 Analyzer，其含义分别如下所示。

（1）Node：即节点。节点是组成 Elasticsearch 集群的基本服务单元，集群中的每个运行中的 Elasticsearch 服务器都可称之为节点。

（2）Cluster：即集群。Elasticsearch 的集群是由具有相同 cluster.name（默认值为 elasticsearch）的一个或多个 Elasticsearch 节点组成的，各个节点协同工作，共享数据。同一个集群内节点的名字不能重复，但集群名称一定要相同。

在实际使用 Elasticsearch 集群时，一般需要给集群起一个合适的名字来替代 cluster.name 的默认值。自定义集群名称的好处是，可以防止一个新启动的节点加入相同网络中的另一个同名的集群中。

在 Elasticsearch 集群中，节点的状态有 Green、Yellow 和 Red 三种，分别如下所述。

① Green：绿色，表示节点运行状态为健康状态。所有的主分片和副本分片都可以正常工作，集群 100% 健康。

② Yellow：黄色，表示节点的运行状态为预警状态。所有的主分片都可以正常工作，但至少有一个副本分片是不能正常工作的。此时集群依然可以正常工作，但集群的高可用性在某种程度上被弱化。

③ Red：红色，表示集群无法正常使用。此时，集群中至少有一个分片的主分片及它的全部副本分片都不可正常工作。虽然集群的查询操作还可以进行，但是也只能返回部分数据（其他正常分片的数据可以返回），而分配到这个有问题分片上的写入请求将会报错，最终导致数据丢失。

（3）Shards：即分片。当索引的数据量太大时，受限于单个节点的内存、磁盘处理能力等，节点无法足够快地响应客户端的请求，此时需要将一个索引上的数据进行水平拆分。拆分出来的每个数据部分称之为一个分片。一般来说，每个分片都会放到不同的服务器上。

进行分片操作之后，索引在规模上进行扩大，性能上也随之水涨船高的有了提升。

Elasticsearch 依赖 Lucene，Elasticsearch 中的每个分片其实都是 Lucene 中的一个索引文件，因此每个分片必须有一个主分片和零到多个副本分片。

当软件开发人员在一个设置有多分片的索引中写入数据时，是通过路由来确定具体写入哪个分片中的，因此在创建索引时需要指定分片的数量，并且分片的数量一旦确定就不能更改。

当软件开发人员在查询索引时，需要在索引对应的多个分片上进行查询。Elasticsearch 会把查询发送给每个相关的分片，并汇总各个分片的查询结果。对上层的应用程序而言，分片是透明的，即应用程序并不知道分片的存在。

在 Elasticsearch 中，默认为一个索引创建 5 个主分片，并分别为每个主分片创建一个副本。

（4）Replicas：即备份，也可称之为副本。副本指的是对主分片的备份，这种备份是精确复制模式。每个主分片可以有零个或多个副本，主分片和备份分片都可以对外提供数据查询服务。当构建索引进行写入操作时，首先在主分片上完成数据的索引，然后数据会从主分片分发到备份分片上进行索引。

当主分片不可用时，Elasticsearch 会在备份分片中选举出一个分片作为主分片，从而避免数据丢失。

一方面，备份分片既可以提升 Elasticsearch 系统的高可用性能，又可以提升搜索时的并发性能；另一方面，备份分片也是一把双刃剑，即如果备份分片数量设置得太多，则在写操作时会增加数据同步的负担。

（5）Index：即索引。在 Elasticsearch 中，索引由一个和多个分片组成。在使用索引时，需要通过索引名称在集群内进行唯一标识。

（6）Type：即类别。类别指的是索引内部的逻辑分区，通过 Type 的名字在索引内进行唯一标识。在查询时如果没有该值，则表示需要在整个索引中查询。

（7）Document：即文档。索引中的每一条数据叫作一个文档，与关系数据库的使用方法类似，一条文档数据通过 _id 在 Type 内进行唯一标识。

（8）Settings：Settings 是对集群中索引的定义信息，比如一个索引默认的分片数、副本数等。

（9）Mapping：Mapping 表示中保存了定义索引中字段（Field）的存储类型、分词方式、是否存储等信息，有点类似于关系数据库（如 MySQL）中的表结构信息。

在 Elasticsearch 中，Mapping 是可以动态识别的。如果没有特殊需求，则不需要手动创建 Mapping，因为 Elasticsearch 会根据数据格式自动识别它的类型。当需要对某些字段添加特殊属性时，如定义使用其他分词器、是否分词、是否存储等，就需要手动设置 Mapping 了。一个索引的 Mapping 一旦创建，若已经存储了数据，就不可修改了。

（10）Analyzer：Analyzer 表示的是字段分词方式的定义。一个 Analyzer 通常由一个 Tokenizer 和零到多个 Filter 组成。在 Elasticsearch 中，默认的标准 Analyzer 包含一个标准的 Tokenizer 和三个 Filter，即 Standard Token Filter、Lower Case Token Filter 和 Stop Token Filter。

3.4 Elasticsearch 的架构设计

Elasticsearch 的架构设计图如图 3-11 所示。

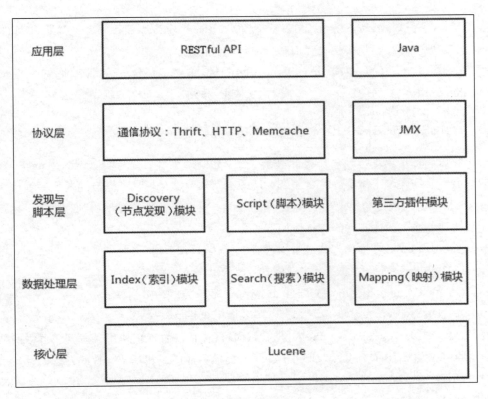

图 3-11

如图 3-11 所示,我们将 Elasticsearch 的架构自底向上分为五层,分别是核心层、数据处理层、发现与脚本层、协议层和应用层。

其中,核心层是指 Lucene 框架——Elasticsearch 是基于 Lucene 框架实现的。

数据处理层主要是指在 Elasticsearch 中对数据的加工处理方式,常见的主要有 Index(索引)模块、Search(搜索)模块和 Mapping(映射)模块。

发现与脚本层主要是 Discovery(节点发现)模块、Script(脚本)模块和第三方插件模块。Discovery 模块是 Elasticsearch 自动发现节点的机制。Script 模块支持脚本的执行,脚本的应用使得我们能很方便的对查询出来的数据进行加工处理,目前 Elasticsearch 支持 JavaScript、Python 等多种语言。第三方插件模块表示 Elasticsearch 支持安装很多第三方的插件,如 elasticsearch-ik 分词插件、elasticsearch-sql 插件等。

协议层是 Elasticsearch 中的数据交互协议。目前 Elasticsearch 支持 Thrift、Memcached 和 HTTP 三种协议,默认的是 HTTP。

应用层指的是 Elasticsearch 的 API 支持模式。Elasticsearch 的特色之一就是 RESTFul 风格的 API，这种 API 接口风格也是当前十分流行的风格之一。

另外，图 3-11 中的 JMX 指的是在 Elasticsearch 中对 Java 的管理框架，用来管理 Elasticsearch 应用。

3.4.1　Elasticsearch 的节点自动发现机制

在 Elasticsearch 内部，通过在集群中配置一个相同的集群名称（即 cluster.name），就能将不同的节点连接到同一个集群。这是怎么实现的呢？本节就来揭晓节点自动发现机制。

Elasticsearch 内嵌自动发现功能，主要提供了 4 种可供选择的发现机制。其中一种是默认实现，其他都是通过插件实现的，具体如下所示。

（1）Azure discovery 插件方式：多播模式。

（2）EC2 discovery 插件方式：多播模式。

（3）Google Compute Engine（GCE）discovery 插件方式：多播模式。

（4）Zen Discovery，默认实现方式，支持多播模式和单播模式。

Zen Discovery 是 Elasticsearch 内置的默认发现模块。发现模块用于发现集群中的节点及选举主节点（又称 master 节点）。Zen Discovery 提供单播模式和基于文件的发现，并且可以扩展为通过插件支持其他形式的发现机制。

在配置前，我们需要了解多播模式和单播模式的配置参数。主要配置参数如下所示：

```
discovery.zen.ping.multicast.enabled: true
discovery.zen.fd.ping_timeout: 100s
discovery.zen.ping.timeout: 100s
discovery.zen.minimum_master_nodes: 2
discovery.zen.ping.unicast.hosts: ["172.31.X.Y"]
discovery.zen.ping.multicast.enabled
```

- discovery.zen.ping.multicast.enabled 表示关闭多播模式的自动发现机制，主要是为了防止其他机器上的节点自动连入。
- discovery.zen.fd.ping_timeout 和 discovery.zen.ping.timeout 表示设置了节点与节点之间连接 ping 命令执行的超时时长。
- discovery.zen.minimum_master_nodes 表示集群中选举主节点时至少需要有多少个节点参与。
- discovery.zen.ping.unicast.hosts 表示在单播模式下，节点应该自动发现哪些节点列表。
  action.auto_create_index: false 表示关闭自动创建索引。

1. 单播模式

Elasticsearch 支持多播模式和单播模式自动两种节点发现机制，不过多播模式已经不被大多数操作系统所支持，加之其安全性不高，所以一般我们会主动关闭多播模式。关闭多播模式的配置如下所示：

```
discovery.zen.ping.multicast.enabled: false  #关闭多播
```

在 Elasticsearch 中，发现机制默认被配置为使用单播模式，以防止节点无意中加入集群。Elasticsearch 支持同一个主机启动多个节点，因此只有在同一台机器上运行的节点才会自动组成集群。当集群的节点运行在不同的机器上时，在单播模式下，我们需要为 Elasticsearch 配置一些它应该去尝试连接的节点列表，配置方式如下所示：

```
#填写集群中的IP地址列表
discovery.zen.ping.unicast.hosts: ["192.168.X1.Y1:9300","192.168.X2.Y2:9300"]
```

因此，单播模式下的配置信息汇总如下：

```
discovery.zen.ping.multicast.enabled: false
discovery.zen.fd.ping_timeout: 100s
discovery.zen.ping.timeout: 100s
discovery.zen.minimum_master_nodes: 2
discovery.zen.ping.unicast.hosts: ["192.168.X1.Y1:9300","192.168.X2.Y2:9300"]
```

配置后，集群构建及主节点选举过程如下：

节点启动后先执行 ping 命令（这里提及的 ping 命令不是 Linux 环境用的 ping 命令，而是 Elasticsearch 的一个 RPC 命令），如果 discovery.zen.ping.unicast.hosts 有设置，则 ping 设置中的 host；否则尝试 ping localhost 的几个端口。

ping 命令的返回结果会包含该节点的基本信息及该节点认为的主节点。

在选举开始时，主节点先从各节点认为的 master 中选。选举规则比较简单，即按照 ID 的字典序排序，取第一个。

如果各节点都没有认为的 master，则从所有节点中选择，规则同上。

需要注意的是，这里有个集群中节点梳理最小值限制条件，即 discovery.zen.minimum_master_nodes。如果节点数达不到最小值的限制，则循环上述过程，直到节点数超过最小限制值，才可以开始选举。

最后选举出一个主节点，如果只有一个本地节点，则主节点就是它自己。

如果当前节点是主节点，则开始等待节点数达到 minimum_master_nodes，再提供服务。如果当前节点不是主节点，则尝试加入主节点所在集群。

2. 多播模式

在多播模式下，我们仅需在每个节点配置好集群名称和节点名称即可。互相通信的节点会根据 Elasticsearch 自定义的服务发现协议，按照多播的方式寻找网络上配置在同样集群内的节点。

3.4.2 节点类型

在 Elasticsearch 中，每个节点可以有多个角色，节点既可以是候选主节点，也可以是数据节点。

节点的角色配置在配置文件 /config/elasticsearch.yml 中设置即可，配置参数如下所示。在 Elasticsearch 中，默认都为 true。

```
node.master: true    //是否为候选主节点
node.data: true      //是否为数据节点
```

其中，数据节点负责数据的存储相关的操作，如对数据进行增、删、改、查和聚合等。正因为如此，数据节点往往对服务器的配置要求比较高，特别是对 CPU、内存和 I/O 的需求很大。此外，数据节点梳理通常随着集群的扩大而弹性增加，以便保持 Elasticsearch 服务的高性能和高可用。

候选主节点是被选举为主节点的节点，在集群中，只有候选主节点才有选举权和被选举权，其他节点不参与选举工作。

一旦候选主节点被选举为主节点，则主节点就要负责创建索引、删除索引、追踪集群中节点的状态，以及跟踪哪些节点是群集的一部分，并决定将哪些分片分配给相关的节点等。

3.4.3 分片和路由

在 Elasticsearch 中，若要进行分片和副本配置，则需要尽早配置。因为当在一个多分片的索引中写入数据时，需要通过路由来确定具体写入哪一个分片中，所以在创建索引时需要指定分片的数量，并且分片的数量一旦确定就不能修改。

分片的数量和副本数量都可以通过创建索引时的 Settings 来配置，Elasticsearch 默认为一个索引创建 5 个主分片，并分别为每个分片创建一个副本。配置的参数如下所示：

```
index.number_of_shards: 5
index.number_of_replicas: 1
```

对文档的新建、索引和删除请求等写操作，必须在主分片上面完成之后才能被复制到相关的副本分片。Elasticsearch 为了加快写入的速度，写入过程往往是并发实施的。为了解决在并发写的过程中出现的数据冲突的问题，Elasticsearch 通过乐观锁进行控制，每个文档都有一个 version（版本号），当文档被修改时版本号递增。

那分片如何使用呢？

当我们向 Elasticsearch 写入数据时，Elasticsearch 根据文档标识符 ID 将文档分配到多个分片上。当查询数据时，Elasticsearch 会查询所有的分片并汇总结果。对用户而言，这个过程是透明的，用户并不知道数据到底存在哪个分片上。

为了避免在查询时部分分片查询失败影响结果的准确性，Elasticsearch 引入了路由功能，即数据在写入时，通过路由将数据写入指定分片；在查询时，可以通过相同的路由指明在哪个分片将数据查出来。在默认情况下，索引数据的分片算法如下所示：

```
shard_num = hash(_routing) % num_primary_shards
```

其中，routing 字段的取值默认是 id 字段或者是 parent 字段。routing 字段在 Hash 分片之后再与有分片的数量取模，最终得到这条数据应该被分配在哪一个分片上。

这样做的目的是通过 Hash 分片来保证在每个分片上数据量的均匀分布，避免各个分片的存储负载不均衡。在做数据检索时，Elasticsearch 默认会搜索所有分片上的数据，最后在主节点上汇总各个分片数据并进行排序处理后，返回最终的结果数据。

3.4.4 数据写入过程

数据写入操作是在 Elasticsearch 的内存中执行的，数据会被分配到特定的分片和副本上，但最终数据是需要存储到磁盘上持久化的。

在 Elasticsearch 中，数据的存储路径在配置文件 ../config/elasticsearch.yml 中进行设置，具体设置如下：

```
path.data: /path/to/data    //索引数据
path.logs: /path/to/logs    //日志记录
```

注：建议不要使用默认值，主要是考虑到当 Elasticsearch 升级时数据的安全性问题，防止因升级 Elasticsearch 而导致数据部分甚至全部丢失。

1. 分段存储

索引数据在磁盘上的是以分段形式存储的。

"段"是 Elasticsearch 从 Lucene 中继承的概念。在索引中，索引文件被拆分为多个子文件，其中每个子文件就叫作段,每个段都是一个倒排索引的小单元。

段具有不变性,一旦索引的数据被写入硬盘,就不能再修改。

为什么要引入分段呢?

可以试想一下,如果我们全部的文档集合仅构建在一个很大的倒排索引文件中,且数据量还在不断增加,当进行修改时,我们需要全量更新当前的倒排索引文件。这会使得数据更新时效性很差、且耗费大量资源,显然这不是我们希望看到的。

其实在 Lucene 中,分段的存储模式可以避免在读写操作时使用锁,从而大大提升 Elasticsearch 的读写性能。这有点类似于 CurrentHashMap 中"分段锁"的概念,二者有异曲同工之妙,都是为了减少锁的使用,提高并发。

当分段被写入磁盘后会生成一个提交点,提交点意味着一个用来记录所有段信息的文件已经生成。因此,一个段一旦拥有了提交点,就表示从此该段仅有读的权限,永远失去了写的权限。

当段在内存中时,此时分段拥有只写的权限,数据还会不断写入,而不具备读数据的权限,意味着这部分数据不能被 Elasticsearch 用户检索到。

那么,既然索引文件分段存储并且不可修改,那么新增、更新和删除如何处理呢?

其实新增是比较容易处理的。既然数据是新的,那么只需在当前文档新增一个段即可。

删除数据时,由于分段不可修改的特性,Elasticsearch 不会把文档从旧的段中移除,因而是新增一个 .del 文件,.del 文件中会记录这些被删除文档的段信息。被标记删除的文档仍然可以被查询匹配到,但它会在最终结果被返回前通过.del 文件将其从结果集中移除。

当更新数据时,由于分段不可修改的特性,Elasticsearch 无法通过修改旧的段来反映文档的更新,于是,更新操作变成了两个操作的结合,即先删除、后新增。Elasticsearch 会将旧的文档从 .del 文件中标记删除,然后将文档的新版本索引到一个新的段中。在查询数据时,两个版本的文档都会被一个查询匹配到,但被删除的旧版本文档在结果集返回前就会被移除。

综上所述,段作为不可修改是具有一定优势的,段的优势主要表现在:不需要锁,从而提升 Elasticsearch 的读写性能。

分段不变性的主要缺点是存储空间占用量大——当删除旧数据时,旧数据不会被马上删除,而是在 .del 文件中被标记为删除。而旧数据只能等到段更新时才能被移除,这样就会导致存储空间的浪费。倘若频繁更新数据,则每次更新都是新增新的数据到新分段,并标记旧的分段中的数据,存储空间的浪费会更多。

在删除和更新数据时,存储空间会浪费;在检索数据时,依然有局限——在查询得到的结果集中会包含所有的结果集,因此主节点需要排除被标记删除的旧数据,随之带来的是查询的负担。

2．延迟写策略

在 Elasticsearch 中，索引写入磁盘的过程是异步的。

因此，为了提升写的性能，Elasticsearch 并没有每新增一条数据就增加一个段到磁盘上，而是采用延迟写策略。延迟写策略的执行过程如下。

每当有新的数据写入时，就将其先写入 JVM 的内存中。在内存和磁盘之间是文件系统缓存，文件缓存空间使用的是操作系统的空间。当达到默认的时间或者内存的数据达到一定量时，会触发一次刷新（Refresh）操作。刷新操作将内存中的数据生成到一个新的分段上并缓存到文件缓存系统，稍后再被刷新到磁盘中并生成提交点。

需要指出的是，由于新的数据会继续写入内存，而内存中的数据并不是以段的形式存储的，因此不能提供检索功能。只有当数据经由内存刷新到文件缓存系统，并生成新的段后，新的段才能供搜索使用，而不需要等到被刷新到磁盘才可以搜索。

在 Elasticsearch 中，写入和打开一个新段的过程叫作刷新。在默认情况下，每个分片会每秒自动刷新一次。这就是 Elasticsearch 能做到近实时搜索的原因，因为文档的变化并不是立即对搜索可见的，但会在一秒之内变为可见。

当然，除自动刷新外，软件开发人员也可以手动触发刷新。

我们还可以在创建索引时，在 Settings 中通过配置 refresh_interval 的值，来调整索引的刷新频率。在设置值时需要注意后面带上时间单位，否则默认是毫秒。当 refresh_interval=-1 时，表示关闭索引的自动刷新。

虽然延迟写策略可以减少数据往磁盘上写的次数，提升 Elasticsearch 的整体写入能力，但文件缓存系统的引入同时也带来了数据丢失的风险，如机房断电等。

为此，Elasticsearch 引入事务日志（Translog）机制。事务日志用于记录所有还没有持久化到磁盘的数据。

于是，在添加了事务日志机制后，数据写入索引的流程如下所示。

（1）新文档被索引之后，先被写入内存中。为了防止数据丢失，Elasticsearch 会追加一份数据到事务日志中。

（2）新的文档持续在被写入内存时，同时也会记录到事务日志中。当然，此时的新数据还不能被检索和查询。

（3）当达到默认的刷新时间或内存中的数据达到一定量后，Elasticsearch 会触发一次刷新，将内存中的数据以一个新段形式刷新到文件缓存系统中并清空内存。这时新段虽未被提交到磁盘，但已经可以对外提供文档的检索功能且不被修改。

（4）随着新文档索引不断被写入，当日志数据大小超过某个值（如 512MB），或者超过一定时间（如 30 min）时，Elasticsearch 会触发一次 Flush。

此时，内存中的数据被写入一个新段，同时被写入文件缓存系统，文件缓存系统中的数据通过 Fsync 刷新到磁盘中，生成提交点。而日志文件被删除，创建一个空的新日志。

3. 段合并

在 Elasticsearch 自动刷新流程中，每秒都会创建一个新的段。这自然会导致短时间内段的数量猛增，而当段数量太多时会带来较大的资源消耗，如对文件句柄、内存和 CPU 的消耗。而在内容搜索阶段，由于搜索请求要检查到每个段，然后合并查询结果，因此段越多，搜索速度越慢。

为此，Elasticsearch 引入段合并机制。段合并机制在后台定期进行，从而小的段被合并到大的段，然后这些大的段再被合并到更大的段。

在段合并过程中，Elasticsearch 会将那些旧的已删除文档从文件系统中清除。被删除的文档不会被拷贝到新的大段中，当然，在合并的过程中不会中断索引和搜索。

段合并是自动进行索引和搜索的，在合并进程中，会选择一小部分大小相似的段，在后台将它们合并到更大的段中，这些段既可以是未提交的，也可以是已提交的。

在合并结束后，老的段会被删除，新的段被 Flush 到磁盘，同时写入一个包含新段且排除旧的和较小的段的新提交点。打开新的段之后，可以用来搜索。

由于段合并的计算量较大，对磁盘 I/O 的消耗也较大，因此段合并会影响正常的数据写入速率，因此 Elasticsearch 不会放任自流，让段合并影响搜索性能。Elasticsearch 在默认情况下会对合并流程进行资源限制，这就是搜索服务仍然有足够的资源仍然可以执行的原因。

3.5 知识点关联

1. 乐观锁

Elasticsearch 引入了乐观锁机制来解决并发写过程中数据冲突的问题，其实乐观锁在多个维度均有应用。

在数据库中，我们用乐观锁来控制表结构，减少长事务中数据库加锁的开销，达到数据表"读多写少"场景下的高性能；

在 Java 中，Java 引入了 CAS（Compare And Swap）乐观锁实现机制实现多线程同步的原子指令，如 AtomicInteger。

命名的艺术

本章重点介绍了 Elasticsearch 的核心概念，这些概念的英文命名方法很值得我们学习借鉴，如 Shard 英文原意为碎片，这个词很形象地解释了倒排索引分解的结果，我们通过这个单词就能见名知意。

其实，命名的学问不仅在 Elasticsearch 中用得很巧，在 Java 中也随处可见。如研发人员经常使用的"<>"操作符，英文原意为 Diamond Operator。这个命名很有想象力，"<>"很像一个菱形，而菱形的英文单词是 Diamond，同时 Diamond 还表示钻石。

2．配置文件格式

前面我们介绍了配置文件格式 YML。该文件格式是由 Clark Evans、Ingy döt Net 和 Oren Ben-Kiki 在 2001 年首次发表的。

YAML 是"YAML Ain't a Markup Language"（YAML 不是一种置标语言）的首字母缩写。有意思的是，在开发这种语言时，YAML 的初衷本是"Yet Another Markup Language"（仍是一种置标语言）。后来为了强调 YAML 语言以数据作为中心，而不是以置标语言为重点，因而采用返璞词来重新命名。

配置文件先后经历了 ini 格式、JSON 格式、XML 格式、Properties 格式和 HOCON 格式。其中，HOCON（Human - Optimized Config Object Notation）格式由 Lightbend 公司开发，它被用于 Sponge，以及利用 SpongeAPI 的独立插件以储存重要的数据，HOCON 文件通常以 .conf 作为后缀名。

在配置文件的格式变迁中，我们能看到配置的方式都在追求语法简单、能继承、支持注释等特性。

如表 3-1 所示，Elasticsearch 中的索引（Index）如果对标关系数据库中的数据库（DataBase）的话，则表（Table）与类型（Type）对应——一个数据库下面可以有多张表（Table），就像 1 个索引（Index）下面有多种类型（Type）一样。

表 3-1

Elasticsearch	关系数据库
索引（Indices）	库（Databases）
类型（Types）	表（Tables）
文档（Documents）	行（Rows）
字段（Fields）	列（Columns）
映射（Mapping）	schema
Put/Post/Delete/Update /Get	增删改查

行（ROW）与文档（Document）对应——一个数据库表（Table）下的数据由多行（ROW）组成，就像 1 个类型 Type 由多个文档（Document）组成一样。

列（column）与字段（Field）对应——数据库表（Table）中一行数据由多列（column）组成，就像 1 个文档（Document）由多个字段（Field）组成一样。

关系数据库中的 schema 与 Elasticsearch 中的映射（Mapping）对应——在关系数据库中，schema 定义了表、表中字段、表和字段之间的关系，就像在 Elasticsearch 中，Mapping 定义了索引下 Type 的字段处理规则，即索引的建立、索引的类型、是否保存原始索引 JSON 文档、是否压缩原始 JSON 文档、是否需要分词处理、如何进行分词处理等。

关系数据库中的增（Insert）删（Delete）改（Update）查（Select）操作可与 Elasticsearch 中的增（Put/Post）删（Delete）改（Update）查（GET）一一对应。

其实，Elasticsearch 中的分片与关系数据库中常用的分库分表方法有异曲同工之妙！

3．副本

副本技术是分布式系统中常见的一种数据组织形式，在日常工作中，"副本"技术也十分常见。比如各级领导都需要指定和培养"二责"人选，当自己出差或请假时，"二责"可以组织团队中的工作。又如在团队中，团队成员之间的工作往往需要多人间相互备份，防止某位成员有事或离职时，相关工作不能继续展开。

在分布式系统中，副本是如何由来的，为什么这么有必要性呢？

副本（Replica 或称 Copy）一般指在分布式系统中为数据或服务提供的冗余。这种冗余设计是提高分布式系统容错率、提高可用性的常用手段。

在服务副本方面，一般指的是在不同服务器中部署同一份代码。如 Tomcat/Jetty 集群部署服务，集群中任意一台服务器都是集群中其他服务器的备份或称副本。

在数据副本方面，一般指的是在不同的节点上持久化同一份数据。当某节点中存储的数据丢失时，系统就可以从副本中读到数据了。

可以说，数据副本是分布式系统解决数据丢失或异常的唯一手段，因此副本协议也成为贯穿整个分布式系统的理论核心。

副本的数据一致性

分布式系统通过副本控制协议，让用户通过一定的方式即可读取分布式系统内部各个副本的数据，这些数据在一定的约束条件下是相同的，即副本数据一致性（Consistency）。副本数据一致性是针对分布式系统中各个节点而言的，不是针对某节点的某个副本而言的。

在分布式系统中，一致性分为强一致性（Strong Consistency）、弱一致性（Week Consistency），还有介于二者之间的会话一致性（Session Consistency）和最终一致性（Eventual Consistency）。

其中，强一致性最难实现。强一致性要求任何时刻用户都可以读到最近一次成功更新的副本数据。弱一致性与强一致性正好相反，数据更新后，用户无法在一定时间内读到最新的值，因此在实际中使用很少。

会话一致性指的是在一次会话内，用户一旦读到某个数据的某个版本的更新数据，则在这个会话中就不会再读到比当前版本更老旧的数据。最终一致性指的是集群中各个副本的数据最终能达到完全一致的状态。

从副本的角度而言，强一致性是最佳的，但对于分布式系统而言，还要考虑其他方面，如分布式系统的整体性能（即系统的吞吐）、系统的可用性、系统的可拓展性等。这也是系统设计要全盘考虑的原因。

副本数据的分布方式

副本的数据是如何分发到位的呢？这就涉及数据的分布方式。

一般来说，数据的分布方式主要有哈希方式、按数据范围分布、按数据量分布和一致性哈希方式（Consistent Hashing）等。

其中哈希方式最为简单，简单是其最大的优势，但缺点同样明显。一方面，可扩展性不高——一旦存储规模需要扩大，则所有数据都需要重新按哈希值分发；另一方面，哈希方式容易导致存储空间的数据分布不均匀。

按数据范围分布也比较常见，一般来说，是将数据按特征值的范围划分为不同的区间，使得集群中不同的服务器处理不同区间的数据。这种方式可以避免哈希值带来的存储空间数据分布不均匀的情况。

按数据量分布和按数据范围分布核心思路比较接近，一般是将数据看作一个顺序增长的，并将数据集按照某一较为固定的大小划分为若干数据块，把不同的数据块分布到不同的服务器上。

而一致性哈希是在工程实践中使用较为广泛的数据分布方式。一致性哈希的基本思路是使用一个哈希函数计算数据的哈希值,而哈希函数的输出值会作为一个封闭的环,我们会根据哈希值将节点随机分布到这个环上,每个节点负责处理从自己开始顺时针至下一个节点的全部哈希值域上的数据。

有了数据分布的方法,那么数据以何种形态进行分布呢?一般来说有两种,一种以服务器为核心,另一种是以数据为核心。

以机器为核心时,机器之间互为副本,副本机器之间的数据完全相同。以机器为核心的策略适用于上述各种数据分布方式,最主要的优点就是简单,容易落地;而缺点也很明显,一旦数据出问题,在数据恢复时就需要恢复多台服务器中的数据,效率很低;而且增加服务器后,会带来可扩展性低的问题。

以数据为核心时,一般将数据拆分为若干个数据段,以数据段为单位去分发。一般来说,每个数据段的大小尽量相等,而且限制数据量大小的上线。在不同的系统中,数据段有很多不同的称谓,如在 Lucene 和 Elasticsearch 中称之为 segment,在 Kafka 中称之为 chunk 和 partition 等。

以数据为核心并不适合所有数据分布方式,一般会采用哈希方式或一致性哈希方式。

将数据拆分为数据段意味着副本的管理将以数据段为单位进行展开,因此副本与机器不再强相关,每台机器都可以负责一定数据段的副本。这带来的好处是当某台服务器中的数据有问题时,我们可以从集群中的任何其他服务器恢复数据,因此数据的恢复效率很高。

副本分发策略

副本分发策略指的是主节点和副本节点之间副本数据同步的方法。一般来说分为两大类:中心化方式和去中心化方式。

中心化方式的基本上线思路是由一个中心节点协调副本数据的更新、维护副本之间的一致性。数据的更新可以是主节点主动向副本节点推送,也可以是副本节点向主节点推送。中心化方式的优点是设计思路较为简单,而缺点也很明显,数据的同步及系统的可用性都有"单点依赖"的风险,即依赖于中心化节点。一旦中心化节点发生异常,则数据同步和系统的可用性都会受到影响。

在去中心化方式中则没有中心节点,所有的节点都是 P2P 形式,地位对等,节点之间通过平等协商达到一致,因此去中心化节点不会因为某个节点的异常而导致系统的可用性受到影响。但有得必有失,去中心化方式的最大的缺点在于各个节点达成共识的过程较长,需要反复进行消息通信来确认内容,实现较为复杂。去中心化方式在区块链中有广泛应用,其共识达成的算法可以参见《区块链底层设计 Java 实战》一书中的"共识算法"部分。

3.6 小结

本章主要介绍了 Elasticsearch 的基本情况及其安装、配置，另外，还介绍了 Elasticsearch 中的核心概念及其架构设计。

在知识点关联部分，先后以乐观锁、命名的艺术、配置文件格式的变迁、副本技术、Elasticsearch 和关系数据库核心概念的类比为核心，进行了关联知识点的介绍。

第 4 章
初级客户端实战

学问勤中得
萤窗万卷书

本章主要介绍在 Elasticsearch 中初级客户端的 API 使用方法。

想要使用 Elasticsearch 服务，则要先获取一个 Elasticsearch 客户端。获取 Elasticsearch 客户端的方法很简单，最常见的就是创建一个可以连接到集群的传输客户端对象。

在 Elasticsearch 中，客户端有初级客户端和高级客户端两种。它们均使用 Elasticsearch 提供了 RESTful 风格的 API，因此，本书中的客户端 API 使用也以 RESTful 风格的 API 为主。在使用 RESTful API 时，一般通过 9200 端口与 Elasticsearch 进行通信。

初级客户端是 Elasticsearch 为用户提供的官方版初级客户端。初级客户端允许通过 HTTP 与 Elasticsearch 集群进行通信，它将请求封装发给 Elasticsearch 集群，将 Elasticsearch 集群的响应封装返回给用户。初级客户端与所有 Elasticsearch 版本都兼容。

高级客户端是用于弹性搜索的高级客户端，它基于初级客户端。高级客户端公开了 API 特定的方法，并负责处理未编组的请求和响应。

4.1 初级客户端初始化

在介绍如何使用初级客户端之前，我们先要了解初级客户端的主要功能，其主要功能包括：

（1）跨所有可用节点的负载平衡。
（2）在节点故障和特定响应代码时的故障转移。
（3）失败连接的惩罚机制。判断一个失败节点是否重试，取决于客户端连接时连续失败的次数；失败的尝试次数越多，客户端再次尝试同一节点之前等待的时间越长。
（4）持久连接。
（5）请求和响应的跟踪日志记录。
（6）自动发现群集节点，该功能可选。

在介绍完初级客户端的主要功能后，下面介绍初级客户端的使用。

1．获取客户端组件

首先，我们需要从 Maven 存储库中获取初级客户端的 jar 包。目前，初级客户端托管在 Maven Central 中央仓库中，其所需的 Java 最低版本为 1.7。

需要指出的是，初级客户端保持了与 Elasticsearch 相同的发布周期。而客户端的版本和客户端可以通信的 Elasticsearch 的版本之间没有关系，即初级客户端可以与所有 Elasticsearch 版本兼容。

从 Maven Central 中央仓库，可以查到当前初级客户端的 Maven 依赖和 Gradle 依赖分别如下。

Maven 依赖：

```
<dependency>
    <groupId>org.elasticsearch.client</groupId>
    <artifactId>elasticsearch-rest-client</artifactId>
    <version>7.2.0</version>
</dependency>
Gradle 依赖：
dependencies {
    compile 'org.elasticsearch.client:elasticsearch-rest-client:7.2.0'
}
```

而 elasticsearch-rest-client 也是有依赖项的，这是因为是初级客户端内部需要使用 Apache HTTP 异步客户端发送 HTTP 请求，因此依赖于以下组件，即异步 HTTP 客户端及其自身的可传递依赖项，具体如下所示：

```
org.apache.httpcomponents:httpasyncclient
org.apache.httpcomponents:httpcore-nio
org.apache.httpcomponents:httpclient
org.apache.httpcomponents:httpcore
```

```
commons-codec:commons-codec
commons-logging:commons-logging
```

此时,工程中的 POM 文件(以 Maven 依赖为例)如下所示:

```xml
<?xml version="1.0" encoding="UTF-8"?>
<project xmlns="http://maven.apache.org/POM/4.0.0" xmlns:xsi="http://www.
            w3.org/2001/XMLSchema-instance"
    xsi:schemaLocation="http://maven.apache.org/POM/4.0.0
                        http://maven.apache.org/xsd/maven-4.0.0.xsd">
    <modelVersion>4.0.0</modelVersion>
    <groupId>com.niudong</groupId>
    <artifactId>esdemo</artifactId>
    <version>0.0.1-SNAPSHOT</version>
    <packaging>jar</packaging>
    <name>esdemo</name>
    <description>ElasticSearch Demo project for Spring Boot</description>
    <parent>
        <groupId>org.springframework.boot</groupId>
        <artifactId>spring-boot-starter-parent</artifactId>
        <version>2.1.6.RELEASE</version>
        <relativePath /> <!-- lookup parent from repository -->
    </parent>
    <properties>
        <project.build.sourceEncoding>UTF-8</project.build.sourceEncoding>
        <project.reporting.outputEncoding>UTF-8</project.reporting.
            outputEncoding>
        <java.version>1.8</java.version>
    </properties>
    <dependencies>
        <dependency>
            <groupId>org.springframework.boot</groupId>
            <artifactId>spring-boot-starter-web</artifactId>
        </dependency>
        <dependency>
            <groupId>org.springframework.boot</groupId>
            <artifactId>spring-boot-starter-test</artifactId>
            <scope>test</scope>
        </dependency>
        <!-- 添加 Elasticsearch 的依赖 -->
        <dependency>
            <groupId>org.elasticsearch.client</groupId>
            <artifactId>elasticsearch-rest-client</artifactId>
            <version>7.2.0</version>
```

```xml
        </dependency>
        <!-- 添加Elasticsearch的依赖 begin-->
        <dependency>
            <groupId>org.apache.httpcomponents</groupId>
            <artifactId>httpasyncclient</artifactId>
            <version>4.1.4</version>
        </dependency>
        <dependency>
            <groupId>org.apache.httpcomponents</groupId>
            <artifactId>httpcore-nio</artifactId>
            <version>4.4.11</version>
        </dependency>
        <dependency>
            <groupId>org.apache.httpcomponents</groupId>
            <artifactId>httpclient</artifactId>
            <version>4.5.9</version>
        </dependency>
        <dependency>
            <groupId>org.apache.httpcomponents</groupId>
            <artifactId>httpcore</artifactId>
            <version>4.4.11</version>
        </dependency>
        <dependency>
            <groupId>commons-codec</groupId>
            <artifactId>commons-codec</artifactId>
            <version>1.12</version>
        </dependency>
        <dependency>
            <groupId>commons-logging</groupId>
            <artifactId>commons-logging</artifactId>
            <version>1.2</version>
        </dependency>
        <!-- 添加Elasticsearch的依赖 end -->
</dependencies>
<build>
    <plugins>
        <plugin>
            <groupId>org.springframework.boot</groupId>
            <artifactId>spring-boot-maven-plugin</artifactId>
        </plugin>
    </plugins>
</build>
</project>
```

2. 客户端初始化

用户可以通过相应的 RestClientBuilder 类来构建 RestClient 实例，该类是通过 RestClient builder（HttpHost...）静态方法创建的。

RestClient 在初始化时，唯一需要的参数是客户端将与之通信的一个或多个主机作为 HttpHost 实例提供的，如下所示：

```
RestClient restClient = RestClient.builder(
    new HttpHost("localhost", 9200, "http"),
    new HttpHost("localhost", 9201, "http")).build();
```

对于 RestClient 类而言，RestClient 类是线程安全的。在理想情况下，它与使用它的应用程序具有相同的生命周期。因此，当不再需要时，应该关闭它，以便释放它使用的所有资源及底层 HTTP 客户机实例及其线程，这一点很重要。关闭的方法如下所示：

```
restClient.close();
```

下面通过一段代码展示初级客户端初始化的使用。代码分为 3 层，分别是 Controller 层、Service 层和 ServiceImpl 实现层。

其中，Controller 层的代码如下所示：

```
package com.niudong.esdemo.controller;
import org.springframework.beans.factory.annotation.Autowired;
import org.springframework.web.bind.annotation.RequestMapping;
import org.springframework.web.bind.annotation.RestController;
import com.niudong.esdemo.service.MeetElasticSearchService;
@RestController
@RequestMapping("/springboot/es")
public class MeetElasticSearchController {
  @Autowired
  private MeetElasticSearchService meetElasticSearchService;

  @RequestMapping("/init")
  public String initElasticSearch() {
    meetElasticSearchService.initEs();
    return "Init ElasticSearch Over!";
  }
}
```

Service 层的代码如下所示：

```
package com.niudong.esdemo.service;
public interface MeetElasticSearchService {
```

```
  public void initEs();
  public void closeEs();
}
```

ServiceImpl 实现层的代码如下所示:

```
package com.niudong.esdemo.service.impl;
import javax.annotation.PostConstruct;
import org.apache.commons.logging.Log;
import org.apache.commons.logging.LogFactory;
import org.apache.http.HttpHost;
import org.elasticsearch.client.RestClient;
import org.springframework.stereotype.Service;
import com.niudong.esdemo.service.MeetElasticSearchService;
@Service
public class MeetElasticSearchServiceImpl implements MeetElasticSearchService {
  private static Log log = LogFactory.getLog (MeetElasticSearchServiceImpl.
      class);
  private RestClient restClient;
  @PostConstruct
  public void initEs() {
    restClient = RestClient
        .builder(new HttpHost("localhost", 9200, "http"), new HttpHost
            ("localhost", 9201, "http"))
        .build();
    log.info("ElasticSearch init in service.");
  }
  public void closeEs() {
    try {
      restClient.close();
    } catch (Exception e) {
      e.printStackTrace();
    }
  }
}
```

在运行代码之前，首先启动 Elasticsearch 服务，启动后，输出内容如下所示。

```
2019-07-30 13:51:57.742  INFO 52088 --- [nio-8080-exec-2] c.n.e.s.i.MeetElasticSearchServiceImpl    : ElasticSearch init in service.
```

然后编译工程，在工程根目录下输入如下命令:

```
mvn clean package
```

接着通过下面的命令启动工程服务:

```
java -jar ./target/esdemo-0.0.1-SNAPSHOT.jar
```

在工程服务启动后,在浏览器中调用如下接口查看 Elasticsearch 客户端的连接情况:

```
http://localhost:8080/springboot/es/init
```

如图 4-1 所示。

图 4-1

可以看到,浏览器页面输出了"Init ElasticSearch Over!",即 Elasticsearch 客户端已经成功连接。

RestClientBuilder 允许在构建 RestClient 时选择性地设置以下配置参数。

(1)请求头配置方法。

```
// 带请求头
public void initEsWithHeader() {
RestClientBuilder builder = RestClient.builder(new HttpHost("localhost",
    9200, "http"));
// 设置每个请求需要发送的默认请求头,以防止在每个请求中指定它们
Header[] defaultHeaders = new Header[] {new BasicHeader("header",
    "value")};
builder.setDefaultHeaders(defaultHeaders);
}
```

(2)配置监听器。

```
// 带失败监听
public void initEsWithFail() {
RestClientBuilder builder = RestClient.builder(new HttpHost("localhost",
    9200, "http"));
// 设置一个监听器,该监听器在每次节点失败时都会收到通知,在启用嗅探时在内部使用
builder.setFailureListener(new RestClient.FailureListener() {
  @Override
  public void onFailure(Node node) {
  }
});
}
```

（3）配置节点选择器。

```java
// 节点选择器
public void initEsWithNodeSelector() {
  RestClientBuilder builder = RestClient.builder(new HttpHost("localhost",
      9200, "http"));
  builder.setNodeSelector(NodeSelector.SKIP_DEDICATED_MASTERS);
}
```

（4）设置超时时间。

```java
// 设置超时时间
public void initEsWithTimeout() {
  RestClientBuilder builder = RestClient.builder(new HttpHost("localhost",
      9200, "http"));
  builder.setRequestConfigCallback(new RestClientBuilder.
      RequestConfigCallback() {
    @Override
    public RequestConfig.Builder customizeRequestConfig(
        RequestConfig.Builder requestConfigBuilder) {
      return requestConfigBuilder.setSocketTimeout(10000);
    }
  });
}
```

4.2 提交请求

在创建好客户端之后，执行请求之前还需构建请求对象。用户可以通过调用客户端的 performRequest 和 performRequestAsync 方法来发送请求。

其中，performRequest 是同步请求方法，它将阻塞调用线程，并在请求成功时返回响应，或者在请求失败时引发异常。

而 performRequestAsync 是异步方法，它接收一个 ResponseListener 对象作为参数。如果请求成功，则该参数使用响应进行调用；如果请求失败，则使用异常进行调用。

1. 构建请求对象 Request

请求对象 Request 的请求方式与 The HTTP 的请求方式相同，如 GET、POST、HEAD 等，代码如下所示：

```
/**
```

```
 * 本部分用于介绍如何构建对Elasticsearch服务的请求
 */
public Request buildRequest() {
  Request request = new Request("GET", // 与The HTTP 的请求方式相,如GET、
                                        // POST、HEAD等
    "/");
  return request;
}
```

2. 请求的执行

在构建请求（Request）后，即可执行请求。请求有同步执行和异步执行两种方式。下面将分别展示如何使用 performRequest 方法和 performRequestAsync 方法发送请求。

同步方式

当以同步方式执行 Request 时，客户端会等待 Elasticsearch 服务器返回的查询结果 Response。在收到 Response 后，客户端继续执行相关的逻辑代码。以同步方式执行的代码如下所示，我们仍然构建三层代码，分别是 Controller 层、Service 层和 ServiceImpl 实现层。

其中，在 Controller 层的 MeetElasticSearchController 类中添加如下代码：

```
@RequestMapping("/buildRequest")
 public String executeRequestForElasticSearch() {
    return meetElasticSearchService.executeRequest();
 }
```

随后，在 Service 层的 MeetElasticSearchService 类中添加如下代码：

```
/**
 * 本部分用于介绍如何构建对Elasticsearch服务的请求
 */
public String executeRequest();
```

在 ServiceImpl 实现层的 MeetElasticSearchServiceImpl 类中添加如下代码：

```
/**
 * 本部分用于介绍如何构建对Elasticsearch服务的请求
 */
public String executeRequest() {
  Request request = new Request("GET",
// 与The HTTP 的请求方式相同,如GET、POST、HEAD等
    "/");
  // 在服务器上请求
  try {
    Response response = restClient.performRequest(request);
```

```
      return response.toString();
    } catch (Exception e) {
      e.printStackTrace();
    }
    try {
      restClient.close();
    } catch (Exception e) {
      e.printStackTrace();
    }

    return "Get result failed!";
  }
```

随后编译工程，在工程根目录下输入如下命令：

```
mvn clean package
```

通过如下命令启动工程服务：

```
java -jar ./target/esdemo-0.0.1-SNAPSHOT.jar
```

在工程服务启动后，在浏览器中通过调用如下接口查看 Elasticsearch 客户端的连接情况：

```
http://localhost:8080/springboot/es/buildRequest
```

在浏览器中可以看到请求发送成功，显示的 Response 信息如下所示。

```
Response{requestLine=GET / HTTP/1.1, host=http://localhost:9200, response=HTTP/1.1 200 OK}
```

异步方式

当以异步方式执行请求时，初级客户端不必同步等待请求结果的返回，可以直接向接口调用方返回异步接口执行成功的结果。

为了处理异步返回的响应信息或处理在请求执行过程中引发的异常信息，用户需要指定监听器。以异步方式调用的代码如下所示。我们在 ServiceImpl 实现层的 MeetElasticSearchServiceImpl 类中添加如下代码：

```
//以异步方式在服务器上执行请求
  public String buildRequestAsync() {
    Request request = new Request("GET",  // 与 The HTTP 的请求方式相同，如 GET、
                                          // POST、HEAD 等
      "/");
    // 在服务器上请求
    restClient.performRequestAsync(request, new ResponseListener() {
      @Override
```

```
        public void onSuccess(Response response) {
        }
        @Override
        public void onFailure(Exception exception) {
        }
    });
    try {
        restClient.close();
    } catch (Exception e) {
        e.printStackTrace();
    }
    return "Get result failed!";
}
```

在异步请求处理后，如果请求执行成功，则调用 ResponseListener 类中的 onResponse 方法进行相关逻辑的处理；如果请求执行失败，则调用 ResponseListener 类中的 onFailure 方法进行相关逻辑的处理。

3．可选参数配置

不论同步请求，还是异步请求，我们都可以在请求中添加参数，添加方法如下所示：

```
request.addParameter("pretty", "true");
```

可以将请求的主体设置为任意 HttpEntity，设置方法如下所示：

```
request.setEntity(new NStringEntity(
        "{\"json\":\"text\"}",
        ContentType.APPLICATION_JSON));
```

还可以将其设置为一个字符串，在 Elasticsearch 中，默认使用 application/json 的内容格式，设置方法如下所示：

```
request.setJsonEntity("{\"json\":\"text\"}");
```

此外，Request 还有一些可选的请求构建选项，通过 RequestOptions 来实现。

需要说明的是，在 RequestOptions 类中保存的请求，可以在同一应用程序的多个请求之间共享。因此用户可以创建一个单实例，然后在所有请求之间共享。具体方法如下所示：

```
private static final RequestOptions COMMON_OPTIONS;
static {
    RequestOptions.Builder builder = RequestOptions.DEFAULT.toBuilder();
    builder.addHeader("Authorization", "Bearer " + TOKEN);
    builder.setHttpAsyncResponseConsumerFactory(
```

```
            new HttpAsyncResponseConsumerFactory
                .HeapBufferedResponseConsumerFactory(30 * 1024 * 1024 * 1024));
    COMMON_OPTIONS = builder.build();
}
```

上述代码中的 TOKEN 表示添加所有请求所需的任何头。

而 AddHeader 用于授权或在 Elasticsearch 前使用代理所需的头信息。在使用时，不需要设置 Content-Type 头，因为客户端将自动在请求的 HttpEntity 中设置 Content-Type 头。

在创建好 RequestOptions 单实例 COMMON_OPTIONS 后，我们就可以在发出请求时使用它了，使用方法如下所示：

```
request.setOptions(COMMON_OPTIONS);
```

Elasticsearch 允许用户根据每个请求定制这些选项，如添加一个额外的标题：

```
RequestOptions.Builder options = COMMON_OPTIONS.toBuilder();
options.addHeader("title","any other things");
request.setOptions(options);
```

RequestOptions 在 ServiceImpl 实现层的 MeetElasticSearchServiceImpl 类中的代码如下所示：

```
//设置全局单实例 RequestOptions
private static final RequestOptions COMMON_OPTIONS;
static {
    RequestOptions.Builder builder = RequestOptions.DEFAULT.toBuilder();
    builder.addHeader("Authorization", "Bearer " + "my-token");
    builder.setHttpAsyncResponseConsumerFactory(
        new HttpAsyncResponseConsumerFactory
            .HeapBufferedResponseConsumerFactory(30 * 1024 * 1024 * 1024));
    COMMON_OPTIONS = builder.build();
}

/**
 * 本部分用于介绍如何构建对Elasticsearch服务的请求
 */
public String buildRequestWithRequestOptions() {
    Request request = new Request("GET",
// 与 The HTTP 的请求方式相同，如 GET、POST、HEAD 等
        "/");
    // 在服务器上请求
    try {
        Response response = restClient.performRequest(request);
```

```
        RequestOptions.Builder options = COMMON_OPTIONS.toBuilder();
        options.addHeader("title", "u r my dear!");
        request.setOptions(COMMON_OPTIONS);

        return response.toString();
    } catch (Exception e) {
        e.printStackTrace();
    }
    try {
        restClient.close();
    } catch (Exception e) {
        e.printStackTrace();
    }
    return "Get result failed!";
}
```

4．多个并行异步操作

除单个操作的执行外，Elasticsearch 的客户端还可以并行执行许多操作。下面通过 ServiceImpl 实现层 MeetElasticSearchServiceImpl 类中的示例，展示如何对多文档进行并行索引。代码如下：

```
//并发处理文档数据
public void multiDocumentProcess(HttpEntity[] documents) {
    final CountDownLatch latch = new CountDownLatch(documents.length);
    for (int i = 0; i < documents.length; i++) {
        Request request = new Request("PUT", "/posts/doc/" + i);
        // 假设 documents 存储在 HttpEntity 数组中
        request.setEntity(documents[i]);
        restClient.performRequestAsync(request, new ResponseListener() {
            @Override
            public void onSuccess(Response response) {
                latch.countDown();
            }
            @Override
            public void onFailure(Exception exception) {
                latch.countDown();
            }
        });
    }
    try {
        latch.await();
    } catch (Exception e) {
        e.printStackTrace();
```

        }
    }

4.3 对请求结果的解析

4.2 节介绍了客户端中请求对象的构建和请求方式，本节介绍对得到的响应结果 Response 的解析。

请求对象有两种请求方式，分别是同步请求和异步请求，因此对于请求的响应结果 Response 的解析也分为两种。

同步请求得到的响应对象是由 performRequest 方法返回的；而异步请求得到的响应对象是通过 ResponseListener 类下 onSuccess(Response)方法中的参数接收的。响应对象中包装 HTTP 客户端返回的响应对象，并公开一些附加信息。

下面通过代码学习对请求结果的解析。以同步请求方式为例，对请求结果的解析代码如下所示。

示例代码共分为三层，分别是 Controller 层、Service 层和 ServiceImpl 实现层。

首先，在 Controller 层的 MeetElasticSearchController 类中添加如下代码：

```
@RequestMapping("/parseEsResponse")
  public String parseElasticSearchResponse() {
    meetElasticSearchService.parseElasticSearchResponse();
    return "Parse ElasticSearch Response Is Over!";
  }
```

然后在 Service 层的 MeetElasticSearchService 类中添加如下代码：

```
  /**
   * 本部分用于介绍如何解析Elasticsearch服务的返回结果
   */
  public void parseElasticSearchResponse();
```

接着在 ServiceImpl 实现层的 MeetElasticSearchServiceImpl 类中添加如下代码：

```
  /**
   * 本部分用于介绍如何解析Elasticsearch服务的返回结果
   */
  public void parseElasticSearchResponse() {
    try {
      Response response = restClient.performRequest(new Request("GET", "/"));
```

```
        //已执行请求的信息
        RequestLine requestLine = response.getRequestLine();
        //Host 返回的信息
        HttpHost host = response.getHost();
        //响应状态行，从中可以解析状态代码
        int statusCode = response.getStatusLine().getStatusCode();
        //响应头，也可以通过 getheader（string）按名称获取
        Header[] headers = response.getHeaders();
        String responseBody = EntityUtils.toString(response.getEntity());

        log.info("parse ElasticSearch Response,responseBody is :" +
            responseBody);
    } catch (Exception e) {
        e.printStackTrace();
    }
}
```

随后编译工程，在工程根目录下输入如下命令：

```
mvn clean package
```

通过如下命令启动工程服务：

```
java -jar ./target/esdemo-0.0.1-SNAPSHOT.jar
```

当工程服务启动后，在浏览器中调用如下接口查看 Elasticsearch 客户端的连接情况：

```
http://localhost:8080/springboot/es/parseEsResponse
```

在服务器控制台中打印 responseBody 的内容，如图 4-2 所示。

```
2019-07-30 20:19:18.213  INFO 22020 --- [nio-8080-exec-1] c.n.e.s.i.MeetElasticSearchServiceImpl   : parse ElasticSearc
Response,responseBody is :{
  "name" : "LAPTOP-1S8BALK3",
  "cluster_name" : "elasticsearch",
  "cluster_uuid" : "6UkGmpPrSwyXlzPHgSaivg",
  "version" : {
    "number" : "7.2.0",
    "build_flavor" : "default",
    "build_type" : "zip",
    "build_hash" : "508c38a",
    "build_date" : "2019-06-20T15:54:18.811730Z",
    "build_snapshot" : false,
    "lucene_version" : "8.0.0",
    "minimum_wire_compatibility_version" : "6.8.0",
    "minimum_index_compatibility_version" : "6.0.0-beta1"
  },
  "tagline" : "You Know, for Search"
```

图 4-2

在浏览器页面输出接口请求成功的内容，如下所示：

```
Parse ElasticSearch Response Is Over!
```

4.4 常见通用设置

除上述客户端 API 外，客户端还支持一些常见通用设置，如超时设置、线程数设置、节点选择器设置和配置嗅探器等。

1. 超时设置

我们可以在构建 RestClient 时提供 requestconfigCallback 的实例来完成超时设置。该接口有一个方法，它接收 org.apache.http.client.config.requestconfig.builder 的实例作为参数，并且具有相同的返回类型。

用户可以修改请求配置生成器 org.apache.http.client.config.requestconfig.builder 的实例，然后返回。

在下面的示例中，增加了连接超时（默认为 1s）和套接字超时（默认为 30s），代码如下所示：

```
.setConnectTimeout(5000)
.setSocketTimeout(60000);
```

在具体使用时，详见 MeetElasticSearchServiceImpl 中的代码，部分代码如下所示：

```
// 设置超时时间
public void initEsWithTimeout() {
  RestClientBuilder builder = RestClient.builder(new HttpHost("localhost",
      9200, "http"));
  builder.setRequestConfigCallback(new RestClientBuilder.
      RequestConfigCallback() {
    @Override
    public RequestConfig.Builder customizeRequestConfig(
        RequestConfig.Builder requestConfigBuilder) {
      return requestConfigBuilder.setSocketTimeout(10000).setSocketTimeout
          (60000);
    }
  });
}
```

2. 线程数设置

Apache HTTP 异步客户端默认启动一个调度程序线程，连接管理器使用的多个工作线程。一般线程数与本地检测到的处理器数量相同，线程数主要取决于 Runtime.getRuntime().availableProcessors()返回的结果。

Elasticsearch 允许用户修改线程数，修改代码如下所示，详见 MeetElasticSearchServiceImpl

类:

```java
/**
 * 本部分用于介绍使用Elasticsearch客户端的通用设置
 */
public void setThreadNumber(int number) {
  RestClientBuilder builder = RestClient.builder(new HttpHost("localhost",
      9200))
    .setHttpClientConfigCallback(new HttpClientConfigCallback() {
      @Override
      public HttpAsyncClientBuilder customizeHttpClient(
          HttpAsyncClientBuilder httpClientBuilder) {
        return httpClientBuilder.setDefaultIOReactorConfig(
            IOReactorConfig.custom().setIoThreadCount(number).build());
      }
    });
}
```

3. 节点选择器设置

在默认情况下,客户端会以轮询的方式将每个请求发送到配置的各个节点中。

Elasticsearch 允许用户自由选择需要连接的节点。一般通过初始化客户端来配置节点选择器,以便筛选节点。

该功能在启用嗅探器时很有用,以防止 HTTP 请求只命中专用的主节点。

配置后,对于每个请求,客户端都通过节点选择器来筛选备选节点。

代码如下所示,详见 MeetElasticSearchServiceImpl 类:

```java
//设置节点选择器
public void setNodeSelector() {
  RestClientBuilder builder = RestClient.builder(new HttpHost("localhost",
      9200, "http"));
  builder.setNodeSelector(new NodeSelector() {
    @Override
    public void select(Iterable<Node> nodes) {
      boolean foundOne = false;
      for (Node node : nodes) {
        String rackId = node.getAttributes().get("rack_id").get(0);
        if ("targetId".equals(rackId)) {
          foundOne = true;
          break;
        }
      }
    }
  });
}
```

```
      if (foundOne) {
        Iterator<Node> nodesIt = nodes.iterator();
        while (nodesIt.hasNext()) {
          Node node = nodesIt.next();
          String rackId = node.getAttributes().get("rack_id").get(0);
          if ("targetId".equals(rackId) == false) {
            nodesIt.remove();
          }
        }
      }
    }
  });
}
```

4. 配置嗅探器

嗅探器允许自动发现运行中的 Elasticsearch 集群中的节点，并将其设置为现有的 RestClient 实例。

在默认情况下，嗅探器使用 nodes info API 检索属于集群的节点，并使用 jackson 解析获得的 JSON 响应。

目前，嗅探器与 Elasticsearch 2.X 及更高版本兼容。

在使用嗅探器之前需添加相关的依赖，代码如下所示：

```
// Maven 版本依赖
<!-- 添加 Elasticsearch 的嗅探器 begin -->
    <dependency>
        <groupId>org.elasticsearch.client</groupId>
        <artifactId>elasticsearch-rest-client-sniffer</artifactId>
        <version>7.2.1</version>
    </dependency>
    <!-- 添加 Elasticsearch 的嗅探器 end -->
// Gradle 依赖
dependencies {
    compile 'org.elasticsearch.client:elasticsearch-rest-client-sniffer:7.2.1'
}
```

在创建好 RestClient 实例（如初始化中代码所示）后，就可以将嗅探器与其进行关联了。嗅探器利用 RestClient 提供的定期机制（在默认情况下定期时间为 5min），从集群中获取当前节点的列表，并通过调用 RestClient 类中的 setNodes 方法来更新它们。

嗅探器的使用代码详见 ServiceImpl 实现层的 MeetElasticSearchServiceImpl 类，部分代码

如下所示：

```
// 配置嗅探器
public void setSniffer() {
  RestClient restClient = RestClient.builder(new HttpHost("localhost", 9200,
      "http")).build();
  Sniffer sniffer = Sniffer.builder(restClient)
      // 嗅探器默认每5min更新一次节点。可以通过setSniffIntervalMillis（以毫秒为单
      // 位）自定义此间隔
      .setSniffIntervalMillis(60000).build();
  // 使用后，结束客户端和嗅探器
  try {
    sniffer.close();
    restClient.close();
  } catch (Exception e) {
    e.printStackTrace();
  }
}
```

当然，除在客户端启动时配置嗅探器外，还可以在失败时启用嗅探器。这意味着在每次失败后，节点列表都会立即更新，而不是在接下来的普通嗅探循环中更新。

在这种情况下，首先需要创建一个 SniffOnFailureListener，然后在创建 RestClient 时配置。在创建嗅探器后，同一个 SniffOnFailureListener 实例会相互关联，以便在每次失败时都通知该实例，并使用嗅探器执行嗅探动作。

嗅探器 SniffOnFailureListener 的使用代码详见 ServiceImpl 实现层的 MeetElasticSearchServiceImpl 类，部分代码如下所示：

```
// 配置嗅探器
public void setSnifferWhenFail(int failTime) {
  SniffOnFailureListener sniffOnFailureListener = new SniffOnFailure
      Listener();
  RestClient restClient = RestClient.builder(new HttpHost("localhost",
      9200))
      .setFailureListener(sniffOnFailureListener).build();
  Sniffer sniffer = Sniffer.builder(restClient).
      setSniffAfterFailureDelayMillis(failTime).build();
  sniffOnFailureListener.setSniffer(sniffer);
  // 使用后，结束客户端和嗅探器
  try {
    sniffer.close();
    restClient.close();
  } catch (Exception e) {
```

```
      e.printStackTrace();
    }
  }
```

由于 Elasticsearch 节点信息 API 不会返回连接到节点时要使用的协议，而是只返回它们的 host:port，因此在默认情况下会使用 HTTP。如果需要使用 HTTPS，则必须手动创建并提供 ElasticSearchNodesNiffer 实例，相关代码如下所示：

```
// 配置基于 HTTPS 的嗅探器
  public void setSnifferWithHTTPS(int failTime) {
    RestClient restClient = RestClient.builder(new HttpHost("localhost", 9200,
        "http")).build();
    NodesSniffer nodesSniffer = new ElasticsearchNodesSniffer(restClient,
        ElasticsearchNodesSniffer.DEFAULT_SNIFF_REQUEST_TIMEOUT,
        ElasticsearchNodesSniffer.Scheme.HTTPS);
    Sniffer sniffer = Sniffer.builder(restClient).setNodesSniffer (nodesSniffer).
        build();
    // 使用后，结束客户端和嗅探器
    try {
      sniffer.close();
      restClient.close();
    } catch (Exception e) {
      e.printStackTrace();
    }
  }
```

4.5 高级客户端初始化

目前，官方计划在 Elasticsearch 7.0 版本中关闭 TransportClient，并且在 8.0 版本中完全删除 TransportClient。作为替代品，我们应该使用高级客户端。高级客户端可以执行 HTTP 请求，而不是序列化 Java 请求。

高级客户端基于初级客户端来实现。

高级客户端的主要目标是公开特定的 API 方法，这些 API 方法将接收请求作为参数并返回响应结果，以便由客户端本身处理请求和响应结果。

与初级客户端一样，高级客户端也有同步、异步两种 API 调用方式。其中，以同步调用方式调用后直接返回响应对象，而异步调用方式则需要配置一个监听器参数才能调用，该参数在收到响应或错误后会得到相应的结果通知。

高级客户端需要 Java 1.8 及其以上的 JDK 环境，而且依赖于 Elasticsearch core 项目，它接收与 TransportClient 相同的请求参数，并返回相同的响应对象。这其实也是为了提高兼容性而做的设计。

高级客户端的版本与 Elasticsearch 版本同步。高级客户端能够与运行着相同主版本和更高版本上的任何 Elasticsearch 节点进行有效通信。高级客户端无须与它通信的 Elasticsearch 节点处于同一个小版本，这是因为向前兼容设计的缘故。这也意味着高级客户端支持与 Elasticsearch 的较新版本进行有效通信。

举例来说，6.0 版本的客户端可以与任何 6.X 版本的 Elasticsearch 节点进行通信，而 6.1 版本的客户端可以确保与 6.1 版本、6.2 版本和任何更高版本的 6.X 节点进行通信，但在与低于 6.0 版本的 Elasticsearch 节点进行通信时可能会出现不兼容问题。

下面介绍高级客户端的使用，首先介绍高级客户端的初始化方法。

在构建高级客户端之前，我们需要在工程中引入必要的依赖配置项。在 Maven 打包和管理依赖时，在 POM 文件中新增的依赖内容如下所示：

```xml
<!-- 添加Elasticsearch的高级客户端 begin -->
<dependency>
    <groupId>org.elasticsearch.client</groupId>
    <artifactId>elasticsearch-rest-high-level-client</artifactId>
    <version>7.2.1</version>
</dependency>
<!-- 添加Elasticsearch的高级客户端 end -->
```

如果基于 Gradle 来管理依赖，则在 Gradle 的配置文件 build.gradle 中添加如下内容：

```
dependencies {
    compile 'org.elasticsearch.client:elasticsearch-rest-high-level-client:7.2.1'
}
```

jar 包 elasticsearch-rest-high-level-client 又依赖以下 jar 包，具体如下所示：

```
org.elasticsearch.client:elasticsearch-rest-client
org.elasticsearch:elasticsearch
```

因此，在 POM 文件中需补充上述二者的依赖。由于在初级客户端的依赖构建过程中已经添加了 elasticsearch-rest-client，所以仅需添加对 Elasticsearch 的依赖即可。此时在 POM 文件中新增内容如下所示：

```xml
<!-- 添加Elasticsearch的高级客户端的依赖 begin -->
<dependency>
    <groupId>org.elasticsearch</groupId>
```

```xml
        <artifactId>elasticsearch</artifactId>
        <version>7.2.1</version>
    </dependency>
    <!-- 添加 Elasticsearch 的高级客户端的依赖 end -->
```

当依赖添加完毕后，即可进入高级客户端的初始化。高级客户端的初始化是基于 RestHighLevelClient 实现的，而 RestHighLevelClient 是基于初级客户端生成器构建的。

高级客户端在内部创建执行请求的初级客户端，该初级客户端会维护一个连接池并启动一些线程，因此当对高级客户端的接口调用完成时，应该关闭它，因为它将同步关闭内部初级客户端，以释放这些资源。

在 Service 层中添加 MeetHighElasticSearchService 类，代码如下所示：

```java
package com.niudong.esdemo.service;
public interface MeetHighElasticSearchService {
  /**
   * 本部分用于介绍如何与 Elasticsearch 构建连接和关闭连接
   */
  public void initEs();
  public void closeEs();
}
```

在 ServiceImpl 实现层中添加 MeetHighElasticSearchServiceImpl 类，代码如下所示：

```java
package com.niudong.esdemo.service.impl;
import javax.annotation.PostConstruct;
import org.apache.commons.logging.Log;
import org.apache.commons.logging.LogFactory;
import org.apache.http.HttpHost;
import org.elasticsearch.client.RestClient;
import org.elasticsearch.client.RestHighLevelClient;
import org.springframework.stereotype.Service;
import com.niudong.esdemo.service.MeetHighElasticSearchService;
@Service
public class MeetHighElasticSearchServiceImpl implements MeetHighElasticSearch
    Service {
  private static Log log = LogFactory.getLog
      (MeetHighElasticSearchServiceImpl.class);
  private RestHighLevelClient restClient;
  /**
   * 本部分用于介绍如何与 ElasticSearch 构建连接和关闭连接
   */
  // 初始化连接
  @PostConstruct
```

```java
public void initEs() {
    restClient = new RestHighLevelClient(RestClient.builder(new HttpHost
        ("localhost", 9200, "http"),
        new HttpHost("localhost", 9201, "http")));

    log.info("ElasticSearch init in service.");
}
// 关闭连接
public void closeEs() {
    try {
        restClient.close();
    } catch (Exception e) {
        e.printStackTrace();
    }
}
```

4.6 创建请求对象模式

在高级客户端中，请求对象 Request、请求结果的解析与初级客户端中的用法相同；与之类似的还有 RequestOptions 的使用及客户端的常见设置，不再赘述。

4.7 知识点关联

初级客户端的同步和异步的思想遍及各类中间件和数据存储平台。如 MySQL 集群、Redis 集群、HBase 集群中各个节点的数据同步方式，甚至区块链中各个节点的数据同步方式，都有同步和异步两种。消息通知机制中也常用到同步和异步两种方式。

那么，什么是同步和异步呢？它们的区别和联系又是什么呢？

和同步、异步相近的一组词是阻塞和非阻塞。

我们可以从线程的维度理解同步和异步、阻塞和非阻塞。

同步和异步，顾名思义是相对的。对同一个线程而言，会有阻塞和非阻塞之分；对不同线程而言，会有同步和异步之分。

换言之，对同一个线程来说，一个时间拍一个快照，这时线程要么处于阻塞状态，要么处于非阻塞状态。因为阻塞状态一般是当前线程同步等待其他线程的处理结果；而非阻塞状态一般是当前线程异步于需配合的其他线程的结果处理过程。

一般来说，同步状态是当前线程发起请求调用后，在被调用的线程处理消息过程中，当前线程必须等被调用的线程处理完才能返回结果；如果被调用的线程还未处理完，则当前线程是不能返回结果的，当前线程必须主动等待所需的结果。

异步状态是当前线程发起请求调用后，被调用的线程直接返回，但是并没有返回给当前线程对应的结果，而是等被调用的线程处理完消息后，通过状态、通知或者回调函数来通知调用线程，当前线程处于被动接收结果的状态。

从上述说明不难发现，同步和异步最主要的区别在于任务或接口调用完成时消息通知的方式。

一般来说，阻塞状态是被调用的线程在返回处理结果前，当前线程会被挂起，不释放 CPU 执行权，当前线程也不能做其他事情，只能等待，直到被调用的线程返回处理结果后，才能接着向下执行。

非阻塞状态是当前线程在没有获取被调用的线程的处理结果前，不是一直等待，而是继续向下执行。如果此时是同步状态，则当前线程可以通过轮询的方式检查被调用线程的处理结果是否返回；如果此时是异步状态，则只有在被调用的线程处理后才会通知当前线程回调。

从上述说明不难发现，阻塞和非阻塞最主要的区别在于当前线程发起任务或接口调用后是否能继续执行。

同步和异步、阻塞和非阻塞还可以两两组合，从而产生 4 种组合，即同步阻塞、同步非阻塞、异步阻塞和异步非阻塞。

前端人员熟悉的 Node.js、运维人员和前后端人员都熟悉的 Nginx 就是异步非阻塞的典型代表。

那么什么是同步阻塞、同步非阻塞和异步阻塞、异步非阻塞呢？如表 4-1 所示。

表 4-1

名 称	说 明
同步阻塞	当前线程在得不到调用结果前不返回，当前线程进入阻塞态等待
同步非阻塞	当前线程在得不到调用结果前不返回，但当前线程不阻塞，一直在 CPU 中运行
异步阻塞	当前线程调用其他线程，当前线程自己并不阻塞，但其他线程会阻塞来等待结果
异步非阻塞	当前线程调用其他线程，其他线程一直在运行，直到得出结果

下面我们通过一个网购的实例来具体解释。

假设现在笔者在电商平台购物，此时会有如下几种选择：

（1）如果笔者付款后，什么事情也不做，仅是坐等快递小哥来送货上门——这是同步阻塞。

（2）如果笔者付款后，什么事情也不做，仅是坐等快递小哥来送货上门；快递小哥到楼下之后会打电话和笔者确认是否在家——这是异步阻塞。

（3）如果笔者付款后，就去锻炼身体了，在休息的间隙，不时地看看App中的送货状态——这就是同步非阻塞。

（4）如果笔者付款后，就去锻炼身体了；在快递小哥到楼下之后会打电话和笔者确认是否在家——这就是异步非阻塞。

4.8 小结

本章主要介绍了在Elasticsearch中初级客户端相关API的使用，例如构建客户端初始化、构建请求对象、解析返回结果及客户端常用的设置属性等。

由于客户端在发出请求时分为同步和异步两种方式，因此在知识点关联部分，详细介绍了同步和异步、阻塞和非阻塞的区别和联系。

第 5 章
高级客户端文档实战一

力学如力耕
勤惰尔自知

在 Elasticsearch 中,高级客户端支持以下文档相关的 API:
(1) Single document APIs——单文档操作 API。
(2) Index API——文档索引 API。
(3) Get API——文档获取 API。
(4) Exists API——文档存在性判断 API。
(5) Delete API——文档删除 API。
(6) Update API——文档更新 API。
(7) Term Vectors API——词向量 API。
(8) Bulk API——批量处理 API。
(9) Multi-Get API——多文档获取 API。
(10) ReIndex API——重新索引 API。
(11) Update By Query API——查询更新 API。
(12) Delete By Query API——查询删除 API。
(13) Multi Term Vectors API——多词条向量 API。
其中,前 7 项较为简单,后 6 项较为复杂,因此我们将用两章介绍这些 API 的使用。

5.1 文档

在 Elasticsearch 中，使用 JavaScript 对象符号（JavaScript Object Notation），作为文档序列化格式，也就是我们常说的 JSON。

JSON 对象由键值对组成，键（key）是字段或属性名称，值（value）可以是多种类型，如字符串、数字、布尔值、一个对象、值数组、日期的字符串、地理位置对象等。

当前，JSON 格式已经被大多数语言所接受并支持，而且已经成为 NoSQL 领域的标准格式。它简洁且容易阅读，常见的 JSON 格式如下所示：

```
{
    "name": "niudong",
    "age": 2,
    "sex": "man",
    "date": "2019-06-01",
    "home": "beijing",
    "company": [
        {
            "company_type": "Electronic Commerce",
            "company_name":  "Alibaba"
        },
        {
            "company_type": "Education",
            "company_name":  "TAL"
        }
    ]
}
```

在 Elasticsearch 中，存储并索引的 JSON 数据被称为文档，文档会以唯一 ID 进行标识并存储于 Elasticsearch 中。文档不仅有自身的键值对数据信息，还包含了元数据，即关于文档的信息。文档一般包含三个必需的元数据信息：

（1）index：文档存储的数据结构，即索引。
（2）type：文档代表的对象类型。
（3）ID：文档的唯一标识。

index 表示的是索引。这有点类似于开发人员常用的关系数据库中的"数据库"的概念，索引表示的是 Elasticsearch 存储和索引关联数据的数据结构。需要指出的是，文档数据最终是被存储和索引在分片中的，索引多个分片存储的逻辑空间。

type 表示的是文档代表的对象类型。在 Java 语言栈中，开发人员对面向对象编程很熟悉。

在面向对象编程时，每个对象都是一个类的实例，这个类定义了对象的多个属性或与对象关联的数据和方法。

就像我们用"数据库"类比"索引_index"，"_type 对象类型"也可以类比"表结构"，这不难理解。在关系数据库中，研发人员经常将相同类的对象存储在同一张表中，因为它们有着相同的结构——在数据库中称之为表结构。

类似的，在 Elasticsearch 中，每个对象类型都有自己的映射结构。所有类型下的文档都被存储在同一个索引下。

在实际使用时，type 的命名是有一定规范的，名字中的字母可以是大写，也可以是小写；但不能包含下画线或逗号。

id 表示的是文档的唯一标识，如 ID，文档标识会以字符串形式出现。当在 Elasticsearch 中创建一个文档时，用户既可以自定义 id，也可以让 Elasticsearch 自动生成。

id、index 和 type 三个元素组合使用时，可以在 Elasticsearch 中唯一标识一个文档。

5.2 文档索引

Elasticsearch 是面向文档的，它可以存储整个文档。

但 Elasticsearch 对文档的操作不仅限于存储，Elasticsearch 还会索引每个文档的内容使之可以被搜索。在 Elasticsearch 中，用户可以对文档数据进行索引、搜索、排序和过滤等操作，而这也是 Elasticsearch 能够执行复杂的全文搜索的原因之一。

下面一一介绍在 Elasticsearch 中对文档操作的各类 API。首先介绍文档索引 API 的使用。

文档索引 API 允许用户将一个类型化的 JSON 文档索引到一个特定的索引中，并且使它可以被搜索到。

在索引 JSON 文档前，首先要构建 JSON 文档。

1．构建 JSON 文档

在 Elasticsearch 中，构建一个 JSON 文档可以用以下几种方式：

（1）手动使用本地 byte[] 或者使用 String 来构建 JSON 文档。

（2）使用 Map，Elasticsearch 会自动把它转换成与其等价的 JSON 文档。

（3）使用如 Jackson 这样的第三方类库来序列化开发人员构建的 Java Bean，以便构建 JSON 文档。

（4）使用内置的帮助类 XContentFactory.jsonBuilder 来构建 JSON 文档。

在 Elasticsearch 内部，每种类型都被转换成 byte[]，即字节数组。因此，用户可以直接使用已经是这种形式的对象；而 JsonBuilder 是高度优化过的 JSON 生成器，它可以直接构建一个 byte[]。

文档通过索引 API 被索引后，意味着文档数据可以被存储和搜索。文档通过其 index、type 和 ID 确定其唯一存储所在。对于 ID，用户可以自己提供一个 ID，或者使用索引 API 自动生成。

下面通过代码展示 JSON 文档的构建方法。

在 ServiceImpl 实现层的 MeetHighElasticSearchServiceImpl 类中添加如下代码：

```java
/**
 * 本部分用于介绍Elasticsearch 索引 API 的使用
 */
// 基于 String 构建 IndexRequest
public void buildIndexRequestWithString(String indexName, String document) {
    // 索引名称
    IndexRequest request = new IndexRequest(indexName);
    // 文档 ID
    request.id(document);
    // String 类型的文档
    String jsonString = "{" + "\"user\":\"niudong\"," + "\"postDate\":\"2019-07-30\","
        + "\"message\":\"Hello Elasticsearch\"" + "}";
    request.source(jsonString, XContentType.JSON);
}
// 基于 Map 构建 IndexRequest
public void buildIndexRequestWithMap(String indexName, String document) {
    Map<String, Object> jsonMap = new HashMap<>();
    jsonMap.put("user", "niudong");
    jsonMap.put("postDate", new Date());
    jsonMap.put("message", "Hello Elasticsearch");
    // 以 Map 形式提供文档源，Elasticsearch 会自动将 Map 形式转换为 JSON 格式
    IndexRequest indexRequest = new IndexRequest(indexName).id(document).
        source(jsonMap);
}
// 基于 XContentBuilder 构建 IndexRequest
public void buildIndexRequestWithXContentBuilder(String indexName, String
    document) {
    try {
        XContentBuilder builder = XContentFactory.jsonBuilder();
        builder.startObject();
```

```
        {
          builder.field("user", "niudong");
          builder.timeField("postDate", new Date());
          builder.field("message", "Hello Elasticsearch");
        }
      builder.endObject();
      // 以 XContentBuilder 对象提供文档源，Elasticsearch 内置的帮助器自动将其生成
      // JSON 格式内容
      IndexRequest indexRequest = new IndexRequest(indexName).id(document).
          source(builder);
    } catch (Exception e) {
      e.printStackTrace();
    }
  }
  // 基于键值对构建 IndexRequest
  public void buildIndexRequestWithKV(String indexName, String document) {
    // 以键值对提供文档源，Elasticsearch 自动将其转换为 JSON 格式
    IndexRequest indexRequest = new IndexRequest(indexName).id(document).
        source("user", "niudong",
      "postDate", new Date(), "message", "Hello Elasticsearch");
  }
```

如上述代码所示，展示了四种构建 JSON 文档的方法，并展示了在 IndexRequest（索引请求）中必选参数的配置方法。

在 IndexRequest 中，除三个必选参数外，还有一些可选参数。可选参数包含路由、超时时间、版本、版本类型和索引管道名称等，代码如下所示：

```
// 配置 IndexRequest 的其他参数
  public void buildIndexRequestWithParam(String indexName, String document) {
    // 将以键值对提供的文档源转换为 JSON 格式
    IndexRequest request = new IndexRequest(indexName).id(document).
        source("user", "niudong",
       "postDate", new Date(), "message", "Hello Elasticsearch");
    request.routing("routing");// 路由值

    // 设置超时时间
    request.timeout(TimeValue.timeValueSeconds(1));
    request.timeout("1s");

    // 设置超时策略
    request.setRefreshPolicy(WriteRequest.RefreshPolicy.WAIT_UNTIL);
    request.setRefreshPolicy("wait_for");
```

```
    // 设置版本
    request.version(2);

    // 设置版本类型
    request.versionType(VersionType.EXTERNAL);

    // 设置操作类型
    request.opType(DocWriteRequest.OpType.CREATE);
    request.opType("create");

    // 在索引文档之前要执行的接收管道的名称
    request.setPipeline("pipeline");
}
```

2. 执行索引文档请求

在 JSON 文档构建后，即可执行索引文档请求。在 Elasticsearch 中，索引文档有两种方式，即同步方式和异步方式。

同步方式

当以同步方式执行 IndexRequest 时，客户端必须等待返回结果 IndexResponse。在收到 IndexResponse 之后，客户端方可继续执行代码。以同步方式执行的代码如下所示：

```
IndexResponse indexResponse = client.index(request, RequestOptions.DEFAULT);
```

当然，在以同步方式执行过程中可能会出现解析 IndexResponse 响应失败、请求超时或没有从服务器返回响应的情况，从而引发 IOException。因此，在程序中需要进行相应的处理。

另外，在 Elasticsearch 服务器返回 4XX 或 5XX 错误代码的情况下，高级客户端会尝试解析 IndexResponse 来响应正文中的错误详细信息，并抛出一个通用的异常 ElasticSearchException。因此，在程序中需要进行相应的处理。

在 ServiceImpl 实现层的 MeetHighElasticSearchServiceImpl 类中添加如下代码：

```
    // 索引文档
public void indexDocuments(String indexName, String document) {
    // 将以键值对提供的文档源转换为 JSON 格式
    IndexRequest indexRequest = new IndexRequest(indexName).id(document).
        source("user", "niudong",
        "postDate", new Date(), "message", "Hello Elasticsearch");
    try {
        restClient.index(indexRequest, RequestOptions.DEFAULT);
    } catch (Exception e) {
        e.printStackTrace();
```

```
    }
  }
```

异步方式

当以异步方式执行 IndexRequest 时,高级客户端不必同步等待请求结果的返回,可以直接向接口调用方返回异步接口执行成功的结果。

为了处理异步返回的响应信息或处理在请求执行过程中引发的异常信息,用户需要指定监听器。以异步方式调用的代码如下所示:

```
client.indexAsync(request, RequestOptions.DEFAULT, listener);
```

其中,request 是要执行的 IndexRequest,listener 是执行完成时使用的 ActionListener。

在异步请求处理后,如果请求执行成功,则调用 ActionListener 类中的 onResponse 方法进行相关逻辑的处理;如果请求执行失败,则调用 ActionListener 类中的 onFailure 方法进行相关逻辑的处理。

当然,在异步调用过程中可能会出现异常或处理失败的情况,因此用户需要在代码中做相应的处理。

在 ServiceImpl 实现层的 MeetHighElasticSearchServiceImpl 类中添加如下代码:

```
// 以异步方式索引文档
  public void indexDocumentsAsync(String indexName, String document) {
// 将以键值对提供的文档源转换为 JSON 格式
IndexRequest indexRequest = new IndexRequest(indexName).id(document).source
    ("user", "niudong",
      "107ostdate", new Date(),"message", "Hello Elasticsearch");
ActionListener listener = new ActionListener<IndexResponse>() {
    @Override
    public void onResponse(IndexResponse indexResponse) {
    }
    @Override
    public void onFailure(Exception e) {
    }
};
try {
    restClient.indexAsync(indexRequest, RequestOptions.DEFAULT, listener);
} catch (Exception e) {
    e.printStackTrace();
}
  }
```

3. 对响应结果的解析

在文档索引的请求发出后，不论同步方式，还是异步方式，客户端均会收到相应的结果 IndexResponse。用户可以通过解析 IndexResponse 来判断文档索引是成功还是失败。

代码分为三层，分别是 Controller 层、Service 层和 ServiceImpl 实现层。其中，Controller 层中 MeetHighElasticSearchController 类的代码如下所示：

```java
package com.niudong.esdemo.controller;
import org.elasticsearch.common.Strings;
import org.springframework.beans.factory.annotation.Autowired;
import org.springframework.web.bind.annotation.RequestMapping;
import org.springframework.web.bind.annotation.RestController;
import com.niudong.esdemo.service.MeetHighElasticSearchService;
@RestController
@RequestMapping("/springboot/es/high")
public class MeetHighElasticSearchController {
  @Autowired
  private MeetHighElasticSearchService meetHighElasticSearchService;

  @RequestMapping("/index/put")
  public String putIndexInHighElasticSearch(String indexName, String document) {
    if(Strings.isEmpty(indexName) || Strings.isEmpty(document)) {
      return "Parameters are error!";
    }
    meetHighElasticSearchService.indexDocuments(indexName, document);
    return "Index High ElasticSearch Client Successed!";
  }
}
```

Service 层中 MeetHighElasticSearchService 类的代码如下所示：

```java
package com.niudong.esdemo.service;
public interface MeetHighElasticSearchService {
  /**
   * 本部分用于介绍如何与Elasticsearch构建连接和关闭连接
   */
  public void initEs();
  public void closeEs();
  // 索引文档
  public void indexDocuments(String indexName, String document);
}
```

ServiceImpl 层中 MeetHighElasticSearch Service Impl 类的代码如下所示：

```java
package com.niudong.esdemo.service.impl;
import java.util.Date;
import java.util.HashMap;
import java.util.Map;
import javax.annotation.PostConstruct;
import org.apache.commons.logging.Log;
import org.apache.commons.logging.LogFactory;
import org.apache.http.HttpHost;
import org.elasticsearch.action.ActionListener;
import org.elasticsearch.action.DocWriteRequest;
import org.elasticsearch.action.index.IndexRequest;
import org.elasticsearch.action.index.IndexResponse;
import org.elasticsearch.action.support.WriteRequest;
import org.elasticsearch.action.support.replication.ReplicationResponse;
import org.elasticsearch.client.RequestOptions;
import org.elasticsearch.client.RestClient;
import org.elasticsearch.client.RestHighLevelClient;
import org.elasticsearch.common.unit.TimeValue;
import org.elasticsearch.common.xcontent.XContentBuilder;
import org.elasticsearch.common.xcontent.XContentFactory;
import org.elasticsearch.common.xcontent.XContentType;
import org.elasticsearch.index.VersionType;
import org.springframework.stereotype.Service;
import org.elasticsearch.action.DocWriteResponse;
import com.niudong.esdemo.service.MeetHighElasticSearchService;
@Service
public class MeetHighElasticSearchServiceImpl implements
    MeetHighElasticSearchService {
  private static Log log = LogFactory.getLog
      (MeetHighElasticSearchServiceImpl.class);
  private RestHighLevelClient restClient;
  /**
   * 本部分介绍如何与Elasticsearch构建连接和关闭连接
   */
  // 初始化连接
  @PostConstruct
  public void initEs() {
    restClient = new RestHighLevelClient(RestClient.builder(new HttpHost
        ("localhost", 9200, "http"),
        new HttpHost("localhost", 9201, "http")));
    log.info("ElasticSearch init in service.");
  }
  // 关闭连接
  public void closeEs() {
```

```java
      try {
        restClient.close();
      } catch (Exception e) {
        e.printStackTrace();
      }
    }
    /**
     * 本部分介绍Elasticsearch 索引API 的使用
     */
    // 基于String 构建IndexRequest
    public void buildIndexRequestWithString(String indexName, String document) {
      // 索引名称
      IndexRequest request = new IndexRequest(indexName);
      // 文档ID
      request.id(document);
      // String 类型的文档
      String jsonString = "{" + "\"user\":\"niudong\"," + "\"postDate\":\"2019-07-30\","
          + "\"message\":\"Hello Elasticsearch\"" + "}";
      request.source(jsonString, XContentType.JSON);
    }
    // 基于Map 构建IndexRequest
    public void buildIndexRequestWithMap(String indexName, String document) {
      Map<String, Object> jsonMap = new HashMap<>();
      jsonMap.put("user", "niudong");
      jsonMap.put("postDate", new Date());
      jsonMap.put("message", "Hello Elasticsearch");
      //将以Map 形式提供的文档源转换为JSON 格式
      IndexRequest indexRequest = new IndexRequest(indexName).id(document).
          source(jsonMap);
    }
    // 基于XContentBuilder 构建IndexRequest
    public void buildIndexRequestWithXContentBuilder(String indexName, String
        document) {
      try {
        XContentBuilder builder = XContentFactory.jsonBuilder();
        builder.startObject();
        {
          builder.field("user", "niudong");
          builder.timeField("postDate", new Date());
          builder.field("message", "Hello Elasticsearch");
        }
        builder.endObject();
        // 将XContentBuilder 对象提供的文档源转换为JSON 格式
```

```java
        IndexRequest indexRequest = new IndexRequest(indexName).id(document).
            source(builder);
    } catch (Exception e) {
        e.printStackTrace();
    }
}
// 基于键值对构建 IndexRequest
public void buildIndexRequestWithKV(String indexName, String document) {
    // 将以键值对形式提供的文档源转换为 JSON 格式
    IndexRequest indexRequest = new IndexRequest(indexName).id(document).
        source("user", "niudong",
        "postDate", new Date(), "message", "Hello Elasticsearch");
}
// 构建 IndexRequest 的其他参数配置
public void buildIndexRequestWithParam(String indexName, String document) {
    //将以键值对形式提供的文档源转换为 JSON 格式
    IndexRequest request = new IndexRequest(indexName).id(document).source
        ("user", "niudong",
        "postDate", new Date(), "message", "Hello Elasticsearch");
    request.routing("routing");// 路由值
    // 设置超时时间
    request.timeout(TimeValue.timeValueSeconds(1));
    request.timeout("1s");
    // 设置超时策略
    request.setRefreshPolicy(WriteRequest.RefreshPolicy.WAIT_UNTIL);
    request.setRefreshPolicy("wait_for");
    // 设置版本
    request.version(2);
    // 设置版本类型
    request.versionType(VersionType.EXTERNAL);
    // 设置操作类型
    request.opType(DocWriteRequest.OpType.CREATE);
    request.opType("create");
    // 在索引文档之前要执行的接收管道的名称
    request.setPipeline("pipeline");
}
// 索引文档
public void indexDocuments(String indexName, String document) {
    // 将以键值对形式提供的文档源转换为 JSON 格式
    IndexRequest indexRequest = new IndexRequest(indexName).id(document).
        source("user", "niudong",
        "postDate", new Date(), "message",
        "Hello Elasticsearch!北京时间 8 月 1 日凌晨 2 点,美联储公布 7 月议息会议结果。一
如市场预期,美联储本次降息 25 个基点,将联邦基金利率的目标范围调至 2.00%~2.25%。此次是 2007—
```

2008年间美国为应对金融危机启动降息周期后，美联储十年多以来首次降息。美联储公布利率决议后，美股下跌，美元上涨，人民币汇率下跌");
```java
        try {
            IndexResponse indexResponse = restClient.index(indexRequest, RequestOptions.
                DEFAULT);
            // 解析索引结果
            processIndexResponse(indexResponse);
        } catch (Exception e) {
            e.printStackTrace();
        }
    }
    // 关闭Elastisearch连接
    closeEs();
}
// 解析索引结果
private void processIndexResponse(IndexResponse indexResponse) {
    String index = indexResponse.getIndex();
    String id = indexResponse.getId();
    log.info("index is " + index + ", id is " + id);
    if (indexResponse.getResult() == DocWriteResponse.Result.CREATED) {
        // 文档创建时
        log.info("Document is created!");
    } else if (indexResponse.getResult() == DocWriteResponse.Result.UPDATED) {
        // 文档更新时
        log.info("Document has updated!");
    }
    ReplicationResponse.ShardInfo shardInfo = indexResponse.getShardInfo();
    if (shardInfo.getTotal() != shardInfo.getSuccessful()) {
        // 处理成功 shards 小于总 shards 的情况
        log.info("Successed shards are not enough!");
    }
    if (shardInfo.getFailed() > 0) {
        for (ReplicationResponse.ShardInfo.Failure failure : shardInfo.
            getFailures()) {
            String reason = failure.reason();
            log.info("Fail reason is " + reason);
        }
    }
}
// 异步索引文档
public void indexDocumentsAsync(String indexName, String document) {
    // 将以键值对形式提供的文档源转换为JSON格式
    IndexRequest indexRequest = new IndexRequest(indexName).id(document).
        source("user", "niudong",
```

```
        "postDate", new Date(), "message", "Hello Elasticsearch");
    ActionListener listener = new ActionListener<IndexResponse>() {
      @Override
      public void onResponse(IndexResponse indexResponse) {
      }
      @Override
      public void onFailure(Exception e) {
      }
    };
    try {
      restClient.indexAsync(indexRequest, RequestOptions.DEFAULT, listener);
    } catch (Exception e) {
      e.printStackTrace();
    }
  }
// 关闭Elasticsearch连接
    closeEs();
}
```

随后编译工程,在工程根目录下输入如下命令:

```
mvn clean package
```

通过如下命令启动工程服务:

```
java -jar ./target/esdemo-0.0.1-SNAPSHOT.jar
```

在工程服务启动后,在浏览器中调用如下接口查看文档索引的情况:

```
http://localhost:8080/springboot/es/high/index/put?indexName=ultraman&document=1
```

调用接口后,服务器输出内容如下所示。

```
2019-08-01 09:44:31.451  INFO 63876 --- [nio-8080-exec-7] n.e.s.i.MeetHighElasticSearchServiceImpl : index is ultraman, id is 1
2019-08-01 09:44:31.453  INFO 63876 --- [nio-8080-exec-7] n.e.s.i.MeetHighElasticSearchServiceImpl : Document is created!
```

更换索引的字符串内容后第二次调用,结果如下所示。

```
2019-08-01 09:45:18.094  INFO 40436 --- [nio-8080-exec-1] n.e.s.i.MeetHighElasticSearchServiceImpl : index is ultraman, id is 1
2019-08-01 09:45:18.096  INFO 40436 --- [nio-8080-exec-1] n.e.s.i.MeetHighElasticSearchServiceImpl : Document has updated!
```

两次调用后浏览器页面均返回:

```
Index High ElasticSearch Client Successed!
```

需要指出的是,在两次索引中,虽然索引内容不同,但由于传入的文档 ID 均为 1,因此,从第二次接口调用中可以看到，在索引后，文档 ID 依然是 1，和预期相同。

5.3 文档索引查询

Elasticsearch 提供了文档索引查询 API，即 Get API。

1. 构建文档索引查询请求

在执行文档索引查询请求前，需要构建文档索引查询请求，即 GetRequest。GetRequest 有两个必选参数，即索引名称和文档 ID。构建 GetRequest 的代码如下所示：

```java
//构建 GetRequest
public void buildGetRequest(String indexName, String document) {
    GetRequest getRequest = new GetRequest(indexName, document);
}
```

在构建 GetRequest 的过程中，可以选择其他可选参数进行相应的配置。可选参数有禁用源检索、为特定字段配置源包含关系、为特定字段配置源排除关系、为特定存储字段配置检索、配置路由值、配置偏好值、配置在检索文档之前执行刷新、配置版本号、配置版本类型等，代码如下所示：

```java
// 构建 GetRequest
public void buildGetRequest(String indexName, String document) {
    GetRequest getRequest = new GetRequest(indexName, document);
    // 可选配置参数
    // 禁用源检索，在默认情况下启用
    getRequest.fetchSourceContext(FetchSourceContext.DO_NOT_FETCH_SOURCE);
    // 为特定字段配置源包含
    String[] includes = new String[] {"message", "*Date"};
    String[] excludes = Strings.EMPTY_ARRAY;
    FetchSourceContext fetchSourceContext = new FetchSourceContext(true,
        includes, excludes);
    getRequest.fetchSourceContext(fetchSourceContext);
    // 为特定字段配置源排除
    includes = Strings.EMPTY_ARRAY;
    excludes = new String[] {"message"};
    fetchSourceContext = new FetchSourceContext(true, includes, excludes);
    getRequest.fetchSourceContext(fetchSourceContext);
getRequest.storedFields("message");  // 为特定存储字段配置检索
//要求在映射中单独存储字段
    try {
```

```java
        GetResponse getResponse = restClient.get(getRequest, RequestOptions.
            DEFAULT);
        String message = getResponse.getField("message").getValue();
// 检索消息存储字段（要求该字段单独存储在映射中）
        log.info("message is " + message);
    } catch (Exception e) {
      e.printStackTrace();
    }
    // 路由值
    getRequest.routing("routing");
    // 偏好值
    getRequest.preference("preference");
    // 将实时标志设置为假（默认为真）
    getRequest.realtime(false);
    // 在检索文档之前执行刷新（默认为 false）
    getRequest.refresh(true);
    // 配置版本号
    getRequest.version(2);
    // 配置版本类型
    getRequest.versionType(VersionType.EXTERNAL);
  }
```

2. 执行文档索引查询请求

在 GetRequest 构建后，即可执行文档索引查询请求。与文档索引请求类似，文档索引查询请求也有同步和异步两种执行方式。

同步方式

当以同步方式执行 GetRequest 时，客户端会等待 Elasticsearch 服务器返回的查询结果 GetResponse。在收到 GetResponse 后，客户端会继续执行相关的逻辑代码。以同步方式执行的代码如下所示：

```java
// 以同步方式执行 GetRequest
  public void getIndexDocuments(String indexName, String document) {
    GetRequest getRequest = new GetRequest(indexName, document);
    try {
      GetResponse getResponse = restClient.get(getRequest, RequestOptions.
          DEFAULT);
    } catch (Exception e) {
      e.printStackTrace();
    }
}
// 关闭 Elasticsearch 连接
closeEs();
  }
```

异步方式

当以异步方式执行 GetRequest 时，高级客户端不必同步等待请求结果的返回，可以直接向接口调用方返回异步接口执行成功的结果。为了处理异步返回的响应信息或处理在请求执行过程中引发的异常信息，用户需要指定监听器 ActionListener。

如果请求执行成功，则调用 ActionListener 中的 onResponse 方法进行相应逻辑的处理。如果请求执行失败，则调用 ActionListener 中的 onFailure 方法进行相应逻辑的处理。

当然，在异步请求执行过程中可能会出现异常，异常的处理与同步方式执行情况相同。

以异步方式执行的代码如下所示：

```java
// 以异步方式执行 GetRequest
public void getIndexDocumentsAsync(String indexName, String document) {
    GetRequest getRequest = new GetRequest(indexName, document);
    ActionListener<GetResponse> listener = new ActionListener<GetResponse>() {
        @Override
        public void onResponse(GetResponse getResponse) {
            String id = getResponse.getId();
            String index = getResponse.getIndex();
            log.info("id is " + id + ", index is " + index);
        }
        @Override
        public void onFailure(Exception e) {
        }
    };
    try {
        restClient.getAsync(getRequest, RequestOptions.DEFAULT, listener);
    } catch (Exception e) {
        e.printStackTrace();
    }
}
```

3．返回结果的解析

不论同步方式，还是异步方式，在 GetRequest 执行后都会收到请求响应结果 GetResponse。用户可以通过解析 GetResponse 查看文档索引的结果。

解析 GetResponse 的代码共分为三层，分别是 Controller 层、Service 层和 ServiceImpl 实现层。

在 Controller 层的 MeetHighElasticSearchController 类中添加如下代码：

```java
// 以同步方式执行 GetRequest
```

```java
@RequestMapping("/index/get")
public String getIndexInHighElasticSearch(String indexName, String
    document) {
    if (Strings.isEmpty(indexName) || Strings.isEmpty(document)) {
        return "Parameters are error!";
    }
    meetHighElasticSearchService.getIndexDocuments(indexName, document);
    return "Get Index High ElasticSearch Client Successed!";
}
```

在 Service 层的 MeetHighElasticSearchService 类中添加如下代码：

```java
// 以同步方式执行 GetRequest
public void getIndexDocuments(String indexName, String document);
```

在 ServiceImpl 实现层的 MeetHighElasticSearchServiceImpl 类中添加如下代码：

```java
// 以同步方式执行 GetRequest
public void getIndexDocuments(String indexName, String document) {
    GetRequest getRequest = new GetRequest(indexName, document);
    try {
        GetResponse getResponse = restClient.get(getRequest, RequestOptions.
            DEFAULT);
        // 处理 GetResponse
        processGetResponse(getResponse);
    } catch (Exception e) {
        e.printStackTrace();
    }
}
// 关闭 Elasticsearch 连接
closeEs();
}
// 处理 GetResponse
private void processGetResponse(GetResponse getResponse) {
    String index = getResponse.getIndex();
    String id = getResponse.getId();
    log.info("id is " + id + ", index is " + index);
    if (getResponse.isExists()) {
        long version = getResponse.getVersion();
// 以字符串形式检索文档
        String sourceAsString = getResponse.getSourceAsString();
// 以 Map<String, Object>形式检索文档
        Map<String, Object> sourceAsMap = getResponse.getSourceAsMap();
// 以 byte[]形式检索文档
        byte[] sourceAsBytes = getResponse.getSourceAsBytes();
        log.info("version is " + version + ", sourceAsString is " +
```

```
            sourceAsString);
    } else {
        // 当找不到文档时在此处处理。注意，尽管返回的响应具有404状态码，但返回的是有效的
        // getResponse，而不是引发异常。这样的响应不包含任何源文档，并且其isexists方法返
        // 回false
    }
}
```

随后编译工程，在工程根目录下输入如下命令：

```
mvn clean package
```

通过如下命令启动工程服务：

```
java -jar ./target/esdemo-0.0.1-SNAPSHOT.jar
```

在工程服务启动后，在浏览器中调用如下接口查看文档索引的查询结果：

```
http://localhost:8080/springboot/es/high/index/get?indexName=ultraman&document=1
```

服务器运行结果如图 5-1 所示。

```
2019-08-01 11:06:50.566  INFO 52128 --- [nio-8080-exec-2] n.e.s.i.MeetHighElasticSearchServiceImpl : id is 1, index is u
ltraman
2019-08-01 11:06:50.674  INFO 52128 --- [nio-8080-exec-2] n.e.s.i.MeetHighElasticSearchServiceImpl : version is 2, sourc
eAsString is {"user":"niudong","postDate":"2019-08-01T01:45:17.741Z","message":"Hello Elasticsearch!北京时间8月1日凌晨2
点，美联储公布7月议息会议结果。一如市场预期，美联储本次降息25个基点，将联邦基金利率的目标范围调至2.00%-2.25%。此次是2007
-2008年间美国为应对金融危机启动降息周期后，美联储十年多以来首次降息。美联储公布利率决议后，美股下跌，美元上涨，人民币汇
率下跌"}
```

图 5-1

对比之前的索引内容，可以看到二者的结果是一样的。

而浏览器端则显示如下内容，表示接口调用成功：

```
Get Index High ElasticSearch Client Successed!
```

5.4 文档存在性校验

Elasticsearch 还提供了校验文档是否存在于某索引中的接口 API，即 Exists API。当调用该接口时，如果被验证的文档存在，则 Exists API 会返回 true，否则返回 false。

这个接口特别适合只检查某文档是否存在的场景。

1. 构建文档存在性校验请求

Exists API 依赖于 GetRequest 的实例，这点很像文档索引查询 API，即 Get API。同理，在

Exists API 中也会支持 GetRequest 的所有可选参数。

由于在 Exists API 的返回结果中只有布尔型的值,即 true 或 false,因此建议用户关闭提取源和任何存储字段,以使请求的构建是轻量级的。

Exists API 的使用代码如下所示:

```
// 以同步方式校验索引文档是否存在
public void checkExistIndexDocuments(String indexName, String document) {
  GetRequest getRequest = new GetRequest(indexName, document);
  // 禁用提取源
  getRequest.fetchSourceContext(new FetchSourceContext(false));
  // 禁用提取存储字段
  getRequest.storedFields("_none_");
}
```

2. 执行文档存在性校验请求

在 GetRequest 构建后,即可执行文档存在性校验请求。与文档索引请求和文档索引查询请求类似,文档存在性校验请求也有同步和异步两种执行方式。

同步方式

当以同步方式执行文档存在性校验请求时,客户端会等待 Elasticsearch 服务器返回的查询结果。在收到查询结果后,客户端会继续执行相关的逻辑代码。以同步方式执行的代码如下所示:

```
// 以同步方式校验索引文档是否存在
public void checkExistIndexDocuments(String indexName, String document) {
  GetRequest getRequest = new GetRequest(indexName, document);
  // 禁用提取源
  getRequest.fetchSourceContext(new FetchSourceContext(false));
  // 禁用提取存储字段
  getRequest.storedFields("_none_");
  try {
    boolean exists = restClient.exists(getRequest, RequestOptions.DEFAULT);
    log.info("索引" + indexName + "下的" + document + "文档的存在性是" +
        exists);
  } catch (Exception e) {
    e.printStackTrace();
  }
  // 关闭Elasticsearch连接
  closeEs();
}
```

异步方式

当以异步方式执行 GetRequest 时,高级客户端不必同步等待请求结果的返回,可以直接向接口调用方返回异步接口执行成功的结果。

为了处理异步返回的响应信息或处理在请求执行过程中引发的异常信息,用户需要指定监听器。以异步方式执行的核心代码如下所示:

```
client.existsAsync(getRequest, RequestOptions.DEFAULT, listener);
```

其中,listener 为监听器。

以异步方式执行后不会阻塞,而是立即返回。在请求执行后,如果请求执行成功,则调用 ActionListener 类中的 onResponse 方法进行相关逻辑的处理;如果请求执行失败,则调用 ActionListener 类中的 onFailure 方法进行相关逻辑的处理。

当然,在异步请求执行过程中可能会出现异常,异常的处理与同步方式执行情况相同。

以异步方式执行的全部代码如下所示:

```java
// 以异步方式校验索引文档是否存在
public void checkExistIndexDocumentsAsync(String indexName, String
    document) {
  GetRequest getRequest = new GetRequest(indexName, document);
  // 禁用提取源
  getRequest.fetchSourceContext(new FetchSourceContext(false));
  // 禁用提取存储字段
  getRequest.storedFields("_none_");
  // 定义监听器
  ActionListener<Boolean> listener = new ActionListener<Boolean>() {
    @Override
    public void onResponse(Boolean exists) {
      log.info("索引" + indexName + "下的" + document + "文档的存在性是" +
        exists);
    }
    @Override
    public void onFailure(Exception e) {
    }
  };
  try {
    restClient.existsAsync(getRequest, RequestOptions.DEFAULT, listener);
  } catch (Exception e) {
    e.printStackTrace();
  }
  // 关闭Elasticsearch连接
  closeEs();
```

随后编译工程，在工程根目录下输入如下命令：

```
mvn clean package
```

通过如下命令启动工程服务：

```
java -jar ./target/esdemo-0.0.1-SNAPSHOT.jar
```

在工程服务启动后，在浏览器中调用如下接口查看特定文档在特定索引下的存在情况：

```
http://localhost:8080/springboot/es/high/index/check?indexName=ultraman&document=1
```

服务器输出的查询结果下所示：

```
2019-08-01 11:47:31.568  INFO 31100 --- [io-8080-exec-10] n.e.s.i.MeetHighElasticSearchServiceImpl : 索引ultraman下的1文档的存在性是true
```

显然上述结果与预期相同，而此时浏览器端会显示接口调用成功，内容如下所示：

```
Check Index High ElasticSearch Client Successed!
```

5.5 删除文档索引

Elasticsearch 提供了在索引中删除文档接口的 API。

1. 构建删除文档索引请求

在执行删除文档索引请求前，需要构建删除文档索引请求，即 DeleteRequest。在 DeleteRequest 中有两个必选参数，即索引名称和文档 ID。

DeleteRequest 的初始化代码如下所示：

```
// 构建DeleteRequest
 public void buildDeleteRequestIndexDocuments (String indexName, String
     document) {
   DeleteRequest request = new DeleteRequest(indexName, document);
 }
}
```

在 DeleteRequest 的构建过程中，可以设置一些可选参数。可选参数有路由设置参数、超时设置参数、刷新策略设置参数、版本设置参数和版本类型设置参数等。各个参数的设置方法如下所示：

```java
// 构建DeleteRequest
public void buildDeleteRequestIndexDocuments (String indexName, String
    document) {
  DeleteRequest request = new DeleteRequest(indexName, document);
  // 设置路由
  request.routing("routing");
  // 设置超时
  request.timeout(TimeValue.timeValueMinutes(2));
  request.timeout("2m");
  // 设置刷新策略
  request.setRefreshPolicy(WriteRequest.RefreshPolicy.WAIT_UNTIL);
  request.setRefreshPolicy("wait_for");
  // 设置版本
  request.version(2);
  // 设置版本类型
  request.versionType(VersionType.EXTERNAL);
}
```

2. 执行删除文档索引请求

在 DeleteRequest 构建后，即可执行删除文档索引请求。与文档索引请求类似，删除文档索引请求也有同步和异步两种执行方式。

同步方式

当以同步方式执行删除文档索引请求时，客户端会等待 Elasticsearch 服务器返回的查询结果 DeleteResponse。在收到 DeleteResponse 后，客户端继续执行相关的逻辑代码。以同步方式执行的代码如下所示：

```java
// 以同步方式删除文档索引请求
public void deleteIndexDocuments(String indexName, String document) {
  DeleteRequest request = new DeleteRequest(indexName, document);
  try {
    DeleteResponse deleteResponse = restClient.delete(request, RequestOptions.
        DEFAULT);
  } catch (Exception e) {
    e.printStackTrace();
  }
  // 关闭Elasticsearch连接
  closeEs();
}
```

异步方式

当以异步方式执行删除文档索引请求时，高级客户端不必同步等待请求结果的返回，可以直接向接口调用方返回异步接口执行成功的结果。

在删除文档索引请求执行后，如果请求执行成功，则调用 ActionListener 类中的 onResponse 方法进行相关逻辑的处理；如果请求执行失败，则调用 ActionListener 类中的 onFailure 方法进行相关逻辑的处理。当然，在异步请求执行过程中可能会出现异常，异常的处理与同步方式执行情况相同。

为了处理异步返回的响应信息或处理在请求执行过程中引发的异常信息，用户需要指定监听器。

以异步方式执行的代码如下所示：

```java
// 以异步方式删除文档索引请求
public void deleteIndexDocumentsAsync(String indexName, String document) {
    DeleteRequest request = new DeleteRequest(indexName, document);
    ActionListener listener = new ActionListener<DeleteResponse>() {
        @Override
        public void onResponse(DeleteResponse deleteResponse) {
            String id = deleteResponse.getId();
            String index = deleteResponse.getIndex();
            long version = deleteResponse.getVersion();
            log.info("delete id is " + id + ", index is " + index + ",version is "
                + version);
        }
        @Override
        public void onFailure(Exception e) {
        }
    };
    try {
        restClient.deleteAsync(request, RequestOptions.DEFAULT, listener);
    } catch (Exception e) {
        e.printStackTrace();
    }
    // 关闭 Elasticsearch 连接
    closeEs();
}
```

3. 对响应结果的解析

不论同步方式，还是异步方式，在请求执行后都会返回删除结果 DeleteResponse。用户可以通过解析 DeleteResponse 来查看文档在索引中的删除信息。

示例代码分为三层，分别是 Controller 层、Service 层和 ServiceImpl 实现层。

在 Controller 层的 MeetHighElasticSearchController 类中增加如下代码：

```java
// 以同步方式删除文档索引请求
@RequestMapping("/index/delete")
public String deleteIndexInHighElasticSearch(String indexName, String
    document) {
  if (Strings.isEmpty(indexName) || Strings.isEmpty(document)) {
    return "Parameters are error!";
  }
  meetHighElasticSearchService.deleteIndexDocuments(indexName, document);
  return "Delete Index High ElasticSearch Client Successed!";
}
```

在 Service 层的 MeetHighElasticSearchService 类中增加如下代码：

```java
// 以同步方式删除文档索引请求
public void deleteIndexDocuments(String indexName, String document);
```

在 ServiceImpl 实现层的 MeetHighElasticSearchServiceImpl 类中增加如下代码：

```java
// 以同步方式删除文档索引请求
public void deleteIndexDocuments(String indexName, String document) {
  DeleteRequest request = new DeleteRequest(indexName, document);
  try {
    DeleteResponse deleteResponse = restClient.delete(request,
        RequestOptions.DEFAULT);

    // 处理 DeleteResponse
    processDeleteRequest(deleteResponse);
  } catch (Exception e) {
    e.printStackTrace();
  }
  // 关闭 Elasticsearch 连接
  closeEs();
}
private void processDeleteRequest(DeleteResponse deleteResponse) {
  String index = deleteResponse.getIndex();
  String id = deleteResponse.getId();
  long version = deleteResponse.getVersion();
  log.info("delete id is " + id + ", index is " + index + ",version is " +
      version);
  ReplicationResponse.ShardInfo shardInfo = deleteResponse.getShardInfo();
  if (shardInfo.getTotal() != shardInfo.getSuccessful()) {
    log.info("Success shards are not enough");
  }
  if (shardInfo.getFailed() > 0) {
    for (ReplicationResponse.ShardInfo.Failure failure : shardInfo.
```

```
        getFailures()) {
      String reason = failure.reason();
      log.info("Fail reason is " + reason);
    }
  }
}
```

随后编译工程，在工程根目录下输入如下命令：

```
mvn clean package
```

通过如下命令启动工程服务：

```
java -jar ./target/esdemo-0.0.1-SNAPSHOT.jar
```

在工程服务启动后，在浏览器中调用如下接口查看删除 ultraman 索引下文档编号为 2 的结果：

```
http://localhost:8080/springboot/es/high/index/delete?indexName=ultraman&document=2
```

在接口调用后，服务器输出内容如下所示：

```
2019-08-01 16:03:50.027  INFO 39848 --- [nio-8080-exec-1] n.e.s.i.MeetHighElasticSearchServiceImpl : delete id is 2, index is ultraman,version is 2
```

从 Elasticsearch 控制台的输出内容可以看到，在文档删除后，文档的版本号增加了 1，变成了 2。此时浏览器端显示的内容如下所示：

```
Delete Index High ElasticSearch Client Successed!
```

5.6 更新文档索引

Elasticsearch 提供了在索引中更新文档接口的 API。

1. 构建更新文档索引请求

在执行更新索引文档请求前，需要构建更新文档索引请求，即 UpdateRequest。UpdateRequest 有两个必选参数，即索引名称和文档 ID。

UpdateRequest 的初始化代码如下所示：

```
// 构建 UpdateRequest
public void buildUpdateRequestIndexDocuments(String indexName, String
    document) {
```

```
    UpdateRequest request = new UpdateRequest(indexName, document);
}
```

在构建 UpdateRequest 的过程中，还需要配置一些可选参数。可选参数有路由设置参数和超时设置参数，另外，还可以配置并发对同一文档更新时的重试次数。此外，用户可以配置启用源检索参数，为特定字段配置源包含关系，为特定字段配置源排除关系。各个参数的配置方法如下所示：

```
// 构建 UpdateRequest
public void buildUpdateRequestIndexDocuments(String indexName, String
    document) {
 UpdateRequest request = new UpdateRequest(indexName, document);
 // 设置路由
 request.routing("routing");
 // 设置超时
 request.timeout(TimeValue.timeValueSeconds(1));
 request.timeout("1s");
 // 设置刷新策略
 request.setRefreshPolicy(WriteRequest.RefreshPolicy.WAIT_UNTIL);
 request.setRefreshPolicy("wait_for");
 // 设置：如果更新的文档在更新时被另一个操作更改，则重试更新操作的次数
 request.retryOnConflict(3);
 // 启用源检索，在默认情况下禁用
 request.fetchSource(true);
 // 为特定字段配置源包含关系
 String[] includes = new String[] {"updated", "r*"};
 String[] excludes = Strings.EMPTY_ARRAY;
 request.fetchSource(new FetchSourceContext(true, includes, excludes));
 // 为特定字段配置源排除关系
 includes = Strings.EMPTY_ARRAY;
 excludes = new String[] {"updated"};
 request.fetchSource(new FetchSourceContext(true, includes, excludes));
}
```

此外，还可以使用其他方式构建 UpdateRequest，例如使用部分文档更新方式。当使用部分文档更新方式时，部分文档将与现有文档合并。

可以用以下几种方式提供被更新的部分文件的内容：

（1）以 JSON 格式提供部分文档源。

（2）以 Map 形式提供部分文档源，Elasticsearch 自动将其转换为 JSON 格式。

（3）以 XContentBuilder 对象形式提供文档源，Elasticsearch 自动将其转换为 JSON 格式。

（4）以键值对形式提供部分文档源，Elasticsearch 自动将其转换为 JSON 格式。

在 ServiceImpl 实现层的 MeetHighElasticSearchServiceImpl 类下的 buildUpdateRequestIndexDocuments 方法中添加如下代码：

```
/*
 * 用其他方式构建 UpdateRequest
 */
  // 方式 2：以 JSON 格式构建文档
  request = new UpdateRequest(indexName, document);
  String jsonString = "{" + "\"updated\":\"2019-12-31\"," +
      "\"reason\":\"Year update! \"" + "}";
  request.doc(jsonString, XContentType.JSON);
  // 方式 3：以 Map 形式提供文档源，Elasticsearch 自动将其转换为 JSON 格式
  Map<String, Object> jsonMap = new HashMap<>();
  jsonMap.put("updated", new Date());
  jsonMap.put("reason", "Year update!");
  request = new UpdateRequest(indexName, document).doc(jsonMap);
  // 方式 4：以 XContentBuilder 对象形式提供文档源，Elasticsearch 内置的帮助器自动将
  // 其生成为 JSON 格式的内容
  try {
    XContentBuilder builder = XContentFactory.jsonBuilder();
    builder.startObject();
    {
      builder.timeField("updated", new Date());
      builder.field("reason", "Year update!");
    }
    builder.endObject();
    request = new UpdateRequest(indexName, document).doc(builder);
  } catch (Exception e) {
    e.printStackTrace();
  }
  // 方式 5：以键值对形式提供文档源，Elasticsearch 自动将其转换为 JSON 格式
  request =
      new UpdateRequest(indexName, document).doc("updated", new Date(),
          "reason", "Year update! ");
```

如果被更新的文档不存在，则可以使用 upsert 方法将某些内容定义为新文档。upsert 方法的代码如下所示。

在 ServiceImpl 实现层的 MeetHighElasticSearchServiceImpl 类下的 buildUpdateRequestIndexDocuments 方法中添加如下代码：

```
/*
 * 下面展示 upsert 的使用
 */
```

```
        jsonString = "{\"created\":\"2019-12-31\"}";
        request.upsert(jsonString, XContentType.JSON);
```

与部分文档更新类似，在使用 upsert 方法时，可以使用字符串、Map 映射、XContentBuilder 对象或键值对定义 upsert 文档的内容。

2．执行更新文档索引请求

在 UpdateRequest 构建后，即可执行更新文档索引请求。与文档索引请求类似，更新文档索引请求也有同步和异步两种执行方式。

同步方式

当以同步方式执行更新文档索引请求时，客户端会等待 Elasticsearch 服务器返回的查询结果 UpdateResponse。在收到 UpdateResponse 后，客户端继续执行相关的逻辑代码。以同步方式执行的代码如下所示：

```
// 以同步方式更新文档索引请求
  public void updateIndexDocuments(String indexName, String document) {
    UpdateRequest request = new UpdateRequest(indexName, document);

    try {
      UpdateResponse updateResponse = restClient.update(request, RequestOptions.
         DEFAULT);
    } catch (Exception e) {
      e.printStackTrace();
    }
    // 关闭Elasticsearch连接
    closeEs();
  }
```

异步方式

当以异步方式执行 UpdateRequest 时，高级客户端不必同步等待请求结果的返回，可以直接向接口调用方返回异步接口执行成功的结果。

在 UpdateRequest 执行后，如果请求执行成功，则调用 ActionListener 类中的 onResponse 方法执行相关逻辑；如果请求执行失败，则调用 ActionListener 类中的 onFailure 方法执行相关逻辑。

当然，在异步请求执行过程中，可能会出现异常，异常的处理与同步方式执行情况相同。

为了处理异步返回的响应信息或处理在请求执行过程中引发的异常信息，用户需要指定监听器。以异步方式执行的代码如下所示：

```
// 以异步方式更新文档索引请求
```

```
public void updateIndexDocumentsAsync(String indexName, String document) {
  UpdateRequest request = new UpdateRequest(indexName, document);
  ActionListener listener = new ActionListener<UpdateResponse>() {
    @Override
    public void onResponse(UpdateResponse updateResponse) {
    }
    @Override
    public void onFailure(Exception e) {
    }
  };
  try {
    restClient.updateAsync(request, RequestOptions.DEFAULT, listener);
  } catch (Exception e) {
    e.printStackTrace();
  }
  // 关闭Elasticsearch连接
  closeEs();
}
```

3. 解析更新文档索引请求的响应结果

不论同步方式，还是异步方式，均会返回更新文档索引请求的响应结果 UpdateResponse。用户可以通过解析 UpdateResponse 查看文档的更新情况。

以同步方式为例，UpdateResponse 的解析代码共分为三层，分别是 Controller 层、Service 层和 ServiceImpl 实现层。

在 Controller 层的 MeetHighElasticSearchController 类中添加如下代码：

```
// 以同步方式更新文档索引请求
@RequestMapping("/index/update")
public String updateIndexInHighElasticSearch(String indexName, String
    document) {
  if (Strings.isEmpty(indexName) || Strings.isEmpty(document)) {
    return "Parameters are error!";
  }
  meetHighElasticSearchService.updateIndexDocuments(indexName, document);
  return "Update Index High ElasticSearch Client Successed!";
}
```

在 Service 层的 MeetHighElasticSearchService 类中添加如下代码：

```
// 以同步方式更新文档索引请求
public void updateIndexDocuments(String indexName, String document);
```

在 ServiceImpl 实现层的 MeetHighElasticSearchServiceImpl 类中添加如下代码：

```java
// 以同步方式更新文档索引请求
public void updateIndexDocuments(String indexName, String document) {
  UpdateRequest request = new UpdateRequest(indexName, document);
  Map<String, Object> jsonMap = new HashMap<>();
  jsonMap.put("updated", new Date());
  jsonMap.put("reason", "Year update!");
  jsonMap.put("content",
      "2015年12月,美联储开启新一轮加息周期,至2018年12月,美联储累计加息9次,每次加息25个基点,其中2018年共加息四次,将联邦基金利率的目标范围调至2.25%-2.50%区间。今年以来,美联储未有一次加息动作,联邦基金利率的目标范围维持不变。\r\n");
  request = new UpdateRequest(indexName, document).doc(jsonMap);
  try {
    UpdateResponse updateResponse = restClient.update(request, RequestOptions.
        DEFAULT);
    // 处理 UpdateResponse
    processUpdateRequest(updateResponse);
  } catch (Exception e) {
    e.printStackTrace();
  }
  // 关闭 Elasticsearch 连接
  closeEs();
}
// 处理 UpdateResponse
private void processUpdateRequest(UpdateResponse updateResponse) {
  String index = updateResponse.getIndex();
  String id = updateResponse.getId();
  long version = updateResponse.getVersion();
  log.info("update id is " + id + ", index is " + index + ",version is " +
      version);
  if (updateResponse.getResult() == DocWriteResponse.Result.CREATED) {
    // 创建文档成功
  } else if (updateResponse.getResult() == DocWriteResponse.Result.UPDATED) {
    // 更新文档成功
    // 查看更新的数据
    log.info(updateResponse.getResult().toString());
  } else if (updateResponse.getResult() == DocWriteResponse.Result.DELETED) {
    // 删除文档成功
  } else if (updateResponse.getResult() == DocWriteResponse.Result.NOOP) {
    // 无文档操作
  }
}
```

随后编译工程,在工程根目录下通过如下命令实现:

```
mvn clean package
```

通过如下命令启动工程服务：

```
java -jar ./target/esdemo-0.0.1-SNAPSHOT.jar
```

在工程服务启动后，在浏览器中通过调用如下接口查看索引的更新情况：

```
http://localhost:8080/springboot/es/high/index/update?indexName=ultraman&document=1
```

此时服务器打印输出的内容如下所示：

```
2019-08-01 17:05:11.831  INFO 69104 --- [nio-8080-exec-1] n.e.s.i.MeetHighElasticSearchServiceImpl : update id is 1, index is ultraman,version is 4
```

当再次运行文档查询接口进行 GET 查询时，服务器打印输出内容如下所示：

```
2019-08-01 17:06:32.180  INFO 67544 --- [nio-8080-exec-7] n.e.s.i.MeetHighElasticSearchServiceImpl : id is 1, index is ultraman
2019-08-01 17:06:32.451  INFO 67544 --- [nio-8080-exec-7] n.e.s.i.MeetHighElasticSearchServiceImpl : version is 4, sourceAsString is {"user":"niudong","postDate":"2019-08-01T01:45:17.741Z","message":"Hello Elasticsearch!北京时间8月1日凌晨2点，美联储公布7月议息会议结果。一如市场预期，美联储本次降息25个基点，将联邦基金利率的目标范围调至2.00%~2.25%。此次是2007-2008年间美国为应对金融危机启动降息周期后，美联储十年多以来首次降息。美联储公布利率决议后，美股下跌，美元上涨，人民币汇率下跌","reason":"Year update!","updated":"2019-08-01T09:05:11.3662","content":"2015年12月，美联储开启新一轮加息周期，至2018年12月，美联储累计加息9次，每次加息25个基点，其中2018年共加息四次，将联邦基金利率的目标范围调至2.25%-2.50%区间。今年以来，美联储未有一次加息动作，联邦基金利率的目标范围维持不变。\r\n"}
```

通过对比数据，可以发现更新的数据和之前的数据进行了合并，即对部分文档使用更新时，部分文档将与现有文档合并。此外，对版本号 version 字段也进行了加 1 处理。

其实，在 Elasticsearch 的索引中处理文档的增、删、改请求时，文档的 version 会随着文档的改变而加 1。Elasticsearch 通过使用这个 version 来保证所有修改都被正确排序。当一个旧版本出现在新版本之后，它会被简单地忽略。

用户可以利用 version 的这一优点确保数据不会因为修改冲突而丢失，因此用户可以指定文档的 version 做想要的更改。当然，如果想要修改的版本号不是最新的，则修改请求会失败。

5.7 获取文档索引的词向量

词向量，在 Elasticsearch 中的英文名称为 "Term Vectors"。那么什么是词向量呢？我们可以通俗地理解为：词向量是关于词的一些统计信息的统称。

因此，词向量相关的 API 接口主要用于返回特定文档中词语的信息和统计信息。需要指出的是，这里说的"文档"既可以是存储在索引中的文档，也可以人工提供。

1. 构建词向量请求

在执行词向量请求前，用户需要构建词向量请求，即 TermVectorsRequest。

TermVectorsRequest 有三个必选参数，即索引名称、检索信息的字段和文档 ID。构建 TermVectorsRequest 的代码如下所示：

```
// 构建 TermVectorsRequest
public void buildTermVectorsRequest(String indexName, String document,
    String field) {
  TermVectorsRequest request = new TermVectorsRequest(indexName, document);
  request.setFields(field);
}
```

人工文档需要基于 XContentBuilder 对象提供，XContentBuilder 对象是生成 JSON 内容的 Elasticsearch 内置帮助器，代码如下所示：

```
// 构建 TermVectorsRequest
public void buildTermVectorsRequest(String indexName, String document,
    String field) {
  //方式1：索引中存在的文档
  TermVectorsRequest request = new TermVectorsRequest(indexName, document);
  request.setFields(field);
  //方式2：索引中不存在的文档，可以人工为文档生成词向量
  try {
    XContentBuilder docBuilder = XContentFactory.jsonBuilder();
    docBuilder.startObject().field("user", "niudong").endObject();
    request = new TermVectorsRequest(indexName, docBuilder);
  } catch (Exception e) {
    e.printStackTrace();
  }
}
```

TermVectorsRequest 除索引名称、检索信息的字段和文档 ID 三个必选参数外，还有一些可选参数需要配置。可选参数有 FieldStatistics 字段、TermStatistics 字段、位置字段、偏移量字段、有效载荷字段、FilterSettings 字段、PerFieldAnalyzer 字段、实时检索字段和路由字段等信息。

我们在 buildTermVectorsRequest 方法中添加 TermVectorsRequest 的可选参数，代码如下所示：

```
/*
 * 可选参数
 */
  // 当把 FieldStatistics 设置为 false（默认为 true）时，可忽略文档计数、文档频率总和
  // 及总术语频率总和
  request.setFieldStatistics(false);
```

```java
// 将 TermStatistics 设置为 true（默认为 false），以显示术语总频率和文档频率
request.setTermStatistics(true);

// 将"位置"设置为"假"（默认为"真"），忽略位置的输出
request.setPositions(false);

// 将"偏移"设置为"假"（默认为"真"），忽略偏移的输出
request.setOffsets(false);

// 将"有效载荷"设置为"假"（默认为"真"），忽略有效载荷的输出
request.setPayloads(false);
Map<String, Integer> filterSettings = new HashMap<>();
filterSettings.put("max_num_terms", 3);
filterSettings.put("min_term_freq", 1);
filterSettings.put("max_term_freq", 10);
filterSettings.put("min_doc_freq", 1);
filterSettings.put("max_doc_freq", 100);
filterSettings.put("min_word_length", 1);
filterSettings.put("max_word_length", 10);
// 设置 FilterSettings，根据 TF-IDF 分数筛选可返回的词条
request.setFilterSettings(filterSettings);
Map<String, String> perFieldAnalyzer = new HashMap<>();
perFieldAnalyzer.put("user", "keyword");
// 设置 PerFieldAnalyzer，指定与字段已有的分析器不同的分析器
request.setPerFieldAnalyzer(perFieldAnalyzer);
// 将 Realtime 设置为 false（默认为 true），以便在 Realtime 附近检索术语向量
request.setRealtime(false);

// 设置路由
request.setRouting("routing");
```

2. 执行词向量请求

当 TermVectorsRequest 构建后，即可执行获取文档词向量请求。与文档索引请求类似，获取文档词向量请求也有同步和异步两种执行方式。

同步方式

当以同步方式执行获取文档词向量请求时，客户端会等待 Elasticsearch 服务器返回的查询结果 TermVectorsResponse。在收到 TermVectorsResponse 后，客户端会继续执行相关的逻辑代码。以同步方式执行的代码如下所示：

```java
// 以同步方式执行获取文档词向量请求
public void exucateTermVectorsRequest(String indexName, String document,
```

```
            String field) {
     TermVectorsRequest request = new TermVectorsRequest(indexName, document);
     request.setFields(field);
     try {
       TermVectorsResponse response = restClient.termvectors(request,
RequestOptions.DEFAULT);
     } catch (Exception e) {
       e.printStackTrace();
     }
     // 关闭Elasticsearch连接
     closeEs();
  }
```

异步方式

当以异步方式执行获取文档词向量请求时，高级客户端不必同步等待请求结果的返回，可以直接向接口调用方返回异步接口执行成功的结果。

为了处理异步返回的响应信息或处理在请求执行过程中引发的异常信息，用户需要指定监听器。以异步方式执行的代码如下所示：

```
client.termvectorsAsync(request, RequestOptions.DEFAULT, listener);
```

其中，listener 为监听器。

当然，在异步请求执行过程中可能会出现异常，异常的处理与同步方式执行情况相同。

在异步请求执行后，如果请求执行成功，则调用 ActionListener 类中的 onResponse 方法进行相关逻辑的处理；如果请求执行失败，则调用 ActionListener 类中的 onFailure 方法进行相关逻辑的处理。

以异步方式调用的代码如下所示：

```
// 以异步方式执行TermVectorsRequest
  public void exucateTermVectorsRequestAsync(String indexName, String
      document, String field) {
     TermVectorsRequest request = new TermVectorsRequest(indexName, document);
     request.setFields(field);
     ActionListener listener = new ActionListener<TermVectorsResponse>() {
       @Override
       public void onResponse(TermVectorsResponse termVectorsResponse) {
       }
       @Override
       public void onFailure(Exception e) {
       }
     };
```

```
    try {
      restClient.termvectorsAsync(request, RequestOptions.DEFAULT, listener);
    } catch (Exception e) {
      e.printStackTrace();
    }
    // 关闭Elasticsearch连接
    closeEs();
}
```

3．解析词向量请求的响应结果

不论同步方式，还是异步方式，Elasticsearch 服务器都会返回词向量请求的响应结果 TermVectorsResponse。用户可以通过解析 TermVectorsResponse 获取相关内容。以同步执行方式为例，获取 TermVectorsResponse 后的解析代码共分为三层，分别是 Controller 层、Service 层和 ServiceImpl 实现层。

在 Controller 层的 MeetHighElasticSearchController 类中添加如下代码：

```
// 以同步方式执行TermVectorsRequest
@RequestMapping("/index/term")
public String termVectorsInHighElasticSearch(String indexName, String document,
    String field) {
  if (Strings.isEmpty(indexName) || Strings.isEmpty(document) || Strings.isEmpty
      (field)) {
    return "Parameters are error!";
  }
  meetHighElasticSearchService.exucateTermVectorsRequest(indexName,
      document, field);
  return "Test TermVectorsRequest High ElasticSearch Client Successed!";
}
```

在 Service 层的 MeetHighElasticSearchService 类中添加如下代码：

```
// 以同步方式执行TermVectorsRequest
public void exucateTermVectorsRequest(String indexName, String document,
    String field);
```

在 ServiceImpl 实现层的 MeetHighElasticSearchServiceImpl 类中添加如下代码：

```
// 以同步方式执行TermVectorsRequest
public void exucateTermVectorsRequest(String indexName, String document,
    String field) {
  TermVectorsRequest request = new TermVectorsRequest(indexName, document);
  request.setFields(field);
  try {
```

```java
        TermVectorsResponse response = restClient.termvectors(request,
            RequestOptions.DEFAULT);
        // 处理TermVectorsResponse
        processTermVectorsResponse(response);
    } catch (Exception e) {
        e.printStackTrace();
    }
    // 关闭Elasticsearch连接
    closeEs();
}
// 处理TermVectorsResponse
private void processTermVectorsResponse(TermVectorsResponse response) {
    String index = response.getIndex();
    String type = response.getType();
    String id = response.getId();
    // 指示是否找到文档
    boolean found = response.getFound();
    log.info("index is " + index + ",id is " + id + ", type is " + type + ",
        found is " + found);
    List<TermVector> list = response.getTermVectorsList();
    log.info("list is " + list.size());
    for (TermVector tv : list) {
        processTermVector(tv);
    }
}
// 处理TermVector
private void processTermVector(TermVector tv) {
    String fieldname = tv.getFieldName();
    int docCount = tv.getFieldStatistics().getDocCount();
    long sumTotalTermFreq = tv.getFieldStatistics().getSumTotalTermFreq();
    long sumDocFreq = tv.getFieldStatistics().getSumDocFreq();
    log.info("fieldname is " + fieldname + "; docCount is " + docCount + ";
        sumTotalTermFreq is "
        + sumTotalTermFreq + ";sumDocFreq is " + sumDocFreq);
    if (tv.getTerms() == null) {
        return;
    }
    List<TermVectorsResponse.TermVector.Term> terms = tv.getTerms();
    for (TermVectorsResponse.TermVector.Term term : terms) {
        String termStr = term.getTerm();
        int termFreq = term.getTermFreq();
        int docFreq = term.getDocFreq() == null ? 0 : term.getDocFreq();
        long totalTermFreq = term.getTotalTermFreq() == null ? 0 :
            term.getTotalTermFreq();
```

```
            float score = term.getScore() == null ? 0 : term.getScore();
            log.info("termStr is " + termStr + "; termFreq is " + termFreq + ";
              docFreq is " + docFreq
              + ";totalTermFreq is " + totalTermFreq + ";score is " + score);
            if (term.getTokens() != null) {
              List<TermVectorsResponse.TermVector.Token> tokens = term.getTokens();
              for (TermVectorsResponse.TermVector.Token token : tokens) {
                int position = token.getPosition() == null ? 0 : token.getPosition();
                int startOffset = token.getStartOffset() == null ? 0 :
                   token.getStartOffset();
                int endOffset = token.getEndOffset() == null ? 0 : token.
                   getEndOffset ();
                String payload = token.getPayload();
                log.info("position is " + position + "; startOffset is " +
                   startOffset + "; endOffset is "
                   + endOffset + ";payload is " + payload);
              }
            }
          }
        }
```

随后编译工程，在工程根目录下输入如下命令：

```
mvn clean package
```

通过如下命令启动工程服务：

```
java -jar ./target/esdemo-0.0.1-SNAPSHOT.jar
```

在工程服务启动后，在浏览器中调用如下接口查看获取文档索引的词向量的情况：

```
http://localhost:8080/springboot/es/high/index/term?indexName=ultraman&document=1&field=content
```

在接口执行后，浏览器中如果显示如下内容，则表明接口运行成功：

```
Test TermVectorsRequest High ElasticSearch Client Successed!
```

在服务器中会输出如下所示内容。

```
2019-08-01 19:08:50.811  INFO 71176 --- [nio-8080-exec-1] n.e.s.i.MeetHighElasticSearchServiceImpl : index is ultraman,i
d is 1, type is _doc, found is true
2019-08-01 19:08:50.813  INFO 71176 --- [nio-8080-exec-1] n.e.s.i.MeetHighElasticSearchServiceImpl : list is 1
2019-08-01 19:08:50.822  INFO 71176 --- [nio-8080-exec-1] n.e.s.i.MeetHighElasticSearchServiceImpl : fieldname is conter
t; docCount is 1; sumTotalTermFreq is 95;sumDocFreq is 57
2019-08-01 19:08:50.832  INFO 71176 --- [nio-8080-exec-1] n.e.s.i.MeetHighElasticSearchServiceImpl : termStr is 12; term
Freq is 2; docFreq is 0;totalTermFreq is 0;score is 0.0
2019-08-01 19:08:50.850  INFO 71176 --- [nio-8080-exec-1] n.e.s.i.MeetHighElasticSearchServiceImpl : position is 2; star
tOffset is 5; endOffset is 7;payload is null
2019-08-01 19:08:50.872  INFO 71176 --- [nio-8080-exec-1] n.e.s.i.MeetHighElasticSearchServiceImpl : position is 19; sta
rtOffset is 28; endOffset is 30;payload is null
2019-08-01 19:08:50.903  INFO 71176 --- [nio-8080-exec-1] n.e.s.i.MeetHighElasticSearchServiceImpl : termStr is 2.25; te
rmFreq is 1; docFreq is 0;totalTermFreq is 0;score is 0.0
2019-08-01 19:08:50.918  INFO 71176 --- [nio-8080-exec-1] n.e.s.i.MeetHighElasticSearchServiceImpl : position is 61; sta
rtOffset is 79; endOffset is 83;payload is null
2019-08-01 19:08:50.932  INFO 71176 --- [nio-8080-exec-1] n.e.s.i.MeetHighElasticSearchServiceImpl : termStr is 2.50; te
rmFreq is 1; docFreq is 0;totalTermFreq is 0;score is 0.0
2019-08-01 19:08:50.943  INFO 71176 --- [nio-8080-exec-1] n.e.s.i.MeetHighElasticSearchServiceImpl : position is 62; sta
rtOffset is 85; endOffset is 89;payload is null
2019-08-01 19:08:50.957  INFO 71176 --- [nio-8080-exec-1] n.e.s.i.MeetHighElasticSearchServiceImpl : termStr is 2015; te
rmFreq is 1; docFreq is 0;totalTermFreq is 0;score is 0.0
2019-08-01 19:08:50.971  INFO 71176 --- [nio-8080-exec-1] n.e.s.i.MeetHighElasticSearchServiceImpl : position is 0; star
tOffset is 0; endOffset is 4;payload is null
2019-08-01 19:08:50.994  INFO 71176 --- [nio-8080-exec-1] n.e.s.i.MeetHighElasticSearchServiceImpl : termStr is 2018; te
rmFreq is 2; docFreq is 0;totalTermFreq is 0;score is 0.0
2019-08-01 19:08:51.002  INFO 71176 --- [nio-8080-exec-1] n.e.s.i.MeetHighElasticSearchServiceImpl : position is 17; sta
rtOffset is 23; endOffset is 27;payload is null
2019-08-01 19:08:51.023  INFO 71176 --- [nio-8080-exec-1] n.e.s.i.MeetHighElasticSearchServiceImpl : position is 40; sta
rtOffset is 54; endOffset is 58;payload is null
2019-08-01 19:08:51.034  INFO 71176 --- [nio-8080-exec-1] n.e.s.i.MeetHighElasticSearchServiceImpl : termStr is 25; term
Freq is 1; docFreq is 0;totalTermFreq is 0;score is 0.0
2019-08-01 19:08:51.052  INFO 71176 --- [nio-8080-exec-1] n.e.s.i.MeetHighElasticSearchServiceImpl : position is 34; sta
rtOffset is 46; endOffset is 48;payload is null
2019-08-01 19:08:51.099  INFO 71176 --- [nio-8080-exec-1] n.e.s.i.MeetHighElasticSearchServiceImpl : termStr is 9; termF
req is 1; docFreq is 0;totalTermFreq is 0;score is 0.0
2019-08-01 19:08:51.115  INFO 71176 --- [nio-8080-exec-1] n.e.s.i.MeetHighElasticSearchServiceImpl : position is 28; sta
rtOffset is 39; endOffset is 40;payload is null
2019-08-01 19:08:51.140  INFO 71176 --- [nio-8080-exec-1] n.e.s.i.MeetHighElasticSearchServiceImpl : termStr is 一; term
Freq is 2; docFreq is 0;totalTermFreq is 0;score is 0.0
2019-08-01 19:08:51.178  INFO 71176 --- [nio-8080-exec-1] n.e.s.i.MeetHighElasticSearchServiceImpl : position is 10; sta
rtOffset is 15; endOffset is 16;payload is null
2019-08-01 19:08:51.186  INFO 71176 --- [nio-8080-exec-1] n.e.s.i.MeetHighElasticSearchServiceImpl : position is 74; sta
rtOffset is 103; endOffset is 104;payload is null
2019-08-01 19:08:51.218  INFO 71176 --- [nio-8080-exec-1] n.e.s.i.MeetHighElasticSearchServiceImpl : termStr is 不; term
2019-08-01 19:08:51.228  INFO 71176 --- [nio-8080-exec-1] n.e.s.i.MeetHighElasticSearchServiceImpl : position is 93; sta
rtOffset is 123; endOffset is 124;payload is null
2019-08-01 19:08:51.265  INFO 71176 --- [nio-8080-exec-1] n.e.s.i.MeetHighElasticSearchServiceImpl : termStr is 个; term
Freq is 1; docFreq is 0;totalTermFreq is 0;score is 0.0
```

索引 ultraman 中编号为 1 的文档下的 content 中词语的统计信息。

5.8 文档处理过程解析

前面介绍了 Elasticsearch 中文档及对文档的一些操作的接口实战，相信读者已经对文档有一定的认知，这是一个很好的开始。基础打好之后，就更容易学习和理解较为复杂的内容。下面介绍在 Elasticsearch 内部是如何处理文档的。

5.8.1 文档的索引过程

首先需要明确一点，写入磁盘的倒排索引是不可变的。Elasticsearch 为什么要这样做，主要是基于以下几个考量：

（1）读写操作轻量级，不需要锁。如果 Elasticsearch 从来不需要更新一个索引，则就不必担心多个程序同时尝试修改索引的情况。

（2）一旦索引被读入文件系统的内存，它就会一直在那儿，因为不会改变。

此外，当文件系统内存有足够大的空间时，大部分的索引读写操作是可以直接访问内存，而不是磁盘就能实现的，显然这有助于提升 Elasticsearch 的性能。

（3）当写入单个大的倒排索引时，Elasticsearch 可以压缩数据，以减少磁盘 I/O 和需要存储索引的内存大小。

当然，倒排索引的不可变性也是一把"双刃剑"，不可变的索引也有它的缺点。首先就是它不可变，用户不能改变它。如果想要搜索一个新文档，则必须重建整个索引。这不仅严重限制了一个索引所能装下的数据，还限制了一个索引可以被更新的频率。

那么 Elasticsearch 是如何在保持倒排索引不可变好处的同时又能更新倒排索引呢？答案是，使用多个索引。

Elasticsearch 不是重写整个倒排索引，而是增加额外的索引反映最近的变化。每个倒排索引都可以按顺序查询，从最"老旧"的索引开始查询，最后把结果聚合起来。

Elasticsearch 的底层依赖于 Lucene，Lucene 中的索引其实是 Elasticsearch 中的分片，Elasticsearch 中的索引是分片的集合。当 Elasticsearch 搜索索引时，它发送查询请求给该索引下的所有分片，然后过滤这些结果，最后聚合成全局的结果。

为了避免混淆，Elasticsearch 引入了 per-segment search 的概念。一个段（segment）就一个是有完整功能的倒排索引。Lucene 中的索引指的是段的集合，再加上提交点（commit point，包括所有段的文件）。新的文档在被写入磁盘的段之前，需要先写入内存区的索引。

一个 per-segment search 的工作流程如下所示：

（1）新的文档首先被写入内存区的索引。

（2）内存中的索引不断被提交，新段不断产生。当新的提交点产生时就将这些新段的数据写入磁盘，包括新段的名称。

写入磁盘是文件同步写入的，也就是说，所有的写操作都需要等待文件系统内存的数据同步到磁盘，确保它们可以被物理写入。

（3）新段被打开，于是它包含的文档就可以被检索到。

（4）内存被清除，等待接收新的文档。

当一个请求被接收，所有段依次被查询时。所有段上的 term 统计信息会被聚合，确保每个 term 和文档的相关性被正确计算。通过这种方式，新的文档就能够以较小的代价加入索引。

段是不可变的，那么 Elasticsearch 是如何删除和更新文档数据的呢？

段的不可变特性，意味着文档既不能从旧的段中移除，旧的段中的文档也不能被更新。于是 Elasticsearch 在每一个提交点都引入一个.del 文件，包含了段上已经被删除的文档。

当一个文档被删除时，它实际上只是在 .del 文件中被标记为删除。在进行文档查询时，被删除的文档依然可以被匹配查询，但是在最终返回之前会从结果中删除。

当一个文档被更新时，旧版本的文档会被标记为删除，新版本的文档在新的段中被索引。当对文档进行查询时，该文档的不同版本都会匹配一个查询请求，但是较旧的版本会从结果中被删除。

被删除的文件越积累越多，每个段消耗的如文件句柄、内存、CPU 等资源越来越大。如果每次搜索请求都需要依次检查每个段，则段越多，查询就越慢。这些势必会影响 Elasticsearch 的性能，那么 Elasticsearch 是如何处理的呢？Elasticsearch 引入了段合并段。在段合并时，我们会展示被删除的文件是如何从文件系统中清除的。

Elasticsearch 通过后台合并段的方式解决了上述问题，在段合并过程中，小段被合并成大段，大段再合并成更大的段。在合并段时，被删除的文档不会被合并到大段中。

在索引过程中，refresh 会创建新的段，并打开它。合并过程是在后台选择一些小的段，把它们合并成大的段。在这个过程中不会中断索引和搜索。当新段合并后，即可打开供搜索；而旧段会被删除。

需要指出的是，合并大的段会消耗很多 I/O 和 CPU。为了不影响 Elasticsearch 的搜索性能，在默认情况下，Elasticsearch 会限制合并过程，这样搜索就可以有足够的资源进行了。

除自动完成段合并外，Elasticsearch 还提供了 optimize API，以便根据需要强制合并段。optimize API 强制分片合并段以达到指定 max_num_segments 参数，这会减少段的数量（通常为 1），达到提高搜索性能的目的。

需要指出的是，不要在动态的索引上使用 optimize API。optimize API 的典型场景是记录日志。日志是按照每天、周、月存入索引的。旧的索引一般是只可读，不可修改的。在这种场景下，用户主动把每个索引的段降至 1 是有效的，因为搜索过程会用到更少的资源，性能更好。

5.8.2 文档在文件系统中的处理过程

前面介绍了文档的索引过程，由于数据最终会持久化在磁盘的文件系统中，那么 Elasticsearch 是如何将文档存储到文件系统中的呢？本节将揭晓。

在 Elasticsearch 的配置文件 elasticsearch-7.2.0\config\elasticsearch.yml 中，有一个配置属性 path.data，该属性包含了 Elasticsearch 中存储的数据的文件夹的路径，下面就从本目录开始讲起。Elasticsearch 根目录的内容如图 5-2 所示。

图 5-2

从图 5-2 中可见，Elasticsearch 根目录中默认有一个 data 目录，data 目录即为 Elasticsearch 中默认的 path.data 属性的值。

在 Elasticsearch 中，是使用 Lucene 来处理分片级别的索引和查询的，因此 data 目录中的文件由 Elasticsearch 和 Lucene 写入。

两者的职责非常明确：Lucene 负责写和维护 Lucene 索引文件，而 Elasticsearch 在 Lucene 之上写与功能相关的元数据，如字段映射、索引设置和其他集群元数据等。

打开 data 目录，如图 5-3 所示。

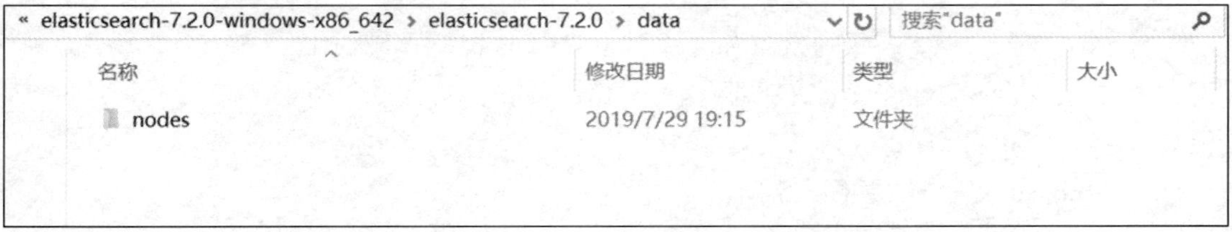

图 5-3

nodes 文件夹用于存储本机中的节点信息。打开 nodes 文件夹，如图 5-4 所示。

图 5-4

在当前的 Elasticsearch 中，因为只有一个主节点分片，所以只有一个编号为 0 的文件夹，打开该文件夹，如图 5-5 所示。

```
« elasticsearch-7.2.0 > data > nodes > 0        搜索"0"
名称                修改日期              类型           大小
_state              2019/9/11 10:02      文件夹
indices             2019/9/6 16:17       文件夹
node.lock           2019/7/29 19:15      LOCK 文件       0 KB
```

图 5-5

图 5-5 中的 node.lock 文件用于确保一次只能从一个数据目录读取、写入一个 Elasticsearch 相关的信息。_state 文件夹用于存放当前节点的索引信息，indices 文件夹用于存放当前节点的索引信息。

打开 state 文件夹，如图 5-6 所示。

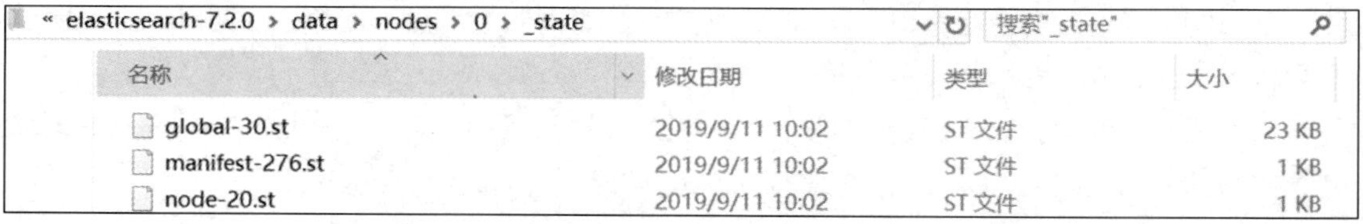

```
« elasticsearch-7.2.0 > data > nodes > 0 > _state     搜索"_state"
名称                修改日期              类型           大小
global-30.st        2019/9/11 10:02      ST 文件        23 KB
manifest-276.st     2019/9/11 10:02      ST 文件        1 KB
node-20.st          2019/9/11 10:02      ST 文件        1 KB
```

图 5-6

图 5-6 中有一个名称为 global-30.st 的文件，其中 global 前缀表示这是一个全局状态文件，而.st 扩展名表示这是一个包含元数据的状态文件。此类二进制文件包含有关用户集群的全局元数据，数字 30 表示集群元数据的版本。

打开 indices 文件夹，如图 5-7 所示。

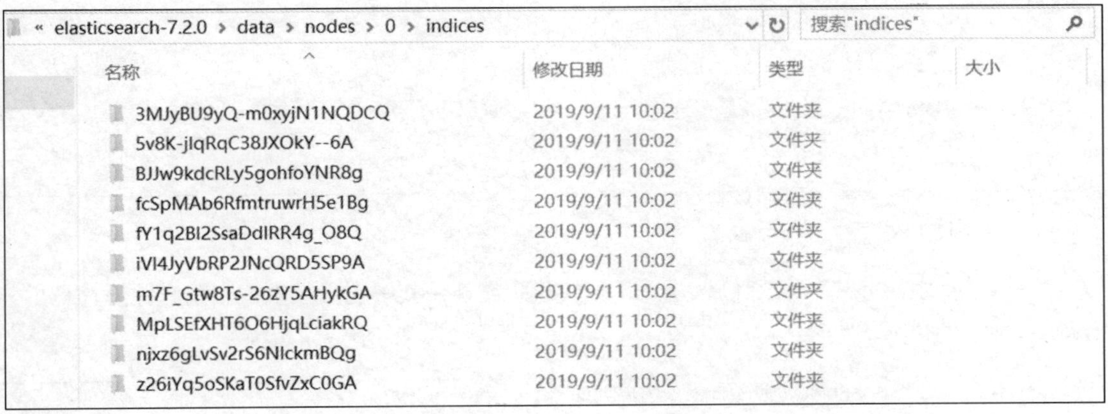

图 5-7

indices 文件夹中的内容为索引信息，每个索引有一个随机字符串的名称，显然该文件夹下的索引数量与本机中的索引数量相同。随机打开一个索引为 3MJyBU9yQ-m0xyjN1NQDCQ 的文件夹，如图 5-8 所示。

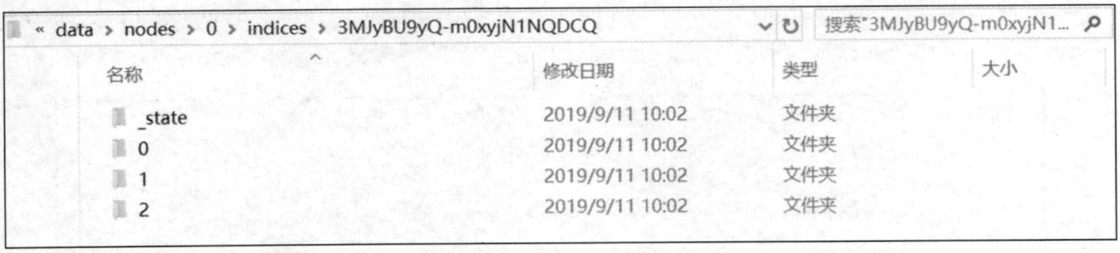

图 5-8

在图 5-8 中，有两类子文件夹：_state 类文件夹和分片类文件夹。其中，0、1、2 为分片文件夹。

_state 文件夹包含了 indices / {index-name} / state / state- {version} .st 路径中的索引状态文件；0、1、2 分片文件夹包含与索引的第一、二、三个分片相关的数据（即分片 0、分片 1 和分片 2）。

打开分片 0 的目录，如图 5-9 所示。

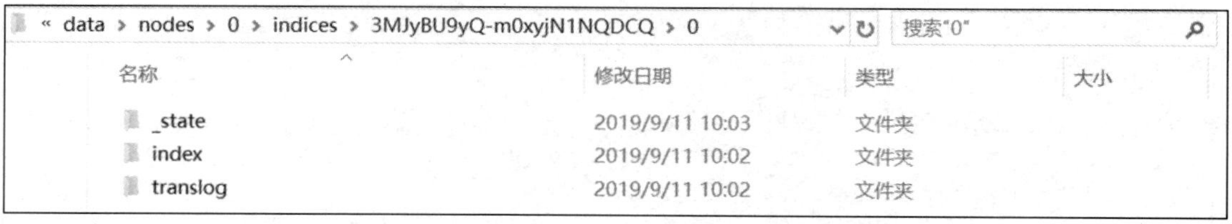

图 5-9

在图 5-9 中，分片 0 目录包含分片相关的状态文件，其中包括版本控制及有关分片是主分片还是副本的信息，如图 5-10 所示。

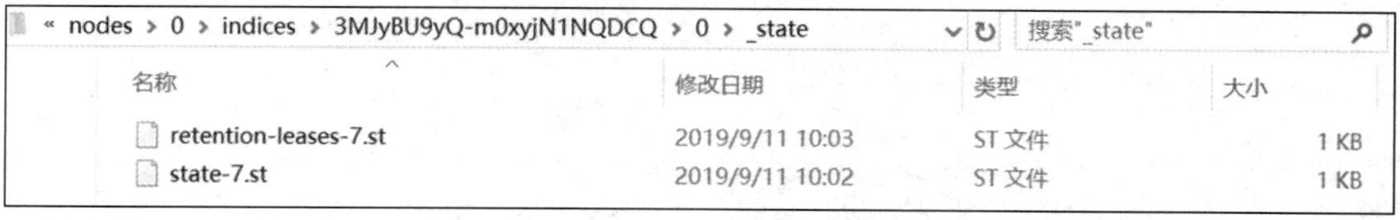

图 5-10

此外，还有本分片下的索引信息和 translog 日志信息。其中，translog 日志信息是 Elasticsearch 的事务日志，在每个分片 translog 目录中的前缀 translog 中存在，如图 5-11 所示。

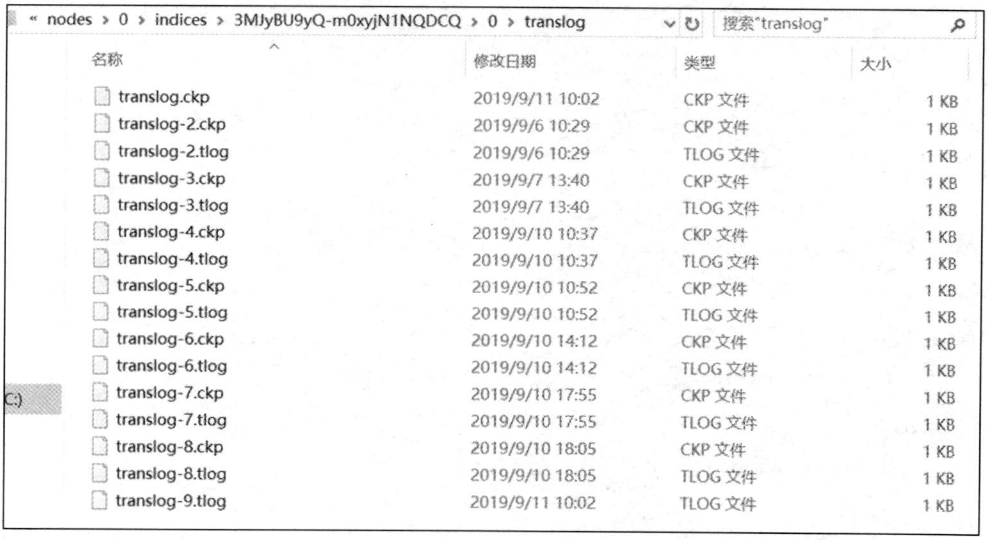

图 5-11

在 Elasticsearch 中，事务日志用于确保安全地将数据索引到 Elasticsearch，而无须为每个文档执行低级 Lucene 提交。当提交 Lucene 索引时，会在 Lucene 级别创建一个新的 segment，即执行 fsync()，会产生大量磁盘 I/O，从而影响性能。

为了能存储索引文档并使其可搜索，而不需要完整地 Lucene 提交，Elasticsearch 将其添加到 Lucene IndexWriter，并将其附加到事务日志中。

这样，在每个 refresh_interval 之后，它将在 Lucene 索引上调用 reopen()，这将使数据可以在不需要提交的情况下就能进行搜索。这是 Lucene 近实时搜索 API 的一部分。当 IndexWriter 最终自动刷新事务日志或由于显式刷新操作而提交时，先前的事务日志将被丢弃，新的事务日志将取代它。

如果需要恢复，则首先恢复在 Lucene 中写入磁盘的 segments，然后重放事务日志，以防止丢失尚未完全提交到磁盘的操作。

5.9　知识点关联

在开发过程中，与 Elasticsearch 相似，且同为文档型存储的中间件是 MongoDB。

MongoDB 是一个基于分布式文件存储的数据库，是基于 C++ 编写的，旨在为 Web 应用提供可扩展的高性能数据存储解决方案。

MongoDB 与关系数据库很像，也是非关系数据库当中功能最丰富的。

那么，Elasticsearch 和 MongoDB 有什么异同之处吗？它们的使用场景有什么区别？

Elasticsearch 和 MongoDB 的相同之处在于都是以 JSON 格式进行数据存储的，都支持对文档数据的增删改查，即 CRUD 操作。

此外，二者都使用了分片和复制技术，都支持处理超大规模数据。

Elasticsearch 和 MongoDB 的不同之处也有很多。

（1）开发语言不同。Elasticsearch 是基于 Java 编写的，而 MongoDB 是基于 C++编写的。

（2）分片方式不同。Elasticsearch 基于 Hash 模式进行分片；而 MongoDB 的分片模式除了 Hash 模式，还有 Range 模式。

（3）集群的配置方式不同。Elasticsearch 天然是分布式的，主副分片自动分配和复制；而 MongoDB 需要手工配置。

（4）全文检索的便捷程度不同。Elasticsearch 全文检索功能强大，字段自动索引；而 MongoDB 仅支持有限的字段检索，且需人工索引。

在使用场景上，Elasticsearch 适用于全文检索场景，而 MongoDB 适用于数据大批量存储的场景。

5.10 小结

本章主要介绍了在 Java 高级客户端中对文档操作 API 的使用，如文档在索引中的增删改查操作，还介绍了文档在索引中存在性检查 API 的使用和文档词向量 API 的使用。

第 6 章
高级客户端文档实战二

> 襟三江而带五湖
> 控蛮荆而引瓯越

在第 5 章中，主要介绍了编号（1）至（7）的高级客户端文档 API 的使用。在本章中，将主要介绍编号（8）至（13）的高级客户端文档 API 的使用。

（1）Single document APIs——单文档操作 API。

（2）Index API——文档索引 API。

（3）Get API——文档获取 API。

（4）Exists API——文档存在性判断 API。

（5）Delete API——文档删除 API。

（6）Update API——文档更新 API。

（7）Term Vectors API——词向量 API。

（8）Bulk API——批量处理 API。

（9）Multi-Get API——多文档获取 API。

（10）ReIndex API——重新索引 API。

（11）Update By Query API——查询更新 API。

（12）Delete By Query API——查询删除 API。

（13）Multi Term Vectors API——多词条向量 API。

6.1 批量请求

为了使用户能够对文档进行高效的增删改查，Elasticsearch 提供了批量请求接口，即 Bulk API。

1．构建批量请求

在使用批量请求前，需要构建批量请求，即 BulkRequest。BulkRequest 可用于通过单次请求执行多个操作请求，如文档索引、文档更新、文档删除等操作。在 BulkRequest 中需要添加至少一个操作，当然，BulkRequest 还支持在其中添加不同类型的操作。

BulkRequest 的构建代码如下所示：

```
// 构建BulkRequest
public void buildBulkRequest(String indexName, String field) {
    /*
     * 方式1：添加同型请求
     */
    BulkRequest request = new BulkRequest();
    // 添加第一个IndexRequest
    request.add(new IndexRequest(indexName).id("1").source(XContentType.JSON,
        field,
        "事实上，自今年年初开始，美联储就已传递出货币政策或将转向的迹象"));
    // 添加第二个IndexRequest
    request.add(new IndexRequest(indexName).id("2").source(XContentType.JSON,
        field,
        "自6月起，市场对于美联储降息的预期愈发强烈"));
    // 添加第三个IndexRequest
    request.add(new IndexRequest(indexName).id("3").source(XContentType.JSON,
        field,
        "从此前美联储降息历程来看，美联储降息将打开全球各国央行的降息窗口"));
    /*
     * 方式2：添加异型请求
     */
    // 添加一个DeleteRequest
    request.add(new DeleteRequest(indexName, "3"));
    // 添加一个UpdateRequest
    request.add(new UpdateRequest(indexName, "2").doc(XContentType.JSON, field,
        "自今年初美联储暂停加息以来，全球范围内的降息大幕就已拉开，不仅包括新兴经济体，发达
        经济体也加入降息阵营，仅7月份一个月内，就有6国央行降息"));
    // 添加一个IndexRequest
    request.add(new IndexRequest(indexName).id("4").source(XContentType.JSON,
        field,
```

```
                "在此次美联储降息后,央行或不会立即跟进降息"));
}
```

BulkRequest 也有一些可选参数供用户进行必要的设置。可选参数有超时时间设置、数据刷新策略、分片副本数量、全局管道标识、全局路由等。

在 buildBulkRequest 方法中添加如下代码:

```
/*
 * 以下是可选参数的配置
 */
    //设置超时时间
    request.timeout(TimeValue.timeValueMinutes(2));
    request.timeout("2m");

    //设置数据刷新策略
    request.setRefreshPolicy(WriteRequest.RefreshPolicy.WAIT_UNTIL);
    request.setRefreshPolicy("wait_for");

    //设置在继续执行索引/更新/删除操作之前必须处于活动状态的分片副本数
    request.waitForActiveShards(2);
    request.waitForActiveShards(ActiveShardCount.ALL);

    //用于所有子请求的全局pipelineid,即全局管道标识
    request.pipeline("pipelineId");

    //用于所有子请求的全局路由 ID
request.routing("routingId");
```

2. 执行批量请求

在 BulkRequest 构建后,即可执行批量请求。与文档索引请求类似,批量请求也有同步和异步两种执行方式。

同步方式

当以同步方式执行批量请求时,客户端会等待 Elasticsearch 服务器返回的查询结果 BulkResponse。在收到 BulkResponse 后,客户端会继续执行相关的逻辑代码。以同步方式执行的代码如下所示:

```
    // 以同步方式执行 BulkRequest
public void executeBulkRequest(String indexName, String field) {
    BulkRequest request = new BulkRequest();
    // 添加第一个 IndexRequest
    request.add(new IndexRequest(indexName).id("1").source(XContentType.JSON,
```

```
        field,
        "事实上,自今年年初开始,美联储就已传递出货币政策或将转向的迹象"));
// 添加第二个 IndexRequest
request.add(new IndexRequest(indexName).id("2").source(XContentType.JSON,
        field,
        "自 6 月起,市场对于美联储降息的预期愈发强烈"));
// 添加第三个 IndexRequest
request.add(new IndexRequest(indexName).id("3").source(XContentType.JSON,
        field,
        "从此前美联储降息历程来看,美联储降息将打开全球各国央行的降息窗口"));
try {
    BulkResponse bulkResponse = restClient.bulk(request, RequestOptions.DEFAULT);
} catch (Exception e) {
    e.printStackTrace();
}
// 关闭 Elasticsearch 连接
closeEs();
}
```

异步方式

当以异步方式执行批量请求时,高级客户端不必同步等待请求结果的返回,可以直接向接口调用方返回异步接口执行成功的结果。

为了处理异步返回的响应信息或处理在请求执行过程中引发的异常信息,用户需要指定监听器。以异步方式执行的代码如下所示:

```
client.bulkAsync(request, RequestOptions.DEFAULT, listener);
```

其中,listener 为监听器。

当然,在异步请求执行过程中可能会出现异常,异常的处理与同步方式执行情况相同。

在异步请求处理后,如果请求执行成功,则调用 ActionListener 类中的 onResponse 方法进行相关逻辑的处理;如果请求执行失败,则调用 ActionListener 类中的 onFailure 方法进行相关逻辑的处理。

以异步方式执行的代码如下所示:

```
// 以异步方式执行 BulkRequest
public void executeBulkRequestAsync(String indexName, String field) {
    BulkRequest request = new BulkRequest();
    // 添加第一个 IndexRequest
    request.add(new IndexRequest(indexName).id("1").source(XContentType.JSON,
            field,
            "事实上,自今年年初开始,美联储就已传递出货币政策或将转向的迹象"));
```

```java
// 添加第二个 IndexRequest
request.add(new IndexRequest(indexName).id("2").source(XContentType.JSON,
    field,
    "自 6 月起，市场对于美联储降息的预期愈发强烈"));
// 添加第三个 IndexRequest
request.add(new IndexRequest(indexName).id("3").source(XContentType.JSON,
    field,
    "从此前美联储降息历程来看，美联储降息将打开全球各国央行的降息窗口"));
// 构建监听器
ActionListener<BulkResponse> listener = new ActionListener<BulkResponse>() {
  @Override
  public void onResponse(BulkResponse bulkResponse) {
  }
  @Override
  public void onFailure(Exception e) {
  }
};
try {
  restClient.bulkAsync(request, RequestOptions.DEFAULT, listener);
} catch (Exception e) {
  e.printStackTrace();
}
// 关闭 Elasticsearch 连接
closeEs();
}
```

3. 解析批量请求的响应结果

不论同步方式，还是异步方式，客户端均会收到批量请求的返回结果 BulkResponse。BulkResponse 中包含有关已执行操作的信息，并允许用户对每个请求的结果进行迭代解析。以同步执行方式为例，BulkResponse 的解析代码共分为三层，分别是 Controller 层、Service 层和 ServiceImpl 实现层。

其中，在 Controller 层的 MeetHighElasticSearchController 类中添加如下代码：

```java
// 以同步方式执行 BulkRequest
@RequestMapping("/index/bulk")
public String bulkGetInHighElasticSearch(String indexName, String field) {
  if (Strings.isEmpty(indexName) || Strings.isEmpty(field)) {
    return "Parameters are error!";
  }
  meetHighElasticSearchService.executeBulkRequest(indexName, field);
  return "Bulk Get In High ElasticSearch Client Successed!";
}
```

在 Service 层的 MeetHighElasticSearchService 类中添加如下代码：

```java
// 以同步方式执行 BulkRequest
public void executeBulkRequest(String indexName, String field);
```

在 ServiceImpl 实现层的 MeetHighElasticSearchServiceImpl 类中添加如下代码：

```java
// 以同步方式执行 BulkRequest
public void executeBulkRequest(String indexName, String field) {
  BulkRequest request = new BulkRequest();
  // 添加第一个 IndexRequest
  request.add(new IndexRequest(indexName).id("1").source(XContentType.JSON,
      field,
      "事实上,自今年年初开始,美联储就已传递出货币政策或将转向的迹象"));
  // 添加第二个 IndexRequest
  request.add(new IndexRequest(indexName).id("2").source(XContentType.JSON,
      field,
      "自 6 月起,市场对于美联储降息的预期愈发强烈"));
  // 添加第三个 IndexRequest
  request.add(new IndexRequest(indexName).id("3").source(XContentType.JSON,
      field,
      "从此前美联储降息历程来看,美联储降息将打开全球各国央行的降息窗口"));
  try {
    BulkResponse bulkResponse = restClient.bulk(request, RequestOptions.
        DEFAULT);
    // 解析 BulkResponse
    processBulkResponse(bulkResponse);
  } catch (Exception e) {
    e.printStackTrace();
  }
  // 关闭 Elasticsearch 连接
  closeEs();
}
// 解析 BulkResponse
private void processBulkResponse(BulkResponse bulkResponse) {
  if (bulkResponse == null) {
    return;
  }
  for (BulkItemResponse bulkItemResponse : bulkResponse) {
    DocWriteResponse itemResponse = bulkItemResponse.getResponse();
    switch (bulkItemResponse.getOpType()) {
      // 索引状态
      case INDEX:
        // 索引生成
```

```java
            case CREATE:
              IndexResponse indexResponse = (IndexResponse) itemResponse;
              String index = indexResponse.getIndex();
              String id = indexResponse.getId();
              long version = indexResponse.getVersion();
              log.info("create id is " + id + ", index is " + index + ",version is
                  " + version);
              break;
            // 索引更新
            case UPDATE:
              UpdateResponse updateResponse = (UpdateResponse) itemResponse;
              break;
            // 索引删除
            case DELETE:
              DeleteResponse deleteResponse = (DeleteResponse) itemResponse;
      }
    }
  }
```

随后编译工程，在工程根目录下输入如下命令：

```
mvn clean package
```

通过如下命令启动工程服务：

```
java -jar ./target/esdemo-0.0.1-SNAPSHOT.jar
```

在工程服务启动后，在浏览器中通过调用如下接口查看批量请求结果的解析情况：

```
http://localhost:8080/springboot/es/high/index/bulk?indexName=ultraman&field=content
```

在请求执行后，浏览器中显示内容如下所示，证明接口执行成功：

```
Bulk Get In High ElasticSearch Client Successed!
```

在服务器中输出的内容如下所示。

```
2019-08-01 19:49:18.811  INFO 72252 --- [nio-8080-exec-2] n.e.s.i.MeetHighElasticSearchServiceImpl : create id is 1, index is ultraman,version is 5
2019-08-01 19:49:18.814  INFO 72252 --- [nio-8080-exec-2] n.e.s.i.MeetHighElasticSearchServiceImpl : create id is 2, index is ultraman,version is 1
2019-08-01 19:49:18.819  INFO 72252 --- [nio-8080-exec-2] n.e.s.i.MeetHighElasticSearchServiceImpl : create id is 3, index is ultraman,version is 1
```

三个请求在同一次批量请求中已经成功执行。因为文档 1 已经进行了 5 次处理，此时 version 为 5；而文档 2、3 刚刚进行索引，因此 version 为 1。

6.2 批量处理器

除批量请求接口 Bulk API 外，Elasticsearch 还提供了 BulkProcessor 进行批量操作处理。BulkProcessor 提供了一个实用程序类，来简化批量 Bulk API 的使用。BulkProcessor 允许将索引、更新、删除文档的操作添加到处理器中透明地执行。

为了执行批量处理请求，BulkProcessor 需要依赖如下组件。

（1）RestHighLevelClient：此客户端用于执行 BulkRequest 和检索 BulkResponse。

（2）BulkProcessor.Listener：在每次执行 BulkRequest 之前和之后，或者当 BulkRequest 失败时，都会调用此监听器。

BulkProcessor 的构建代码如下所示：

```
// 构建 BulkProcessor
 public void buildBulkRequestWithBulkProcessor(String indexName, String
        field) {
   BulkProcessor.Listener listener = new BulkProcessor.Listener() {
     @Override
     public void beforeBulk(long executionId, BulkRequest request) {
       // 批量处理前的动作
     }
     @Override
     public void afterBulk(long executionId, BulkRequest request,
         BulkResponse response) {
       // 批量处理后的动作
     }
     @Override
     public void afterBulk(long executionId, BulkRequest request, Throwable
         failure) {
       // 批量处理后的动作
     }
 };
 BulkProcessor bulkProcessor = BulkProcessor.builder((request, bulkListener)
     -> restClient
         .bulkAsync(request, RequestOptions.DEFAULT, bulkListener),
         listener).build();
 }
```

在 BulkProcessor 中，用户还可以根据当前添加的操作数设置刷新批量请求的时间、根据当前添加的操作大小设置刷新批量请求的时间、设置允许执行的并发请求数、设置刷新间隔，以及设置后退策略等。如上述代码所述，BulkProcessor.Builder 提供了配置 BulkProcessor 批量处理器处理请求执行的方法，我们在上述方法 buildBulkRequestWithBulkProcessor 中添加如下代码：

```java
/*
 * BulkProcessor 的配置
 */
    BulkProcessor.Builder builder = BulkProcessor.builder((request, bulkListener)
        -> restClient
            .bulkAsync(request, RequestOptions.DEFAULT, bulkListener), listener);

    // 根据当前添加的操作数，设置刷新批量请求的时间（默认值为1000，使用-1表示禁用）
    builder.setBulkActions(500);

    // 根据当前添加的操作大小，设置刷新批量请求的时间（默认为5MB，使用-1表示禁用）
    builder.setBulkSize(new ByteSizeValue(1L, ByteSizeUnit.MB));

    // 设置允许执行的并发请求数（默认为1，当使用0时表示仅允许执行单个请求）
    builder.setConcurrentRequests(0);

    // 设置刷新间隔（默认为未设置）
    builder.setFlushInterval(TimeValue.timeValueSeconds(10L));

    // 设置一个恒定的后退策略，该策略最初等待1s，最多重试3次
    builder.setBackoffPolicy(BackoffPolicy.constantBackoff(TimeValue.
        timeValueSeconds(1L), 3));
```

BulkProcessor 类提供了一个简单的接口，它可以根据请求数量或在指定的时间段后自动地刷新批量操作。

在创建完 BulkProcessor 后，用户就可以向其添加请求了。我们在上述方法 buildBulkRequestWithBulkProcessor 中添加如下代码：

```java
/**
 * 添加索引请求
 */
    IndexRequest one = new IndexRequest(indexName).id("6").source (XContentType.
        JSON, "title",
        "8月1日，中国空军发布强军宣传片《初心伴我去战斗》，通过歼-20、轰-6K等新型战机练
        兵备战的震撼场景，展现新时代空军发展的新气象，彰显中国空军维护国家主权、保卫国家安
        全、保障和平发展的意志和能力。");
    IndexRequest two = new IndexRequest(indexName).id("7").source (XContentType.
        JSON, "title",
        "在2分钟的宣传片中，中国空军现役先进战机悉数亮相，包括歼-20、歼-16、歼-11、歼-
        10B/C、苏-35、苏-27、轰-6K等机型");
    IndexRequest three = new IndexRequest(indexName).id("8").source (XContentType.
```

```
            JSON, "title",
            "宣传片发布正逢八一建军节,而今年是新中国成立70周年,也是人民空军成立70周年。70
            年来,中国空军在各领域取得全面发展,战略打击、战略预警、空天防御和战略投送等能力得
            到显著进步。");
    bulkProcessor.add(one);
    bulkProcessor.add(two);
bulkProcessor.add(three);
```

上述代码中的三个文档索引请求将由 BulkProcessor 执行,它负责为每个批量请求调用 BulkProcessor.Listener,配置的监听器提供访问 BulkRequest 和 BulkResponse 的方法。代码为分三层,分别是 Controller 层、Service 层和 ServiceImpl 实现层。

在 Controller 层的 MeetHighElasticSearchController 类中添加如下代码:

```
// 以同步方式执行 BulkRequest
@RequestMapping("/index/bulkProcessor")
public String bulkProcessorGetInHighElasticSearch(String indexName, String
    field) {
  if (Strings.isEmpty(indexName) || Strings.isEmpty(field)) {
    return "Parameters are error!";
  }
  meetHighElasticSearchService.executeBulkRequestWithBulkProcessor
    (indexName, field);
  return "BulkProcessor Get In High ElasticSearch Client Successed!";
}
```

在 Service 层的 MeetHighElasticSearchService 类中添加如下代码:

```
//以同步方式执行 BulkRequest
public void executeBulkRequestWithBulkProcessor(String indexName, String
    field);
```

在 ServiceImpl 实现层的 MeetHighElasticSearchServiceImpl 类中添加如下代码:

```
//以同步方式执行 BulkRequest
  public void executeBulkRequestWithBulkProcessor(String indexName, String
    field) {
  BulkProcessor.Listener listener = new BulkProcessor.Listener() {
    @Override
    public void beforeBulk(long executionId, BulkRequest request) {
      // 批量处理前的动作
      int numberOfActions = request.numberOfActions();
      log.info("Executing bulk " + executionId + " with " + numberOfActions
        + " requests");
    }
```

```java
    @Override
    public void afterBulk(long executionId, BulkRequest request,
      BulkResponse response) {
      // 批量处理后的动作
      if (response.hasFailures()) {
        log.info("Bulk " + executionId + " executed with failures");
      } else {
        log.info("Bulk " + executionId + " completed in " +
          response.getTook().getMillis()
          + " milliseconds");
      }
    }
    @Override
    public void afterBulk(long executionId, BulkRequest request, Throwable
        failure) {
      // 批量处理后的动作
      log.error("Failed to execute bulk", failure);
    }
};
BulkProcessor bulkProcessor = BulkProcessor.builder((request, bulkListener) ->
    restClient
      .bulkAsync(request, RequestOptions.DEFAULT, bulkListener), listener).build();
/**
 * 添加索引请求
 */
IndexRequest one = new IndexRequest(indexName).id ("6").source (XContentType.
    JSON, "title",
    "8月1日,中国空军发布强军宣传片《初心伴我去战斗》,通过歼-20、轰-6K等新型战机练
    兵备战的震撼场景,展现新时代空军发展的新气象,彰显中国空军维护国家主权、保卫国家安
    全、保障和平发展的意志和能力。");
IndexRequest two = new IndexRequest(indexName).id("7").source (XContentType.
    JSON, "title",
    "在2分钟的宣传片中,中国空军现役先进战机悉数亮相,包括歼-20、歼-16、歼-11、歼-
    10B/C、苏-35、苏-27、轰-6K等机型");
IndexRequest three = new IndexRequest(indexName).id("8").source (XContentType.
    JSON, "title",
    "宣传片发布正逢八一建军节,而今年是新中国成立70周年,也是人民空军成立70周年。70
    年来,中国空军在各领域取得全面发展,战略打击、战略预警、空天防御和战略投送等能力得
    到显著进步。");
bulkProcessor.add(one);
bulkProcessor.add(two);
bulkProcessor.add(three);
}
```

随后编译工程，在工程根目录下输入如下命令：

```
mvn clean package
```

通过如下命令启动工程服务：

```
java -jar ./target/esdemo-0.0.1-SNAPSHOT.jar
```

在工程服务启动后，在浏览器中调用如下接口查看批量处理器对多请求的处理情况：

```
http://localhost:8080/springboot/es/high/index/bulkProcessor?indexName=ultra
man&field=content
```

在接口调用成功后，即可完成上述三个文档索引的批量写入。随后，调用文档查询接口查看文档的索引情况，如选择索引名称为 ultraman、文档编号为 6 的文档的索引，接口调用情况如下所示：

```
localhost:8080/springboot/es/high/index/get?indexName=ultraman&document=6
```

服务器的输出内容如下所示，可见批量操作的请求已经执行成功：

```
2019-08-02 11:06:49.909  INFO 80140 --- [nio-8080-exec-6] n.e.s.i.MeetHighElasticSearchServiceImpl : id is 6, index is u
ltraman
```

6.3 MultiGet 批量处理实战

Elasticsearch 不仅提供了批量请求接口 Bulk API，还提供了批量获取 API。批量获取 API 可以合并多个请求，以达到减少每个请求单独处理所需的网络开销。

如果用户需要从 Elasticsearch 中检索多个文档，则与一个一个的检索相比，更快的方法是在一个请求中使用 MultiGet 或者 MGet API。

顾名思义，MultiGet API 是在单个 HTTP 请求中并行执行多个 Get 请求。

1. 构建 MultiGet 批量处理请求

在执行 MultiGet 批量处理请求前，需要构建 MultiGet 批量处理请求，即 MultiGetRequest。MultiGetRequest 在初始化时为空，需要添加 MultiGetRequest.Item 以配置要提取的内容。MultiGetRequest.Item 有两个必选参数，即索引名称和文档 ID。构建 MultiGetRequest 的代码如下所示：

在 MeetHighElasticSearchServiceImpl 类中添加如下代码：

```
// 构建 MultiGetRequest
```

```java
public void buildMultiGetRequest(String indexName, String[] documentIds) {
    if (documentIds == null || documentIds.length <= 0) {
        return;
    }
    MultiGetRequest request = new MultiGetRequest();
    for (String documentId : documentIds) {
        //添加请求
        request.add(new MultiGetRequest.Item(indexName, documentId));
    }
}
```

类似地，MultiGetRequest 也有一些可选参数供用户配置。可选参数主要有禁用源检索、为特定字段配置源包含关系、为特定字段配置源排除关系、为特定存储字段配置检索、配置路由、配置版本和版本类型、配置偏好值、配置实时标志、配置在检索文档之前执行刷新策略等。

由此可见，MultiGet API 的可选参数列表与 Get API 的可选参数列表相同。

MultiGetRequest 的可选参数配置代码如下所示，在 buildMultiGetRequest 方法中添加如下代码：

```java
/*
 * 可选参数使用介绍
 */
    // 禁用源检索，在默认情况下启用
    request.add(new MultiGetRequest.Item(indexName, documentIds[0])
        .fetchSourceContext(FetchSourceContext.DO_NOT_FETCH_SOURCE));
    // 为特定字段配置源包含关系
    String[] excludes = Strings.EMPTY_ARRAY;
    String[] includes = {"title", "content"};
    FetchSourceContext fetchSourceContext = new FetchSourceContext(true,
        includes, excludes);
    request.add(
        new MultiGetRequest.Item(indexName, documentIds[0]). fetchSourceContext
            (fetchSourceContext));
    // 为特定字段配置源排除关系
    fetchSourceContext = new FetchSourceContext(true, includes, excludes);
    request.add(
        new MultiGetRequest.Item(indexName, documentIds[0]).
            fetchSourceContext(fetchSourceContext));
    // 为特定存储字段配置检索（要求字段在索引中单独存储字段）
    try {
        request.add(new MultiGetRequest.Item(indexName, documentIds
            [0]).storedFields("title"));
        MultiGetResponse response = restClient.mget(request, RequestOptions.
```

```
            DEFAULT);
    MultiGetItemResponse item = response.getResponses()[0];
    // 检索title存储字段(要求该字段单独存储在索引中)
    String value = item.getResponse().getField("title").getValue();
    log.info("value is " + value);
} catch (Exception e) {
    e.printStackTrace();
} finally {
    // 关闭Elasticsearch连接
    closeEs();
}
// 配置路由
request.add(new MultiGetRequest.Item(indexName, documentIds [0]).routing
    ("routing"));
// 配置版本和版本类型
request.add(new MultiGetRequest.Item(indexName, documentIds[0])
    .versionType(VersionType.EXTERNAL).version(10123L));
// 配置偏好值
request.preference("title");
// 将实时标志设置为假(默认为真)
request.realtime(false);
// 在检索文档之前执行刷新(默认为false)
request.refresh(true);
```

2. 执行 MultiGet 批量处理请求

在 MultiGetRequest 构建后，即可执行 MultiGet 批量处理请求。与文档索引请求类似，MultiGet 批量处理请求也有同步和异步两种执行方式。

同步方式

当以同步方式执行 MultiGet 批量处理请求时，客户端会等待 Elasticsearch 服务器返回的查询结果 MultiGetResponse。在收到 MultiGetResponse 后，客户端会继续执行相关的逻辑代码。以同步方式执行的核心代码如下所示：

```
//以同步方式执行MultiGetRequest
public void executeMultiGetRequest(String indexName, String[] documentIds)
{
    if (documentIds == null || documentIds.length <= 0) {
        return;
    }
    MultiGetRequest request = new MultiGetRequest();
    for (String documentId : documentIds) {
        // 添加请求
        request.add(new MultiGetRequest.Item(indexName, documentId));
```

```
    }
    try {
      MultiGetResponse response = restClient.mget(request,RequestOptions.
         DEFAULT);
    } catch (Exception e) {
      e.printStackTrace();
    } finally {
      // 关闭Elasticsearch连接
      closeEs();
    }
  }
```

异步方式

当以异步方式执行 MultiGet 批量处理请求时，高级客户端不必同步等待请求结果的返回，可以直接向接口调用方返回异步接口执行成功的结果。

为了处理异步返回的响应信息或处理在请求执行过程中引发的异常信息，用户需要指定监听器。

在异步请求处理后，如果请求执行成功，则调用 ActionListener 类中的 onResponse 方法进行相关逻辑的处理；如果请求执行失败，则调用 ActionListener 类中的 onFailure 方法进行相关逻辑的处理。

以异步方式执行的代码如下所示：

```
  // 以异步方式执行MultiGetRequest
  public void executeMultiGetRequestAsync(String indexName, String[]
      documentIds) {
    if (documentIds == null || documentIds.length <= 0) {
      return;
    }
    MultiGetRequest request = new MultiGetRequest();
    for (String documentId : documentIds) {
      // 添加请求
      request.add(new MultiGetRequest.Item(indexName, documentId));
    }
    // 添加ActionListener
    ActionListener listener = new ActionListener<MultiGetResponse>() {
      @Override
      public void onResponse(MultiGetResponse response) {
      }
      @Override
      public void onFailure(Exception e) {
      }
```

```
        };
        // 执行批量获取
        try {
         MultiGetResponse response = restClient.mget(request, RequestOptions.
            DEFAULT);
        } catch (Exception e) {
         e.printStackTrace();
        } finally {
         // 关闭Elasticsearch连接
         closeEs();
        }
       }
```

当然,在异步请求执行过程中可能会出现异常,对异常的处理与同步方式执行情况相同。

3. 解析 MultiGet 批量处理请求的响应结果

不论同步方式,还是异步方式,在执行 MultiGet 批量处理后,客户端均会对返回结果 MultiGetResponse 进行处理和解析。

MultiGetResponse 中包含了 MultiGetItemResponse 的列表,列表中的元素顺序与请求的顺序相同。如果 Get 请求执行成功,则包含 GetResponse;如果 Get 请求执行失败,则 MultiGetItemResponse 中包含 MultiGetResponse.failure。

以同步执行方式为例,MultiGetResponse 的解析代码分为三层,分别是 Controller 层、Service 层和 ServiceImpl 实现层。

在 Controller 层的 MeetHighElasticSearchController 类中添加如下代码:

```
// 以同步方式执行MultiGetreques
@RequestMapping("/index/multiget")
public String multigetInHighElasticSearch(String indexName, String
    documentId) {
 if (Strings.isEmpty(indexName) || Strings.isEmpty(documentId)) {
  return "Parameters are error!";
 }
 // 将field(英文逗号分隔的)转化成String[]
 List<String> documentIds = Splitter.on(",").splitToList(documentId);
 meetHighElasticSearchService.executeMultiGetRequest(indexName,
    documentIds.toArray(new String[documentIds.size()]));
 return "MultiGet In High ElasticSearch Client Successed!";
}
```

在 Service 层的 MeetHighElasticSearchService 类中添加如下代码:

```
// 以同步方式执行 MultiGetRequest
public void executeMultiGetRequest(String indexName, String[] documentIds);
```

在 ServiceImpl 实现层的 MeetHighElasticSearchServiceImpl 类中添加如下代码：

```
// 以同步方式执行 MultiGetRequest
public void executeMultiGetRequest(String indexName, String[] documentIds)
  {
  if (documentIds == null || documentIds.length <= 0) {
    return;
  }
  MultiGetRequest request = new MultiGetRequest();
  for (String documentId : documentIds) {
    // 添加请求
    request.add(new MultiGetRequest.Item(indexName, documentId));
  }
  try {
    MultiGetResponse response = restClient.mget(request, RequestOptions.
        DEFAULT);
    // 解析 MultiGetResponse
    processMultiGetResponse(response);
  } catch (Exception e) {
    e.printStackTrace();
  } finally {
    // 关闭 Elasticsearch 连接
    closeEs();
  }
}
// 解析 MultiGetResponse
private void processMultiGetResponse(MultiGetResponse multiResponse) {
  if (multiResponse == null) {
    return;
  }
  MultiGetItemResponse[] responses = multiResponse.getResponses();
  log.info("responses is " + responses.length);
  for (MultiGetItemResponse response : responses) {
    GetResponse getResponse = response.getResponse();
    String index = response.getIndex();
    String id = response.getId();
    log.info("index is " + index + ";id is " + id);
    if (getResponse.isExists()) {
      long version = getResponse.getVersion();
      // 按字符串方式获取内容
      String sourceAsString = getResponse.getSourceAsString();
```

```
        // 按Map方式获取内容
        Map<String, Object> sourceAsMap = getResponse.getSourceAsMap();
        // 按字节数组方式获取内容
        byte[] sourceAsBytes = getResponse.getSourceAsBytes();
        log.info("version is " + version + ";sourceAsString is " +
            sourceAsString);
      }
    }
  }
```

随后编译工程，在工程根目录下输入如下命令：

```
mvn clean package
```

通过如下命令启动工程服务：

```
java -jar ./target/esdemo-0.0.1-SNAPSHOT.jar
```

在工程服务启动后，在浏览器中调用如下接口查看 MultiGETRequest 请求的执行情况：

```
http://localhost:8080/springboot/es/high/index/multiget?indexName=ultraman&documentId=6,7,8
```

服务器中的输出结果如下所示：

```
2019-08-02 13:51:33.624  INFO 59948 --- [nio-8080-exec-2] n.e.s.i.MeetHighElasticSearchServiceImpl : responses is 3
2019-08-02 13:51:33.626  INFO 59948 --- [nio-8080-exec-2] n.e.s.i.MeetHighElasticSearchServiceImpl : index is ultraman;id is 6
2019-08-02 13:51:33.627  INFO 59948 --- [nio-8080-exec-2] n.e.s.i.MeetHighElasticSearchServiceImpl : index is ultraman;id is 7
2019-08-02 13:51:33.628  INFO 59948 --- [nio-8080-exec-2] n.e.s.i.MeetHighElasticSearchServiceImpl : index is ultraman;id is 8
```

从输出结果可以看出，三个文档的索引查询请求均执行成功。

而在发出请求的浏览器页面上会显示如下内容，表明接口执行成功：

```
MultiGet In High ElasticSearch Client Successed!
```

6.4 文档 ReIndex 实战

文档 ReIndex，可以译为文档重新索引，用于从一个或更多的索引中复制相关的文档到一个新的索引中进行索引重建。因此，文档 ReIndex 请求需要一个现有的源索引和一个可能存在或不存在的目标索引。需要指出的是，文档 ReIndex 不尝试设置目标索引，它不会复制源索引的设置。因此，用户需要在运行文档 ReIndex 操作之前设置目标索引，包括设置映射、分片计数、副本等。

1. 构建文档 ReIndex 请求

在执行文档 ReIndex 请求前，需要构建文档 ReIndex 请求，即 ReindexRequest。ReindexRequest 的构建代码如下所示：

```
// 构建 ReindexRequest
public void bulidReindexRequest(String fromIndex, String toIndex) {
    ReindexRequest request = new ReindexRequest();
    // 添加要从中复制的源的列表
    request.setSourceIndices("source1", "source2",fromIndex);
    // 添加目标索引
    request.setDestIndex(toIndex);
}
```

类似地，ReindexRequest 也有一些可选参数供用户进行配置。可选参数主要有设置目标索引的版本类型、设置目标索引的操作类型、设置版本冲突时的处理策略、添加查询来限制文档的策略、设置大小来限制已处理文档的数量、设置 ReIndex 的批次、设置管道模式、设置排序策略、分片设置、滚动设置、超时时间设置、ReIndex 后刷新索引的策略设置等。

在 bulidReindexRequest 方法中添加可选参数配置的代码如下：

```
/*
 * ReindexRequest 的参数配置
 */
    // 设置目标索引的版本类型
    request.setDestVersionType(VersionType.EXTERNAL);
    // 设置目标索引的操作类型为创建类型
    request.setDestOpType("create");
    // 在默认情况下，版本冲突会中止重新索引进程，我们可以用以下方法计算它们
    request.setConflicts("proceed");
    // 通过添加查询限制文档。下面仅复制用户字段设置为 kimchy 的文档
    request.setSourceQuery(new TermQueryBuilder("user", "kimchy"));
    // 通过设置大小限制已处理文档的数量
    request.setSize(10);

    // 在默认情况下，ReIndex 使用 1000 个批次。可以使用 sourceBatchSize 更改批大小
    request.setSourceBatchSize(100);
    // 指定管道模式
    request.setDestPipeline("my_pipeline");
    // 如果需要用到源索引中的一组特定文档，则需要使用 sort。建议最好选择更具选择性的查询，
    // 而不是进行大小和排序
    request.addSortField("field1", SortOrder.DESC);
    request.addSortField("field2", SortOrder.ASC);
    // 使用切片滚动对 uid 进行切片。使用 setslices 指定要使用的切片数
```

```
request.setSlices(2);
// 使用 scroll 参数控制"search context",保持活动的时间
request.setScroll(TimeValue.timeValueMinutes(10));
// 设置超时时间
request.setTimeout(TimeValue.timeValueMinutes(2));
// 调用 reindex 后刷新索引
request.setRefresh(true);
```

在上述可选参数的配置中,versionType 的配置可以像索引 API 一样配置 dest 元素来控制乐观并发控制。如果省略 versionType 或将其设置为 Internal,就会导致 Elasticsearch 盲目地将文档转存到目标中。如果将 versionType 设置为 external,则 Elasticsearch 会保留源文件中的版本,并更新目标索引中版本比源文件索引中版本旧的所有文档。

当 opType 设置为 cause_reindex 时,会在目标索引中创建缺少的文档。所有现有文档都将导致版本冲突。默认 opType 是 index。

2. 执行文档 ReIndex 请求

在 ReindexRequest 构建后,即可执行文档重新索引请求。与文档索引请求类似,文档重新索引请求也有同步和异步两种执行方式。

同步方式

当以同步方式执行文档重新索引请求时,客户端会等待 Elasticsearch 服务器返回的查询结果 BulkByScrollResponse。在收到 BulkByScrollResponse 后,客户端会继续执行相关的逻辑代码。以同步方式执行的代码如下所示:

```
// 以同步方式执行 ReindexRequest
public void executeReindexRequest(String fromIndex, String toIndex) {
  ReindexRequest request = new ReindexRequest();
  // 添加要从中复制的源的列表
  request.setSourceIndices(fromIndex);
  // 添加目标索引
  request.setDestIndex(toIndex);
  try {
    BulkByScrollResponse bulkResponse = restClient.reindex(request,
        RequestOptions.DEFAULT);
  } catch (Exception e) {
    e.printStackTrace();
  } finally {
    // 关闭 Elasticsearch 连接
    closeEs();
  }
}
```

异步方式

当以异步方式执行文档重新索引请求时,高级客户端不必同步等待请求结果的返回,可以直接向接口调用方返回异步接口执行成功的结果。

为了处理异步返回的响应信息或处理在请求执行过程中引发的异常信息,用户需要指定监听器。以异步方式执行的代码如下所示:

```
client.reindexAsync(request, RequestOptions.DEFAULT, listener);
```

其中,listener 为监听器。

在异步请求处理后,如果请求执行成功,则调用 ActionListener 类中的 onResponse 方法进行相关逻辑的处理;如果请求执行失败,则调用 ActionListener 类中的 onFailure 方法进行相关逻辑的处理。

当然,在异步请求执行过程中可能会出现异常,异常的处理与同步方式执行情况相同。

以异步方式执行的完整代码如下所示:

```java
// 以异步方式执行ReindexRequest
public void executeReindexRequestAsync(String fromIndex, String toIndex) {
    ReindexRequest request = new ReindexRequest();
    // 添加要从中复制的源的列表
    request.setSourceIndices(fromIndex);
    // 添加目标索引
    request.setDestIndex(toIndex);
    // 构建监听器
    ActionListener listener = new ActionListener<BulkByScrollResponse>() {
        @Override
        public void onResponse(BulkByScrollResponse bulkResponse) {
        }
        @Override
        public void onFailure(Exception e) {
        }
    };
    try {
        restClient.reindexAsync(request, RequestOptions.DEFAULT, listener);
    } catch (Exception e) {
        e.printStackTrace();
    } finally {
        // 关闭Elasticsearch连接
        closeEs();
    }
}
```

3. 解析 ReIndex 请求的响应结果

不论同步请求，还是异步请求，客户端均需对返回结果 BulkByScrollResponse 进行处理和解析。返回的 BulkByScrollResponse 中包含有关已执行操作的信息，并允许按序对每个结果进行迭代解析。

BulkByScrollResponse 的解析代码如下所示，以同步请求方式为例，代码分为三层，分别是 Controller 层、Service 层和 ServiceImpl 实现层。

在 Controller 层的 MeetHighElasticSearchController 类中添加如下代码：

```java
// 以同步方式执行 ReindexRequest
@RequestMapping("/index/reindex")
public String reindexInHighElasticSearch(String fromIndexName, String
    toIndexName) {
  if (Strings.isEmpty(fromIndexName) || Strings.isEmpty(toIndexName)) {
    return "Parameters are error!";
  }
  meetHighElasticSearchService.executeReindexRequest(fromIndexName,
      toIndexName);
  return "Reindex In High ElasticSearch Client Successed!";
}
```

在 Service 层的 MeetHighElasticSearchService 类中添加如下代码：

```java
    // 以同步方式执行 ReindexRequest
public void executeReindexRequest(String fromIndex, String toIndex);
```

在 ServiceImpl 实现层的 MeetHighElasticSearchServiceImpl 类中添加如下代码：

```java
    // 以同步方式执行 ReindexRequest
public void executeReindexRequest(String fromIndex, String toIndex) {
  ReindexRequest request = new ReindexRequest();
  // 添加要从中复制的源的列表
  request.setSourceIndices(fromIndex);
  // 添加目标索引
  request.setDestIndex(toIndex);
  try {
    BulkByScrollResponse bulkResponse = restClient.reindex(request,
        RequestOptions.DEFAULT);
    // 解析 BulkByScrollResponse
    processBulkByScrollResponse(bulkResponse);
  } catch (Exception e) {
    e.printStackTrace();
  } finally {
    // 关闭 Elasticsearch 连接
```

```java
        closeEs();
    }
}
// 解析BulkByScrollResponse
private void processBulkByScrollResponse(BulkByScrollResponse bulkResponse) {
    if (bulkResponse == null) {
        return;
    }
    // 获取总耗时
    TimeValue timeTaken = bulkResponse.getTook();
    log.info("time is " + timeTaken.getMillis());
    // 检查请求是否超时
    boolean timedOut = bulkResponse.isTimedOut();
    log.info("timedOut is " + timedOut);
    // 获取已处理的文档总数
    long totalDocs = bulkResponse.getTotal();
    log.info("totalDocs is " + totalDocs);
    // 已更新的文档数
    long updatedDocs = bulkResponse.getUpdated();
    log.info("updatedDocs is " + updatedDocs);
    // 已创建的文档数
    long createdDocs = bulkResponse.getCreated();
    log.info("createdDocs is " + createdDocs);
    // 已删除的文档数
    long deletedDocs = bulkResponse.getDeleted();
    log.info("deletedDocs is " + deletedDocs);
    // 已执行的批次数
    long batches = bulkResponse.getBatches();
    log.info("batches is " + batches);
    // 跳过的文档数
    long noops = bulkResponse.getNoops();
    log.info("noops is " + noops);
    // 版本冲突数
    long versionConflicts = bulkResponse.getVersionConflicts();
    log.info("versionConflicts is " + versionConflicts);
    // 重试批量索引操作的次数
    long bulkRetries = bulkResponse.getBulkRetries();
    log.info("bulkRetries is " + bulkRetries);
    // 重试搜索操作的次数
    long searchRetries = bulkResponse.getSearchRetries();
    log.info("searchRetries is " + searchRetries);
    // 请求阻塞的总时间,不包括当前处于休眠状态的限制时间
    TimeValue throttledMillis = bulkResponse.getStatus().getThrottled();
    log.info("throttledMillis is " + throttledMillis.getMillis());
```

```
    // 查询失败数量
    List<ScrollableHitSource.SearchFailure> searchFailures = bulkResponse.
        getSearchFailures();
    log.info("searchFailures is " + searchFailures.size());
    // 批量操作失败数量
    List<BulkItemResponse.Failure> bulkFailures = bulkResponse.
        getBulkFailures();
    log.info("bulkFailures is " + bulkFailures.size());
}
```

随后编译工程，在工程根目录下输入如下命令：

```
mvn clean package
```

通过如下命令启动工程服务：

```
java -jar ./target/esdemo-0.0.1-SNAPSHOT.jar
```

在工程服务启动后，在浏览器中调用如下接口查看文档查询索引的情况：

```
http://localhost:8080/springboot/es/high/index/reindex?fromIndexName=ultraman&toIndexName=ultraman1
```

在接口调用成功后，如果浏览器中显示内容如下，则证明接口调用成功：

```
Reindex In High ElasticSearch Client Successed!
```

服务器中展示内容如下所示：

```
2019-08-02 14:41:11.193  INFO 80924 --- [nio-8080-exec-1] n.e.s.i.MeetHighElasticSearchServiceImpl : time is 1778
2019-08-02 14:41:11.195  INFO 80924 --- [nio-8080-exec-1] n.e.s.i.MeetHighElasticSearchServiceImpl : timedOut is false
2019-08-02 14:41:11.197  INFO 80924 --- [nio-8080-exec-1] n.e.s.i.MeetHighElasticSearchServiceImpl : totalDocs is 3
2019-08-02 14:41:11.198  INFO 80924 --- [nio-8080-exec-1] n.e.s.i.MeetHighElasticSearchServiceImpl : updatedDocs is 0
2019-08-02 14:41:11.199  INFO 80924 --- [nio-8080-exec-1] n.e.s.i.MeetHighElasticSearchServiceImpl : createdDocs is 3
2019-08-02 14:41:11.201  INFO 80924 --- [nio-8080-exec-1] n.e.s.i.MeetHighElasticSearchServiceImpl : deletedDocs is 0
2019-08-02 14:41:11.204  INFO 80924 --- [nio-8080-exec-1] n.e.s.i.MeetHighElasticSearchServiceImpl : batches is 1
2019-08-02 14:41:11.206  INFO 80924 --- [nio-8080-exec-1] n.e.s.i.MeetHighElasticSearchServiceImpl : noops is 0
2019-08-02 14:41:11.208  INFO 80924 --- [nio-8080-exec-1] n.e.s.i.MeetHighElasticSearchServiceImpl : versionConflicts is 0
2019-08-02 14:41:11.212  INFO 80924 --- [nio-8080-exec-1] n.e.s.i.MeetHighElasticSearchServiceImpl : bulkRetries is 0
2019-08-02 14:41:11.213  INFO 80924 --- [nio-8080-exec-1] n.e.s.i.MeetHighElasticSearchServiceImpl : searchRetries is 0
2019-08-02 14:41:11.215  INFO 80924 --- [nio-8080-exec-1] n.e.s.i.MeetHighElasticSearchServiceImpl : throttledMillis is 0
2019-08-02 14:41:11.217  INFO 80924 --- [nio-8080-exec-1] n.e.s.i.MeetHighElasticSearchServiceImpl : searchFailures is 0
2019-08-02 14:41:11.219  INFO 80924 --- [nio-8080-exec-1] n.e.s.i.MeetHighElasticSearchServiceImpl : bulkFailures is 0
```

从展示内容可以看出，文档重建索引操作已经成功，名称为 ultraman1 的索引中已经复制了索引名称为 ultraman 的数据。

这时，在浏览器中通过下面的接口查看名称为 ultraman1 的索引中的文档信息：

```
http://localhost:8080/springboot/es/high/index/get?indexName=ultraman1&document=1
```

服务器中展示内容如下所示：

```
2019-08-02 14:42:51.300 INFO 65956 --- [nio-8080-exec-2] n.e.s.i.MeetHighElasticSearchServiceImpl : id is 1, index is u
ltraman1
2019-08-02 14:42:51.383 INFO 65956 --- [nio-8080-exec-2] n.e.s.i.MeetHighElasticSearchServiceImpl : version is 1, sourc
eAsString is {"content":"事实上，自今年年初开始，美联储就已传递出货币政策或将转向的迹象"}
```

可见，ultraman 的重建索引已经成功。此时在新索引 ultraman1 中，文档 1 的版本号码为 1，为新建状态。而旧索引 ultraman 中的版本与此不同。我们可以通过下面的接口来查看：

```
http://localhost:8080/springboot/es/high/index/get?indexName=ultraman&document=1
```

此时，服务器中展示内容如下所示。

```
2019-08-02 14:46:03.580 INFO 80476 --- [nio-8080-exec-9] n.e.s.i.MeetHighElasticSearchServiceImpl : id is 1, index is u
ltraman
2019-08-02 14:46:03.653 INFO 80476 --- [nio-8080-exec-9] n.e.s.i.MeetHighElasticSearchServiceImpl : version is 5, sourc
eAsString is {"content":"事实上，自今年年初开始，美联储就已传递出货币政策或将转向的迹象"}
```

编号为 1 的文档在旧索引 ultraman 中的版本号为 5。

6.5 文档查询时更新实战

在文档查询时更新接口，以更新索引中的文档。与 update 的更新方式不同，查询时更新（update_by_query）是在不更改源的情况下对索引中的每个文档进行更新，用可以在文档查询时更新接口来修改字段或新增字段。

1．构建文档查询时更新请求

在执行文档查询时更新操作前，需要构建文档查询时更新的请求，即 UpdateByQueryRequest。UpdateByQueryRequest 的必选参数只有一个，即需要执行查询时更新操作的索引名称。

构建 UpdateByQueryRequest 的代码如下所示，我们在 MeetHighElasticSearchServiceImpl 类中添加如下代码：

```
// 构建 UpdateByQueryRequest
public void buildUpdateByQueryRequest(String indexName) {
  UpdateByQueryRequest request = new UpdateByQueryRequest(indexName);
}
```

类似地，UpdateByQueryRequest 也有可选参数供用户进行配置。可选参数有版本冲突处理策略、限制查询条件设置、设置大小来限制已处理文档的数量、设置批量处理的大小、设置管道名称、设置滚动模式、设置全局路由、设置超时时间、设置操作后的刷新策略和设置索引选项等。

UpdateByQueryRequest 的可选参数配置代码如下所示，我们将这部分代码添加在

MeetHighElasticSearchServiceImpl 类的 buildUpdateByQueryRequest 方法中：

```java
// 构建 UpdateByQueryRequest
public void buildUpdateByQueryRequest(String indexName) {
    UpdateByQueryRequest request = new UpdateByQueryRequest(indexName);

    /*
     * 配置 UpdateByQueryRequest
     */

    //在默认情况下，版本冲突将中止 UpdateByQueryRequest 进程，但我们可以使用以下方法来计
    //算它们
    request.setConflicts("proceed");

    //通过添加查询条件来限制。下面仅更新字段设置为 niudong 的文档
    request.setQuery(new TermQueryBuilder("user", "niudong"));

    //设置大小来限制已处理文档的数量
    request.setSize(10);

    //在默认情况下，UpdateByQueryRequest 使用的批数为 1000。可以使用 setBatchSize 更
    //改批大小
    request.setBatchSize(100);

    //指定管道模式
    request.setPipeline("my_pipeline");

    //设置分片滚动来并行化
    request.setSlices(2);

    //使用滚动参数控制"搜索上下文"，保持连接的时间
    request.setScroll(TimeValue.timeValueMinutes(10));

    //如果提供路由，那么路由将被复制到滚动查询，从而限制与该路由值匹配的分片处理
    request.setRouting("=cat");

    //设置等待请求的超时时间
    request.setTimeout(TimeValue.timeValueMinutes(2));

    //调用 update by query 后刷新索引
    request.setRefresh(true);

    //设置索引选项
    request.setIndicesOptions(IndicesOptions.LENIENT_EXPAND_OPEN);
```

    }

2. 执行文档查询时更新请求

在 UpdateByQueryRequest 构建后，即可执行文档查询时更新请求。与文档索引请求类似，文档查询时更新请求也有同步和异步两种执行方式。

同步方式

当以同步方式执行文档查询时更新请求时，客户端会等待 Elasticsearch 服务器返回的查询结果 UpdateByQueryResponse。在收到 UpdateByQueryResponse 后，客户端会继续执行相关的逻辑代码。以同步方式执行的代码如下所示：

```
// 以同步方式执行 UpdateByQueryRequest
public void executeUpdateByQueryRequest(String indexName) {
  UpdateByQueryRequest request = new UpdateByQueryRequest(indexName);
  try {
    BulkByScrollResponse bulkResponse = restClient.updateByQuery(request,
        RequestOptions.DEFAULT);
  } catch (Exception e) {
    e.printStackTrace();
  } finally {
    // 关闭 Elasticsearch 连接
    closeEs();
  }
}
```

异步方式

当以异步方式执行文档查询时更新请求时，高级客户端不必同步等待请求结果的返回，可以直接向接口调用方返回异步接口执行成功的结果。

为了处理异步返回的响应信息或处理在请求执行过程中引发的异常信息，用户需要指定监听器。

以异步方式执行的代码如下所示：

```
// 以异步方式执行 UpdateByQueryRequest
public void executeUpdateByQueryRequestAsync(String indexName) {
  UpdateByQueryRequest request = new UpdateByQueryRequest(indexName);
  // 添加监听器
  ActionListener listener = new ActionListener<BulkByScrollResponse>() {
    @Override
    public void onResponse(BulkByScrollResponse bulkResponse) {
    }
```

```
    @Override
    public void onFailure(Exception e) {
    }
};
try {
    restClient.updateByQueryAsync(request, RequestOptions.DEFAULT,
        listener);
} catch (Exception e) {
    e.printStackTrace();
} finally {
    // 关闭Elasticsearch连接
    closeEs();
}
```

在异步处理完成后,如果请求执行成功,则调用 ActionListener 类中的 onResponse 方法进行相关逻辑的处理;如果请求执行失败,则调用 ActionListener 类中的 onFailure 方法进行相关逻辑的处理。

3. 解析文档查询时更新请求的响应结果

不论同步方式,还是异步方式,在文档查询时更新请求执行后,客户端均需要对返回结果 UpdateByQueryResponse 进行处理和解析。UpdateByQueryResponse 中包含了有关已执行操作的信息,并允许用户按序对每个结果进行迭代解析。

下面以同步方式执行为例,展示对文档查询时更新请求的返回结果 UpdateByQueryResponse 的解析。代码共分为三层,分别是 Controller 层、Service 层和 ServiceImpl 实现层。

在 Controller 层的 MeetHighElasticSearchController 类中添加如下代码:

```
// 以同步方式执行UpdateByQueryResponse
@RequestMapping("/index/updateByQuery")
public String updateByQueryInHighElasticSearch(String indexName) {
    if (Strings.isEmpty(indexName)) {
        return "Parameters are error!";
    }
    meetHighElasticSearchService.executeUpdateByQueryRequest(indexName);
    return "UpdateByQuery In High ElasticSearch Client Successed!";
}
```

在 Service 层的 MeetHighElasticSearchService 类中添加如下代码:

```
// 以同步方式执行 UpdateByQueryRequest
  public void executeUpdateByQueryRequest(String indexName);
```

在 ServiceImpl 实现层的 MeetHighElasticSearchServiceImpl 类中添加如下代码:

```
// 以同步方式执行 UpdateByQueryRequest
  public void executeUpdateByQueryRequest(String indexName) {
    UpdateByQueryRequest request = new UpdateByQueryRequest(indexName);
    try {
      BulkByScrollResponse bulkResponse = restClient.updateByQuery(request,
          RequestOptions.DEFAULT);
      // 处理 BulkByScrollResponse
      processBulkByScrollResponse(bulkResponse);
    } catch (Exception e) {
      e.printStackTrace();
    } finally {
      // 关闭 Elasticsearch 连接
      closeEs();
    }
  }
```

其中，processBulkByScrollResponse 方法与文档 ReIndex 接口中的同名方法相同。

随后编译工程，在工程根目录下输入如下命令:

```
mvn clean package
```

通过如下命令启动工程服务:

```
java -jar ./target/esdemo-0.0.1-SNAPSHOT.jar
```

在工程服务启动后，在浏览器中调用如下接口查看文档查询时更新请求的执行情况:

```
http://localhost:8080/springboot/es/high/index/updateByQuery?indexName=ultraman1
```

在接口执行后，我们可以在服务器中看到如下所示内容:

```
2019-08-02 15:43:26.988 INFO 80032 --- [nio-8080-exec-3] n.e.s.i.MeetHighElasticSearchServiceImpl : time is 201
2019-08-02 15:43:26.991 INFO 80032 --- [nio-8080-exec-3] n.e.s.i.MeetHighElasticSearchServiceImpl : timedOut is false
2019-08-02 15:43:26.994 INFO 80032 --- [nio-8080-exec-3] n.e.s.i.MeetHighElasticSearchServiceImpl : totalDocs is 3
2019-08-02 15:43:26.997 INFO 80032 --- [nio-8080-exec-3] n.e.s.i.MeetHighElasticSearchServiceImpl : updatedDocs is 3
2019-08-02 15:43:27.000 INFO 80032 --- [nio-8080-exec-3] n.e.s.i.MeetHighElasticSearchServiceImpl : createdDocs is 0
2019-08-02 15:43:27.003 INFO 80032 --- [nio-8080-exec-3] n.e.s.i.MeetHighElasticSearchServiceImpl : deletedDocs is 0
2019-08-02 15:43:27.005 INFO 80032 --- [nio-8080-exec-3] n.e.s.i.MeetHighElasticSearchServiceImpl : batches is 1
2019-08-02 15:43:27.009 INFO 80032 --- [nio-8080-exec-3] n.e.s.i.MeetHighElasticSearchServiceImpl : noops is 0
2019-08-02 15:43:27.013 INFO 80032 --- [nio-8080-exec-3] n.e.s.i.MeetHighElasticSearchServiceImpl : versionConflicts is 0
2019-08-02 15:43:27.017 INFO 80032 --- [nio-8080-exec-3] n.e.s.i.MeetHighElasticSearchServiceImpl : bulkRetries is 0
2019-08-02 15:43:27.019 INFO 80032 --- [nio-8080-exec-3] n.e.s.i.MeetHighElasticSearchServiceImpl : searchRetries is 0
2019-08-02 15:43:27.022 INFO 80032 --- [nio-8080-exec-3] n.e.s.i.MeetHighElasticSearchServiceImpl : throttledMillis is 0
2019-08-02 15:43:27.026 INFO 80032 --- [nio-8080-exec-3] n.e.s.i.MeetHighElasticSearchServiceImpl : searchFailures is 0
2019-08-02 15:43:27.031 INFO 80032 --- [nio-8080-exec-3] n.e.s.i.MeetHighElasticSearchServiceImpl : bulkFailures is 0
```

从中可以看到，索引名称为 ultraman1 的 3 个文档进行了 3 次更新。

此时，如果浏览器中显示内容如下所示，则证明接口执行成功：

```
UpdateByQuery In High ElasticSearch Client Successed!
```

6.6 文档查询时删除实战

除文档查询时更新外，Elasticsearch 还提供了在查询时删除的接口。

1．构建文档查询时删除请求

在执行文档查询时删除请求前，需要先构建查询时删除请求，即 DeleteByQueryRequest。DeleteByQueryRequest 有一个必选参数，即执行查询时删除操作的现有索引名称。

构建 DeleteByQueryRequest 的代码如下所示：

```
// 构建 DeleteByQueryRequest
public void buildDeleteByQueryRequest (String indexName) {
    DeleteByQueryRequest  request = new DeleteByQueryRequest (indexName);
}
```

类似地，DeleteByQueryRequest 也有一些可选参数供用户进行配置。可选参数有版本冲突处理策略、限制查询条件设置、设置大小来限制已处理文档的数量、设置批量处理的大小、设置管道名称、设置滚动模式、设置全局路由、设置超时时间、设置操作后的刷新策略、设置索引选项等。

DeleteByQueryRequest 的可选参数设置代码如下所示，我们将这部分代码添加到 MeetHighElasticSearchServiceImpl 类的 buildDeleteByQueryRequest 方法中：

```
// 构建 DeleteByQueryRequest
 public void buildDeleteByQueryRequest (String indexName) {
    DeleteByQueryRequest  request = new DeleteByQueryRequest (indexName);
    /*
     * 配置 DeleteByQueryRequest
     */
    //在默认情况下，版本冲突将中止 DeleteByQueryRequest 进程，但我们可以使用以下方法来
    //计算它们
    request.setConflicts("proceed");
    // 通过添加查询条件进行限制。例如仅删除用户字段设置为 niudong 的文档
    request.setQuery(new TermQueryBuilder("user", "niudong"));
    // 设置大小，限制已处理文档的数量
    request.setSize(10);
    //在默认情况下，UpdateByQueryRequest 使用的批数为 1000。可以使用 setBatchSize 更
```

```
    //改批大小
    request.setBatchSize(100);
    // 设置分片滚动，以实现并行化
    request.setSlices(2);
    // 使用滚动参数控制"搜索上下文"，保持连接的时间
    request.setScroll(TimeValue.timeValueMinutes(10));
    // 如果提供路由，那么路由将被复制到滚动查询，从而限制与该路由值匹配的分片处理
    request.setRouting("=cat");
    //设置等待请求的超时时间
    request.setTimeout(TimeValue.timeValueMinutes(2));
    // 调用 update by query 后刷新索引
    request.setRefresh(true);
    // 设置索引选项
    request.setIndicesOptions(IndicesOptions.LENIENT_EXPAND_OPEN);
}
```

2．执行文档查询时删除请求

在 DeleteByQueryRequest 构建后，即可执行文档查询时删除请求。与文档索引请求类似，文档查询时删除请求也有同步和异步两种执行方式。

同步方式

当以同步方式执行文档查询时删除请求时，客户端会等待 Elasticsearch 服务器返回的查询结果 DeleteByQueryResponse。在收到 DeleteByQueryResponse 后，客户端会继续执行相关的逻辑代码。以同步方式执行的代码如下所示：

```
// 以同步方式执行 DeleteByQueryRequest
public void executeDeleteByQueryRequest(String indexName) {
    DeleteByQueryRequest request = new DeleteByQueryRequest(indexName);
    try {
        BulkByScrollResponse bulkResponse = restClient.deleteByQuery(request,
            RequestOptions.DEFAULT);
    } catch (Exception e) {
        e.printStackTrace();
    } finally {
        // 关闭 Elasticsearch 连接
        closeEs();
    }
}
```

异步方式

当以异步方式执行 DeleteByQueryRequest 时，高级客户端不必同步等待请求结果的返回，可以直接向接口调用方返回异步接口执行成功的结果。

为了处理异步返回的响应信息或处理在请求执行过程中引发的异常信息，用户需要指定监听器。以异步方式执行的代码如下所示：

```java
//以异步方式执行 DeleteByQueryRequest
public void executeDeleteByQueryRequestAsync(String indexName) {
    DeleteByQueryRequest request = new DeleteByQueryRequest(indexName);
    // 构建监听器
    ActionListener listener = new ActionListener<BulkByScrollResponse>() {
      @Override
      public void onResponse(BulkByScrollResponse bulkResponse) {
      }
      @Override
      public void onFailure(Exception e) {
      }
    };
    // 执行 DeleteByQuery
    try {
      restClient.deleteByQueryAsync(request, RequestOptions.DEFAULT, listener);
    } catch (Exception e) {
      e.printStackTrace();
    } finally {
      // 关闭Elasticsearch连接
      closeEs();
    }
}
```

在异步请求处理后，如果请求执行成功，则调用 ActionListener 类中的 onResponse 方法进行相关逻辑的处理；如果请求执行失败，则调用 ActionListener 类中的 onFailure 方法进行相关逻辑的处理。

当然，在异步请求执行过程中可能会出现异常，异常的处理与同步方式执行情况相同。

3．解析文档查询时删除请求的响应结果

不论同步方式，还是异步方式，客户端均需要对文档查询时删除请求的响应结果 DeleteByQueryResponse 进行处理和解析。DeleteByQueryResponse 中包含有关已执行操作的信息，并允许用户按序对每个结果进行迭代解析。

以同步方式执行文档查询时删除请求为例，响应结果 DeleteByQueryResponse 的解析示例代码如下所示。代码共分为三层，分别是 Controller 层、Service 层和 ServiceImpl 实现层。

以同步方式执行的代码如下所示：

在 Controller 层的 MeetHighElasticSearchController 类中添加如下代码：

```
//以同步方式执行 DeleteByQueryRequest
@RequestMapping("/index/deleteByQuery")
public String deleteByQueryInHighElasticSearch(String indexName) {
  if (Strings.isEmpty(indexName)) {
    return "Parameters are error!";
  }
  meetHighElasticSearchService.executeDeleteByQueryRequest(indexName);
  return "DeleteByQuery In High ElasticSearch Client Successed!";
}
```

在 Service 层的 MeetHighElasticSearchService 类中添加如下代码：

```
// 以同步方式执行 DeleteByQueryRequest
public void executeDeleteByQueryRequest(String indexName);
```

在 ServiceImpl 实现层的 MeetHighElasticSearchServiceImpl 类中添加如下代码：

```
//以同步方式执行 DeleteByQueryRequest
public void executeDeleteByQueryRequest(String indexName) {
  DeleteByQueryRequest request = new DeleteByQueryRequest(indexName);
  // 通过添加查询条件进行限制。例如，仅删除字段 content 设置为 niudong 的文档
  request.setQuery(new TermQueryBuilder("content", "niudong"));
  try {
    BulkByScrollResponse bulkResponse = restClient.deleteByQuery(request,
        RequestOptions.DEFAULT);
    // 处理 BulkByScrollResponse
    processBulkByScrollResponse(bulkResponse);
  } catch (Exception e) {
    e.printStackTrace();
  } finally {
    // 关闭 Elasticsearch 连接
    closeEs();
  }
}
```

随后编译工程，在工程根目录下输入如下命令：

```
mvn clean package
```

通过如下命令启动工程服务：

```
java -jar ./target/esdemo-0.0.1-SNAPSHOT.jar
```

在工程服务启动后，在浏览器中调用如下接口查看文档查询时删除请求的执行情况：

```
http://localhost:8080/springboot/es/high/index/deleteByQuery?indexName=ultraman1
```

在接口执行后，如果浏览器中显示内容如下所示，则表明接口执行成功：

```
DeleteByQuery In High ElasticSearch Client Successed!
```

此时，在服务器中输出内容如下所示：

```
2019-08-02 16:01:07.911  INFO 82916 --- [nio-8080-exec-1] n.e.s.i.MeetHighElasticSearchServiceImpl : time is 51
2019-08-02 16:01:07.913  INFO 82916 --- [nio-8080-exec-1] n.e.s.i.MeetHighElasticSearchServiceImpl : timedOut is false
2019-08-02 16:01:07.918  INFO 82916 --- [nio-8080-exec-1] n.e.s.i.MeetHighElasticSearchServiceImpl : totalDocs is 0
2019-08-02 16:01:07.920  INFO 82916 --- [nio-8080-exec-1] n.e.s.i.MeetHighElasticSearchServiceImpl : updatedDocs is 0
2019-08-02 16:01:07.923  INFO 82916 --- [nio-8080-exec-1] n.e.s.i.MeetHighElasticSearchServiceImpl : createdDocs is 0
2019-08-02 16:01:07.924  INFO 82916 --- [nio-8080-exec-1] n.e.s.i.MeetHighElasticSearchServiceImpl : deletedDocs is 0
2019-08-02 16:01:07.926  INFO 82916 --- [nio-8080-exec-1] n.e.s.i.MeetHighElasticSearchServiceImpl : batches is 0
2019-08-02 16:01:07.932  INFO 82916 --- [nio-8080-exec-1] n.e.s.i.MeetHighElasticSearchServiceImpl : noops is 0
2019-08-02 16:01:07.936  INFO 82916 --- [nio-8080-exec-1] n.e.s.i.MeetHighElasticSearchServiceImpl : versionConflicts is 0
2019-08-02 16:01:07.940  INFO 82916 --- [nio-8080-exec-1] n.e.s.i.MeetHighElasticSearchServiceImpl : bulkRetries is 0
2019-08-02 16:01:07.944  INFO 82916 --- [nio-8080-exec-1] n.e.s.i.MeetHighElasticSearchServiceImpl : searchRetries is 0
2019-08-02 16:01:07.954  INFO 82916 --- [nio-8080-exec-1] n.e.s.i.MeetHighElasticSearchServiceImpl : throttledMillis is
2019-08-02 16:01:07.960  INFO 82916 --- [nio-8080-exec-1] n.e.s.i.MeetHighElasticSearchServiceImpl : searchFailures is 0
2019-08-02 16:01:07.962  INFO 82916 --- [nio-8080-exec-1] n.e.s.i.MeetHighElasticSearchServiceImpl : bulkFailures is 0
```

由于没有 content 为 niudong 的文档，所以查询时删除的文档为 0。

此时，我们执行文档编号为 6、7、8 的三个文档查询，验证索引名称为 ultraman1 的索引中仅有的这三个文档是否依然存在，可以通过如下接口验证：

```
http://localhost:8080/springboot/es/high/index/multiget?indexName=ultraman1&
documentId=6,7,8
```

在请求执行后，我们可以在服务器中看到如下所示的内容输出，表明文档编号为 6、7、8 的三个文档依然存在：

```
2019-08-02 16:04:40.120  INFO 82568 --- [nio-8080-exec-1] n.e.s.i.MeetHighElasticSearchServiceImpl : responses is 3
2019-08-02 16:04:40.123  INFO 82568 --- [nio-8080-exec-1] n.e.s.i.MeetHighElasticSearchServiceImpl : index is ultraman1;
id is 6
2019-08-02 16:04:40.137  INFO 82568 --- [nio-8080-exec-1] n.e.s.i.MeetHighElasticSearchServiceImpl : index is ultraman1;
id is 7
2019-08-02 16:04:40.143  INFO 82568 --- [nio-8080-exec-1] n.e.s.i.MeetHighElasticSearchServiceImpl : index is ultraman1;
id is 8
```

6.7 获取文档索引的多词向量

与其他接口类似，词向量接口也有批量实现的方式，即多词向量接口。多词向量接口允许用户一次获取多个词向量信息。

1. 构建多词向量请求

在执行多词向量请求前，我们需要构建多词向量请求，即 MultitemVectorsRequest。Elasticsearch 提供了两种构建方法，分别是：

（1）创建一个空的 MultiTermVectorsRequest，然后向其添加单个 Term Vectors 请求。

（2）在所有词向量请求共享相同参数（如索引和其他设置）时，可以使用所有必要的设置创建模板 TermVectorsRequest，并且可以将此模板请求连同执行这些请求的所有文档 ID，传递给 MultitemVectorsRequest 对象。

两种构建多终端向量请求的代码如下所示：

```
// 构建 MultiTermVectorsRequest
public void buildMultiTermVectorsRequest(String indexName, String[]
   documentIds, String field) {
 // 方法1：创建一个空的 MultiTermVectorsRequest，向其添加单个 Term Vectors 请求
 MultiTermVectorsRequest request = new MultiTermVectorsRequest();
 for (String documentId : documentIds) {
   TermVectorsRequest tvrequest = new TermVectorsRequest(indexName,
       documentId);
   tvrequest.setFields(field);
   request.add(tvrequest);
 }

 // 方法2：所有词向量请求共享相同参数（如索引和其他设置）
 TermVectorsRequest tvrequestTemplate = new TermVectorsRequest(indexName,
     "1");
 tvrequestTemplate.setFields(field);
 String[] ids = {"1", "2"};
 request = new MultiTermVectorsRequest(ids, tvrequestTemplate);
}
```

2. 执行多词向量请求

在 MultitemVectorsRequest 构建后，即可执行多词向量请求。与文档索引请求类似，多词向量请求也有同步和异步两种执行方式。

同步方式

当以同步方式执行多词向量请求时，客户端会等待 Elasticsearch 服务器返回的查询结果 MultiTermVectorsResponse。在收到 MultiTermVectorsResponse 后，客户端会继续执行相关的逻辑代码。以同步方式执行的代码如下所示：

```
// 以同步方式执行 MultiTermVectorsRequest
public void executeMultiTermVectorsRequest(String indexName, String[]
documentIds, String field) {
 // 方法1：创建一个空的 MultiTermVectorsRequest，向其添加单个 Term Vectors 请求
 MultiTermVectorsRequest request = new MultiTermVectorsRequest();
 for (String documentId : documentIds) {
   TermVectorsRequest tvrequest = new TermVectorsRequest(indexName,
       documentId);
```

```
            tvrequest.setFields(field);
            request.add(tvrequest);
        }
        try {
            MultiTermVectorsResponse response = restClient.mtermvectors(request,
RequestOptions.DEFAULT);
        } catch (Exception e) {
            e.printStackTrace();
        } finally {
            // 关闭Elasticsearch连接
            closeEs();
        }
    }
```

异步方式

当以异步方式执行多词向量请求时,高级客户端不必同步等待请求结果的返回,可以直接向接口调用方返回异步接口执行成功的结果。

为了处理异步返回的响应信息或处理在请求执行过程中引发的异常信息,用户需要指定监听器。以异步方式执行的代码如下所示:

```
//以异步方式执行MultiTermVectorsRequest
  public void executeMultiTermVectorsRequestAsync(String indexName, String[]
    documentIds,
    String field) {
    // 方法1:创建一个空的MultiTermVectorsRequest,向其添加单个Term Vectors请求
    MultiTermVectorsRequest request = new MultiTermVectorsRequest();
    for (String documentId : documentIds) {
      TermVectorsRequest tvrequest = new TermVectorsRequest(indexName,
          documentId);
      tvrequest.setFields(field);
      request.add(tvrequest);
    }
    // 构建监听器
    ActionListener listener = new ActionListener<MultiTermVectorsResponse>()
{
        @Override
        public void onResponse(MultiTermVectorsResponse mtvResponse) {
        }
        @Override
        public void onFailure(Exception e) {
        }
    };
```

```
    try {
        restClient.mtermvectorsAsync(request, RequestOptions.DEFAULT, listener);
    } catch (Exception e) {
        e.printStackTrace();
    } finally {
        // 关闭Elasticsearch连接
        closeEs();
    }
}
```

在异步请求处理后，如果请求执行成功，则调用 ActionListener 类中的 onResponse 方法进行相关逻辑的处理；如果请求执行失败，则调用 ActionListener 类中的 onFailure 方法进行相关逻辑的处理。

当然，在异步请求执行过程中可能会出现异常，异常的处理与同步方式执行情况相同。

3. 解析多词向量请求的响应结果

不论同步方式，还是异步方式，客户端均需要对多词向量请求的查询结果 MultiTermVectorsResponse 进行处理和解析。MultitermVectorsResponse 允许用户获取术语向量响应列表，每个响应都可以按照词向量 API 中的描述进行检查。

下面以同步方式执行多词向量请求为例，展示返回结果 MultiTermVectorsResponse 的解析方法。解析 MultiTermVectorsResponse 的代码。代码共分为三层，分别是 Controller 层、Service 层和 ServiceImpl 实现层。

其中，在 Controller 层的 MeetHighElasticSearchController 类中添加如下代码：

```
// 以同步方式执行 MultiTermVectorsRequest
@RequestMapping("/index/multiterm")
public String multitermInHighElasticSearch(String indexName, String
    documentId, String field) {
    if (Strings.isEmpty(indexName) || Strings.isEmpty(documentId) ||
      Strings.isEmpty(field)) {
        return "Parameters are error!";
    }
    // 将 field(英文逗号分隔的)转化成 String[]
    List<String> documentIds = Splitter.on(",").splitToList(documentId);
    meetHighElasticSearchService.executeMultiTermVectorsRequest(indexName,
        documentIds.toArray(new String[documentIds.size()]), field);
    return "MultiTermVectorsRequest In High ElasticSearch Client Successed!";
}
```

在 Service 层的 MeetHighElasticSearchService 类中添加如下代码：

```java
// 以同步方式执行 MultiTermVectorsRequest
public void executeMultiTermVectorsRequest(String indexName, String[] documentIds, String field);
```

在 ServiceImpl 实现层的 MeetHighElasticSearchServiceImpl 类中添加如下代码：

```java
// 以同步方式执行 MultiTermVectorsRequest
public void executeMultiTermVectorsRequest(String indexName, String[]
    documentIds, String field) {
  // 方法1：创建一个空的 MultiTermVectorsRequest，向其添加单个 term vectors 请求
  MultiTermVectorsRequest request = new MultiTermVectorsRequest();
  for (String documentId : documentIds) {
    TermVectorsRequest tvrequest = new TermVectorsRequest(indexName,
        documentId);
    tvrequest.setFields(field);
    request.add(tvrequest);
  }
  try {
    MultiTermVectorsResponse response = restClient.mtermvectors(request,
        RequestOptions.DEFAULT);
    // 解析 MultiTermVectorsResponse
    processMultiTermVectorsResponse(response);
  } catch (Exception e) {
    e.printStackTrace();
  } finally {
    // 关闭 Elasticsearch 连接
    closeEs();
  }
}
// 解析 MultiTermVectorsResponse
private void processMultiTermVectorsResponse(MultiTermVectorsResponse
    response) {
  if (response == null) {
    return;
  }
  List<TermVectorsResponse> tvresponseList = response.
      getTermVectorsResponses();
  if (tvresponseList == null) {
    return;
  }
  log.info("tvresponseList size is " + tvresponseList.size());
```

```
    for (TermVectorsResponse tvresponse : tvresponseList) {
      String id = tvresponse.getId();
      String index = tvresponse.getIndex();
      log.info("id size is " + id + "; index is " + index);
    }
  }
```

随后编译工程，在工程根目录下输入如下命令：

```
mvn clean package
```

通过如下命令启动工程服务：

```
java -jar ./target/esdemo-0.0.1-SNAPSHOT.jar
```

在工程服务启动后，在浏览器中调用如下接口获取多词向量请求的执行情况：

```
http://localhost:8080/springboot/es/high/index/multiterm?indexName=ultraman1&documentId=1&field=content
```

此时，在服务器中输出的内容如下所示：

```
2019-08-02 16:28:00.535  INFO 81804 --- [io-8080-exec-10] n.e.s.i.MeetHighElasticSearchServiceImpl : tvresponseList size is 1
2019-08-02 16:28:00.540  INFO 81804 --- [io-8080-exec-10] n.e.s.i.MeetHighElasticSearchServiceImpl : id size is 1; index is ultraman1
```

Elasticsearch 服务器返回了一个命中的文档。

如果在浏览器中输出如下内容，则表明接口执行成功：

```
MultiTermVectorsRequest In High ElasticSearch Client Successed!
```

6.8 文档处理过程解析

下面介绍在 Elasticsearch 内部是如何将文档分片存储的。

6.8.1 Elasticsearch 文档分片存储

一个索引一般由多个分片构成，当用户执行添加、删除、修改文档操作时，Elasticsearch 需要决定把这个文档存储在哪个分片上，这个过程就称为数据路由。

当把文档路由到分片上时需要使用路由算法，Elasticsearch 中的路由算法如下所示：

```
shard = hash(routing) % number_of_primary_shards
```

下面通过示例展示文档路由到分片的过程。

假设某个索引由 3 个主分片组成，用户每次对文档进行增删改查时，都有一个 routing 值，默认是该文档 ID 的值。随后对这个 routing 值使用 Hash 函数进行计算，计算出的值再和主分片个数取余数，余数的取值范围永远是（0 ~ number_of_primary_shards - 1）之间，文档知道应该存储在哪个对应的分片上。

需要指出的是，虽然 routing 值默认是文档 ID 的值，但 Elasticsearch 也支持用户手动指定一个值。手动指定对于负载均衡及提升批量读取的性能有一定的帮助。

正是 Elasticsearch 的这种路由机制，主分片的个数在索引建立之后不能修改。因为修改索引主分片数目会直接导致路由规则出现严重问题，部分数据将无法被检索。

那么读者每次对文档进行增删改查时，主分片和副本分片是如何工作的呢？下面进行说明。

假设读者本机有三个节点的集群，该集群包含一个名叫作 niudong 的索引，并拥有两个主分片，每个主分片有两个副本分片。

在 Elasticsearch 中，出于数据安全和容灾等因素考虑，相同的分片不会放在同一个节点上，所以此时的集群如图 6-1 所示。

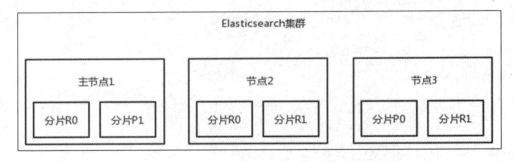

图 6-1

在图 6-1 中，有一个标记为主节点的节点，也称之为请求节点。一般来说，用户能够发送请求给集群中的任意一个节点。每个节点都有能力处理用户提交的任意请求，每个节点都知道任意文档所在的节点，所以也可以将请求转发到需要的节点。

当用户执行新建索引、更新和删除请求等写操作时，文档必须在主分片上成功完成请求，才能复制到相关的副本分片上，分片数据同步过程如图 6-2 所示。

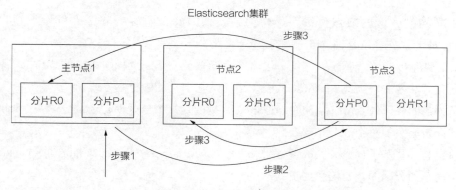

图 6-2 分片数据同步过程

客户端发送了一个索引或者删除的请求给主节点 1。此时，主节点会进行如下处理：

（1）主节点 1 通过请求中文档的 ID 值判断出该文档应该被存储在哪个分片上，如现在已经判断出需要分片编号为 0 的分片在节点 3 中，则主节点 1 会把这个请求转发到节点 3 处。

（2）节点 3 在分片编号为 0 的主分片上执行请求。如果请求执行成功，则节点 3 将并行地将该请求发给分片编号为 0 的所有副本上，即图 6-2 中位于主节点 1 和节点 2 中的副本分片。

如果所有的副本分片都成功地执行了请求，那么将向节点 3 回复一个请求执行成功的确认消息。当节点 3 收到所有副本节点的确认信息后，会向客户端返回一个成功的响应消息。

（3）当客户端收到成功的响应信息时，文档的操作就已经被应用于主分片和所有的副本分片上，此时操作就会生效。

6.8.2　Elasticsearch 的数据分区

一般来说，搜索引擎有两种数据分区方式，即基于文档的分区方式和基于词条的分区方式。Elasticsearch 使用的是基于文档的分区方式。

基于文档的分区（Document Based Partitioning）指的是每个文档只存一个分区，每个分区持有整个文档集的一个子集。这里说的分区是指一个功能完整的倒排索引。

基于文档的分区的优点汇总如下：

（1）每个分区都可以独立地处理查询。

（2）可以非常方便地添加以文档为单位的索引信息。

（3）在搜索过程中网络开销很小，每个节点可以分别独立地执行搜索，执行完之后只需返回文档的 ID 和评分信息即可。而呈现给用户的结果集是在执行分布式搜索的节点上执行合并操作实现的。

基于文档的分区的缺点也很明显，如果查询需要在所有的分区上执行，则它将执行 $O(K×N)$ 次磁盘操作（K 是词条 term 的数量，N 是分区的数量）。

从实用性角度来看，基于文档的分区方式已经被证明是一个构建大型的分布式信息检索系统的行之有效的方法，因此 Elasticsearch 使用的是基于文档的分区方式。

基于词条的分区（Term Based Partitioning）指的是每个分区拥有一部分词条，词条里面包含了与该词条相关的整个 index 的文档数据。

目前也有一些基于词条分区的搜索引擎系统，如 Riak Search、Lucandra 和 Solandra。

基于词条的分区的优点汇总如下：

（1）部分分区执行查询。一般来说，用户只需要在很小的部分分区上执行查询就可以了。举个例子，假如用户有 3 个 term 词条的查询，则 Elasticsearch 在搜索时将至多命中 3 个分区。如果足够幸运，这 3 个 term 词条都保存同一个分区中，那么用户只需访问一个分区即可。在搜索的背后，用户无须知道 Elasticsearch 中实际的分区数量。

（2）时间复杂度低。一般而言，当对应 K 个 term 词条的查询时，用户只需执行 $O(K)$ 次磁盘查找即可。

基于词条的分区的缺点汇总如下：

（1）最主要的问题是 Lucene Segment 概念里面很多固有的结构都将失去。

对于比较复杂的查询，搜索过程中的网络开销将变得非常高，并且可能使得系统可用性大大降低，特别是注入前置搜索或模糊搜索的场景。

（2）获取每个文档的信息将会变得非常困难。由于是按词条分区存储的，如果用户想获取文档的一部分数据做进一步的控制，或获取每个文档的这些数据，都将变得非常困难，因为这种分区方式使得文档的数据被分散到了不同的地方，所以实现注入评分、自定义评分等都将变得难以实现。

6.9 知识点关联

如前文所述，Elasticsearch 使用了乐观锁来解决数据一致性问题。当用户对文档进行操作时，并不需要对文档作加锁、解锁的操作，只需指定要操作的版本即可。当版本号一致时，Elasticsearch 会允许该操作顺利进行；当版本号冲突时，Elasticsearch 会提示冲突并抛出异常 VersionConflictEngineException。在 Elasticsearch 中，文档的版本号的取值范围为 1 到 $2^{63}-1$。

其实，乐观锁不仅是一种锁的类型，更是一种设计思想，这种设计思想在软件开发过程中十分常见。

这里先简要介绍一下乐观锁。在乐观锁的思想中，会认为数据一般不会引发冲突，因此在数据更新时，才会检测是否存在数据冲突。在检测时，如果发现数据冲突，则返回冲突结果，以便读者自主决定如何去做。

和乐观锁对应的是悲观锁，在悲观锁的思想中，会认为数据一般会引发冲突。也就是说，在读数据时写数据操作往往也正在进行，因此在读数据前需要上锁，没有拿到锁的读者或进程只能等待锁的释放。悲观锁在关系数据库中有大量应用，我们常见的行锁、表锁、读锁、写锁等都是悲观锁。

乐观锁除在 Elasticsearch 中有应用外，在关系数据库和 Java 中都有应用。

在关系数据库中，如果某张表对应的应用场景是读多写少的场景，则可以使用乐观锁。采用乐观锁控制数据库表后，表中会新增一列字段，一般称之为 version，version 用于记录行数据的版本。当数据初始写入时，版本默认为 1；每当对数据有修改操作时，version 都会加 1。在修改操作过程中，数据库会对比数据地区的 version 和当前数据库的 version，如果相同，则新数据方可写入。

在 Java 中，乐观锁思想的实现就是 CAS 技术，即 Compare and Swap。在 JDK 1.5 中新增的 java.util.concurrent 就是建立在 CAS 之上的，开发人员常用的 AtomicInteger 就是其中之一。

在 Java 中，当多个线程尝试使用 CAS 同时更新同一个变量时，只有一个线程能成功更新变量的值，而其他线程都会失败。失败的线程并不会被挂起，而是被告知失败，并可再次尝试。

在具体实现上，CAS 操作包含三个操作数，即内存位置（V）、预期原值（A）和新值(B)。如果内存位置的值与预期原值相同，则处理器会自动将该位置值更新为新值；否则，处理器不做任何操作。

6.10 小结

本章主要介绍了文档高级 API 的使用，均为批量操作接口，如文档批量请求接口、批量处理器的使用、MultiGet 批量处理方法的使用、文档 ReIndex 接口的使用、文档查询时更新接口的使用、文档查询时删除接口的使用和获取文档索引多词向量的接口的使用。

第 7 章
搜索实战

> 众里寻他千百度
> 蓦然回首
> 那人却在，灯火阑珊处

前面介绍了在高级客户端中对文档的各种操作，总体来说，这些接口都是围绕着文档索引展开的，下面我们将围绕文档搜索这个核心过程展开。

在 Elasticsearch 中，高级客户端支持以下搜索 API：

（1）Search API

（2）Search Scroll API

（3）Clear Scroll API

（4）Search Template API

（5）Multi-Search-Template API

（6）Multi-Search API

（7）Field Capabilities API

（8）Ranking Evaluation API

（9）Explain API

（10）Count API

这些搜索 API 允许读者执行搜索查询并返回匹配查询的搜索命中结果，它们可以跨一个或多个索引，以及跨一个或多个类型来执行。

7.1 搜索 API

搜索 API，即 Search API，允许用户执行一个搜索查询并返回与查询匹配的搜索点击，用户可以使用简单的查询字符串作为参数或使用请求主体提供查询。

最基本的搜索 API 是空搜索，空搜索不指定任何的查询条件，只返回集群索引中的所有文档。在实际开发中，基本不会用到空搜索。

在常见的搜索请求发出前，用户需要构建搜索请求，即 SearchRequest。

1．构建搜索请求

SearchRequest 可用于与搜索文档、聚合、Suggest 有关的任何操作，另外还提供了请求高亮显示结果文档的方法。

SearchRequest 的构建需要依赖 SearchSourceBuilder。在基本查询中，用户需要构建查询 Query，并把它添加到查询 Request 中。SearchRequest 的构建代码如下所示：

```
// 构建 SearchRequest
public void buildSearchRequest() {
  SearchRequest searchRequest = new SearchRequest();

  // 大多数搜索参数都添加到 SearchSourceBuilder 中，它为进入搜索请求主体的所有内容提供
  // setter
  SearchSourceBuilder searchSourceBuilder = new SearchSourceBuilder();

  // 向 searchSourceBuilder 中添加"全部匹配"查询
  searchSourceBuilder.query(QueryBuilders.matchAllQuery());

  // 将 searchSourceBuilder 添加到 searchRequest 中
  searchRequest.source(searchSourceBuilder);
}
```

类似地，SearchRequest 也有可选参数供用户进行配置。SearchRequest 的可选参数有限制请求类型、设置路由参数、设置 IndicesOptions 和使用首选参数等。

在 SearchRequest 中配置可选参数的代码如下所示：

```
// 构建 SearchRequest
public void buildSearchRequest() {
  SearchRequest searchRequest = new SearchRequest();
  // 大多数搜索参数都添加到 SearchSourceBuilder 中，它为进入搜索请求主体的所有内容提供
  // setter
  SearchSourceBuilder searchSourceBuilder = new SearchSourceBuilder();
  // 向 searchSourceBuilder 中添加"全部匹配"查询
```

```
    searchSourceBuilder.query(QueryBuilders.matchAllQuery());
    // 将 searchSourceBuilder 添加到 serchRequest 中
    searchRequest.source(searchSourceBuilder);
    /*
     * 可选参数配置
     */
    // 在索引上限制请求
    searchRequest = new SearchRequest("posts");

    // 设置路由参数
    searchRequest.routing("routing");

    // 设置 IndicesOptions 控制方法
    searchRequest.indicesOptions(IndicesOptions.lenientExpandOpen());

    // 使用首选参数，例如，执行搜索以首选本地分片。默认值是在分片之间随机的
    searchRequest.preference("_local");
}
```

使用 SearchSourceBuilder

除 SearchRequest 的构建需要依赖 SearchSourceBuilder 外，事实上，大多数控制搜索行为的选项都可以在 SearchSourceBuilder 上进行设置。SearchSourceBuilder 包含了与 REST API 搜索请求正文中相同的选项。

SearchSourceBuilder 常用的搜索选项有设置查询条件、设置搜索结果索引的起始地址（默认为 0）、设置要返回的搜索命中数的大小（默认为 10）、设置一个可选的超时时间（以便控制允许搜索的时间）等。

在 SearchSourceBuilder 中配置常见搜索选项的代码如下所示：

```
/*
 * 使用 SearchSourceBuilder
 */
    SearchSourceBuilder sourceBuilder = new SearchSourceBuilder();
    // 设置查询条件
    sourceBuilder.query(QueryBuilders.termQuery("content", "货币"));
    // 设置搜索结果索引的起始地址，默认为 0
    sourceBuilder.from(0);
    // 设置要返回的搜索命中数的大小，默认为 10
    sourceBuilder.size(5);
    // 设置一个可选的超时时间，控制允许搜索的时间
    sourceBuilder.timeout(new TimeValue(60, TimeUnit.SECONDS));
```

```
//将 searchSourceBuilder 添加到 searchRequest 中
searchRequest.source(sourceBuilder);
```

生成查询

搜索查询是基于 QueryBuilder 创建的。在 Elasticsearch 中，每个搜索查询都需要用到 QueryBuilder。QueryBuilder 是基于 Elasticsearch DSL 实现的。

QueryBuilder 既可以使用其构建函数来创建，也可以使用 QueryBuilders 工具类来创建。QueryBuilders 工具类提供了流式编程格式来创建 QueryBuilder。

QueryBuilder 的构建代码如下所示，我们在 buildSearchRequest 方法中添加如下代码：

```
/*
 * 搜索查询 MatchQueryBuilder 的使用
 */
//方法 1
    MatchQueryBuilder matchQueryBuilder = new MatchQueryBuilder("content",
        "货币");
    //创建 QueryBuilder，提供配置搜索查询选项的方法
    //对匹配查询启用模糊匹配
    matchQueryBuilder.fuzziness(Fuzziness.AUTO);

    //在匹配查询上设置前缀长度
    matchQueryBuilder.prefixLength(3);

    //设置最大扩展选项以控制查询的模糊过程
    matchQueryBuilder.maxExpansions(10);
// 方法 2
    matchQueryBuilder = QueryBuilders.matchQuery("content", "货币").fuzziness
(Fuzziness.AUTO)
        .prefixLength(3).maxExpansions(10);
```

上述代码展示了构建函数的方式，以及如何使用 QueryBuilders 工具类创建 QueryBuilder。在 QueryBuilder 中以 content 字段中的货币为查询对象。

无论用于创建 QueryBuilder 的方法是什么，用户都必须将 matchQueryBuilder 添加到 searchSourceBuilder 中，代码如下所示：

```
//添加 matchQueryBuilder 到 searchSourceBuilder 中
searchSourceBuilder.query(matchQueryBuilder);
```

设定排序策略

在构建搜索请求时，Elasticsearch 还支持用户配置搜索结果的排序策略。一般我们在使用搜索引擎时，对搜索结果常见的排序策略有按时间排序和按相关性排序两种。按时间排序的搜

索结果如图 7-1 所示。

```
2019世界人工智能大会的最新相关信息
...百度智能云提出人工智能工业化概念;2019世界人工...  凤凰科技    53分钟前
本周(2019年8月26日-9月1日)本周,人工智能领域大事不断。2019上海人工智能大会召开,第
四届百度云智峰会在京召开。阿里旗下平头哥发布SoC芯片设计平...
2019世界人工智能大会"汇言"政协论坛在徐汇滨江举行  东方网     5小时前
...行业变革原力 小机器人多项创新成果亮相2019世...  新民晚报    2小时前
2019世界人工智能大会圆满落幕33个重磅项目集中签约  央视网新闻  14小时前
2019世界人工智能大会闭幕 上海将建设人工智能高地  中国广播网  3小时前
```

图 7-1

SearchSourceBuilder 允许用户添加一个或多个 SortBuilder 实例。有 4 种特殊的实现类，分别是 FieldSortBuilder、ScoreSortBuilder、GeoDistanceSortBuilder 和 ScriptSortBuilder。

在 SearchSourceBuilder 中配置 SortBuilder 实例的代码如下所示，代码添加在前文中使用的 buildSearchRequest 方法中：

```
/*
 * 指定排序方法
 */
    // 按分数降序排序（默认）
    sourceBuilder.sort(new ScoreSortBuilder().order(SortOrder.DESC));
    // 按 ID 字段升序排序
    sourceBuilder.sort(new FieldSortBuilder("_id").order(SortOrder.ASC));
```

源筛选

在默认情况下，搜索请求一般会返回文档源的内容。不过，与 REST API 一样，用户可以覆盖此行为，即用户可以完全关闭源检索。

在 SearchSourceBuilder 实例中配置源搜索开关的代码如下所示，代码添加在前文中的 buildSearchRequest 方法中：

```
/*
 * 源筛选的方法使用
 */
    sourceBuilder.fetchSource(false);
```

该方法还接受一个或多个通配符模式的数组，以更细粒度的方式控制哪些字段被包括或被排除。相关代码如下所示，代码添加在前文中使用的 buildSearchRequest 方法中：

```
/*
 * 源筛选的方法使用
```

```
    */
    sourceBuilder.fetchSource(false);

    //该方法还接受一个或多个通配符模式的数组,以更细的粒度控制哪些字段被包括或排除
    String[] includeFields = new String[] {"title", "innerObject.*"};
    String[] excludeFields = new String[] {"user"};
sourceBuilder.fetchSource(includeFields, excludeFields);
```

请求高亮显示

在搜索查询请求中,用户还可以设置请求的高亮显示。

用户可以在 SearchSourceBuilder 上设置 HighlightBuilder 来达到高亮显示搜索结果的目的。

通过向 HighlightBuilder 中添加一个或多个 HighlightBuilder.Field 实例,就可以为每个字段定义不同的高亮显示行为。

相关代码如下所示,代码添加在前文中使用的 buildSearchRequest 方法中:

```
/*
 * 配置请求高亮显示
 */
HighlightBuilder highlightBuilder = new HighlightBuilder();
// 为 title 字段创建字段高亮
HighlightBuilder.Field highlightTitle = new HighlightBuilder.Field("title");
// 设置字段高亮类型
highlightTitle.highlighterType("unified");
// 将 highlightTitle 添加到 highlightBuilder 中
highlightBuilder.field(highlightTitle);
// 添加第二个高亮显示字段
HighlightBuilder.Field highlightUser = new HighlightBuilder.Field("user");
highlightBuilder.field(highlightUser);
searchSourceBuilder.highlighter(highlightBuilder);
```

请求聚合

在搜索查询请求中,用户还可以设置请求聚合结果。

在获取请求聚合前,需要先创建聚合构建器 AggregationBuilder,然后将 AggregationBuilder 添加到 SearchSourceBuilder 中。

在下面的代码中,我们创建了公司名称的术语聚合,并对公司中员工的平均年龄进行了子聚合,相关代码如下所示。代码添加在前文中使用的 buildSearchRequest 方法中:

```
/*
 * 聚合请求的使用
 */
```

```
TermsAggregationBuilder aggregation =
    AggregationBuilders.terms("by_company").field("company.keyword");
aggregation.subAggregation(AggregationBuilders.avg("average_age").field
("age"));
searchSourceBuilder.aggregation(aggregation);
```

建议请求

在搜索查询请求中，用户还可以设置请求结果。请求在搜索引擎中非常常见，例如，我们在搜索框中输入 2019，则会给出 10 条请求提示结果，如图 7-2 所示。

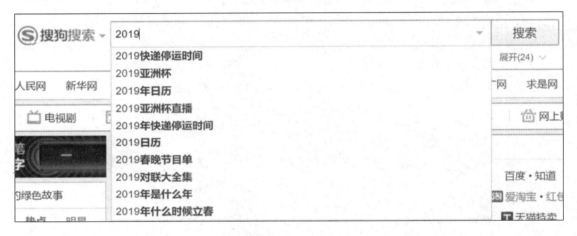

图 7-2

在 Elasticsearch 中，想要在搜索请求中添加请求，则需要使用 SuggestBuilder 工厂类。SuggestBuilder 工厂类是 SuggestionBuilder 类的实现类之一，它的特性是简单易用。

SuggestBuilder 工厂类需要添加到顶级 SuggestBuilder 中，并将顶级 SuggestBuilder 添加到 SearchSourceBuilder 中。

相关代码如下所示，代码添加在前文中使用的 buildSearchRequest 方法中：

```
/*
 * Suggestions 建议请求的使用
 */
   //在 TermSuggestionBuilder 中为 content 字段添加货币的 Suggestions
   SuggestionBuilder termSuggestionBuilder =
       SuggestBuilders.termSuggestion("content").text("货币");
   SuggestBuilder suggestBuilder = new SuggestBuilder();

   // 添加建议生成器并命名
```

```
    suggestBuilder.addSuggestion("suggest_user", termSuggestionBuilder);

    // 将 suggestBuilder 添加到 searchSourceBuilder 中
    searchSourceBuilder.suggest(suggestBuilder);
```

分析查询和聚合

Elasticsearch 从 2.2 版本开始提供 Profile API，以供用户检索、聚合、过滤执行时间和其他细节信息，帮助读者分析每次检索各个环节所用的时间。

在使用 Profile API 时，用户必须在 SearchSourceBuilder 实例中将配置标志设置为 true。相关代码如下所示，代码添加在前文中使用的 buildSearchRequest 方法中：

```
/*
 * 分析查询和聚合 API 的使用
 */
    searchSourceBuilder.profile(true);
```

2. 执行搜索请求

在 SearchRequest 实例构建后，即可执行搜索查询请求。与文档索引请求类似，搜索的查询请求也有同步和异步两种执行方式。

同步方式

当以同步方式执行搜索查询请求时，客户端会等待 Elasticsearch 服务器返回的查询结果 SearchResponse。在收到 SearchResponse 后，客户端会继续执行相关的逻辑代码。以同步方式执行的代码如下所示：

```
    // 参数化构建 SearchRequest
    public SearchRequest buildSearchRequest(String filed, String text) {
    SearchRequest searchRequest = new SearchRequest();
    // 大多数搜索参数都添加到 SearchSourceBuilder 中，它为进入搜索请求主体的所有内容提供
    //setter
    SearchSourceBuilder searchSourceBuilder = new SearchSourceBuilder();
    // 向 searchSourceBuilder 中添加 "全部匹配" 查询
    searchSourceBuilder.query(QueryBuilders.matchAllQuery());
    // 将 searchSourceBuilder 添加到 searchRequest 中
    searchRequest.source(searchSourceBuilder);
    /*
     * 使用 SearchSourceBuilder
     */
    // 设置查询条件
    searchSourceBuilder.query(QueryBuilders.termQuery(filed, text));
    // 设置搜索结果索引的起始地址，默认为 0
```

```java
    searchSourceBuilder.from(0);
    // 设置要返回的搜索命中数的大小，默认为10
    searchSourceBuilder.size(5);
    // 设置一个可选的超时时间，控制允许搜索的时间
    searchSourceBuilder.timeout(new TimeValue(60, TimeUnit.SECONDS));
    // 将SearchSourceBuilder添加到SearchRequest中
    searchRequest.source(searchSourceBuilder);
    /*
     * 配置请求高亮显示
     */
    HighlightBuilder highlightBuilder = new HighlightBuilder();
    // 为title字段创建字段高亮
    HighlightBuilder.Field highlightTitle = new HighlightBuilder.Field(filed);
    // 设置字段高亮类型
    highlightTitle.highlighterType("unified");
    // 将highlightTitle添加到highlightBuilder中
    highlightBuilder.field(highlightTitle);
    searchSourceBuilder.highlighter(highlightBuilder);
    /*
     * 建议请求的使用
     */
    //在TermSuggestionBuilder 中为content字段添加货币的Suggestions
    SuggestionBuilder termSuggestionBuilder = SuggestBuilders.termSuggestion
        (filed).text(text);
    SuggestBuilder suggestBuilder = new SuggestBuilder();
    // 添加建议生成器并命名
    suggestBuilder.addSuggestion("suggest_user", termSuggestionBuilder);
    // 将suggestBuilder添加到searchSourceBuilder中
    searchSourceBuilder.suggest(suggestBuilder);
    return searchRequest;
}
//以同步方式执行SearchRequest
public void executeSearchRequest() {
    SearchRequest searchRequest = buildSearchRequest("content", "货币");
    // 运行
    try {
        SearchResponse searchResponse = restClient.search(searchRequest,
            RequestOptions. DEFAULT);
        log.info(searchResponse.toString());
    } catch (Exception e) {
        e.printStackTrace();
    } finally {
        // 关闭Elasticsearch连接
```

```
        closeEs();
    }
}
```

在 Service 层的 SearchService 类中添加如下代码：

```
// 以同步方式执行 SearchRequest
public void executeSearchRequest();
```

异步方式

当以异步方式执行 SearchRequest 时，高级客户端不必同步等待请求结果的返回，可以直接向接口调用方返回异步接口执行成功的结果。

为了处理异步返回的响应信息或处理在请求执行过程中引发的异常信息，读者需要指定监听器。以异步方式执行的代码如下所示，在 ServiceImpl 实现层的 SearchServiceImpl 类中添加如下代码：

```
//以异步方式执行 SearchRequest
public void executeSearchRequestAsync() {
    SearchRequest searchRequest = buildSearchRequest("content", "货币");
    // 构建监听器
    ActionListener<SearchResponse> listener = new ActionListener
        <SearchResponse>() {
        @Override
        public void onResponse(SearchResponse searchResponse) {
            log.info("response is " + searchResponse.toString());
        }
        @Override
        public void onFailure(Exception e) {
        }
    };
    // 运行
    try {
        restClient.searchAsync(searchRequest, RequestOptions.DEFAULT, listener);
    } catch (Exception e) {
        e.printStackTrace();
    } finally {
        // 关闭 Elasticsearch 连接
        closeEs();
    }
}
```

在执行搜索请求后，如果执行成功，则调用 ActionListener 类中的 onResponse 方法进行相关逻辑的处理；如果执行失败，则调用 ActionListener 类中的 onFailure 方法进行相关逻辑的处理。

当然，在异步请求执行过程中可能会出现异常，异常的处理与同步方式执行情况相同。

3. 解析搜索请求的响应结果

不论同步方式，还是异步方式，客户端均需对搜索请求执行的返回结果 SearchResponse 进行处理和解析。SearchResponse 中提供了有关搜索执行本身及对返回文档访问的详细信息，如 HTTP 状态代码、执行时间，或者请求是提前终止还是超时等。

相关代码如下所示，代码添加在 SearchServiceImpl 类中：

```java
// 解析 SearchResponse
private void processSearchResponse(SearchResponse searchResponse) {
  if (searchResponse == null) {
    return;
  }
  // 获取HTTP状态代码
  RestStatus status = searchResponse.status();
  // 获取请求执行时间
  TimeValue took = searchResponse.getTook();
  // 获取请求是否提前终止
  Boolean terminatedEarly = searchResponse.isTerminatedEarly();
  // 获取请求是否超时
  boolean timedOut = searchResponse.isTimedOut();
  log.info("status is " + status + ";took is " + took + ";terminatedEarly
     is " + terminatedEarly
     + ";timedOut is " + timedOut);
}
```

搜索请求相关的分片

响应结果 SearchResponse 提供了有关搜索影响的分片总数，以及成功与失败分片的统计信息，并提供了有关分片级别执行的信息。对于请求执行失败的信息，可以通过解析 ShardSearchFailure 实例数组元素来处理。

该部分的代码如下所示，代码添加在 ServiceImpl 实现层 SearchServiceImpl 类的 processSearchResponse 方法内：

```java
// 查看搜索影响的分片总数
  int totalShards = searchResponse.getTotalShards();
  //搜索成功的分片的统计信息
  int successfulShards = searchResponse.getSuccessfulShards();
  //搜索失败的分片的统计信息
```

```
    int failedShards = searchResponse.getFailedShards();
    log.info("totalShards is " + totalShards + ";successfulShards is " +
       successfulShards
       + ";failedShards is " + failedShards);
    for (ShardSearchFailure failure : searchResponse.getShardFailures()) {
      log.info("fail is " + failure.toString());
    }
```

获取搜索结果

获取结果响应中包含的搜索结果 SearchHits，获取代码如下所示：

```
SearchHits hits = searchResponse.getHits();
```

SearchHits 中提供了有关所有搜索结果的全部信息，如点击总数或最高分数等，这些内容的获取代码如下所示：

```
TotalHits totalHits = hits.getTotalHits();
// 搜索结果的总量数
long numHits = totalHits.value;
// 搜索结果的相关性数据
TotalHits.Relation relation = totalHits.relation;
float maxScore = hits.getMaxScore();
```

SearchHits 的解析代码如下所示，代码添加在 ServiceImpl 实现层 SearchServiceImpl 类中的 processSearchResponse 方法内：

```
// 获取响应中包含的搜索结果
    SearchHits hits = searchResponse.getHits();
    // SearchHits 提供了有关所有结果的全部信息，如点击总数或最高分数
    TotalHits totalHits = hits.getTotalHits();
    // 点击总数
    long numHits = totalHits.value;
    // 最高分数
    float maxScore = hits.getMaxScore();
    log.info("numHits is " + numHits + ";maxScore is " + maxScore);
```

对 SearchHits 的解析还可以通过遍历 SearchHit 数组实现，代码如下所示：

```
SearchHit[] searchHits = hits.getHits();
    for (SearchHit hit : searchHits) {
        // 解析 SearchHit
}
```

SearchHits 提供了对文档基本信息的访问，如索引名称、文档 ID 和每次搜索的得分，文

档基本信息可以通过如下代码获取：

```java
//获取索引名称
String index = hit.getIndex();
//获取文档ID
String id = hit.getId();
//获取搜索的得分
float score = hit.getScore();
```

通过遍历 SearchHit 数组来解析 SearchHits 的全部代码如下所示，代码添加在 ServiceImpl 实现层 SearchServiceImpl 类中的 processSearchResponse 方法内：

```java
//嵌套在SearchHits中的是可以迭代的单个搜索结果
SearchHit[] searchHits = hits.getHits();
for (SearchHit hit : searchHits) {
    // SearchHits 提供对基本信息的访问，如索引、文档ID和每次搜索的得分
    String index = hit.getIndex();
    String id = hit.getId();
    float score = hit.getScore();
    log.info("docId is " + id + ";docIndex is " + index + ";docScore is " +
        score);
}
```

此外，Elasticsearch 允许以简单的 JSON 字符串或键值对形式返回文档源。在这个映射中，常规字段由字段名作为键值，包含字段值；而多值字段作为对象列表返回；嵌套对象作为另一个键值返回。这部分内容的代码如下所示，代码添加在 ServiceImpl 实现层 SearchServiceImpl 类中的 processSearchResponse 方法内：

```java
// 以JSON字符串形式返回文档源
String sourceAsString = hit.getSourceAsString();
// 以键值对的形式返回文档源
Map<String, Object> sourceAsMap = hit.getSourceAsMap();
String documentTitle = (String) sourceAsMap.get("title");
List<Object> users = (List<Object>) sourceAsMap.get("user");
Map<String, Object> innerObject = (Map<String, Object>) sourceAsMap.get
    ("innerObject");
log.info(
    "sourceAsString is " + sourceAsString + ";sourceAsMap size is " +
        sourceAsMap.size());
```

搜索结果高亮显示

如果对搜索结果展示有需求，则用户可以从返回结果中的每条数据结果，即 SearchHits，

自己获取搜索高亮显示的文本片段。

SearchHits 提供了对 HighlightField 实例中字段名称的映射访问，每个实例包含一个或多个高亮显示的文本片段。

代码如下所示，代码添加在 ServiceImpl 实现层 SearchServiceImpl 类中的 processSearchResponse 方法内：

```java
// 高亮显示
    Map<String, HighlightField> highlightFields = hit.getHighlightFields();
    HighlightField highlight = highlightFields.get("content");
    // 获取包含高亮显示字段内容的一个或多个片段
    Text[] fragments = highlight.fragments();
    String fragmentString = fragments[0].string();
    log.info("fragmentString is " + fragmentString);
```

搜索聚合结果

用户可以通过 SearchResponse 实例获取搜索的聚合结果，首先获取聚合树的根，即聚合对象，然后按照名称获取搜索聚合结果。

获取搜索聚合结果的代码如下所示，代码添加在 ServiceImpl 实现层 SearchServiceImpl 类中的 processSearchResponse 方法内：

```java
// 聚合搜索
    Aggregations aggregations = searchResponse.getAggregations();
    // 按 content 聚合
    Terms byCompanyAggregation = aggregations.get("by_content");
    // 获取 Elastic 为关键词的 buckets
    Bucket elasticBucket = byCompanyAggregation.getBucketByKey("Elastic");
    // 获取平均年龄的子聚合
    Avg averageAge = elasticBucket.getAggregations().get("average_age");
    double avg = averageAge.getValue();
    log.info("avg is " + avg);
```

解析 Suggestions 结果

首先在 SearchResponse 实例中使用 Suggest 对象作为入口点，然后检索嵌套的 Suggest 对象。

对 Suggestions 结果的解析代码如下所示，代码添加在 ServiceImpl 实现层 SearchServiceImpl 类中的 processSearchResponse 方法内：

```java
// Suggest 搜索
    Suggest suggest = searchResponse.getSuggest();
    // 按 content 搜索 Suggest
```

```
    TermSuggestion termSuggestion = suggest.getSuggestion("content");
    for (TermSuggestion.Entry entry : termSuggestion.getEntries()) {
      for (TermSuggestion.Entry.Option option : entry) {
        String suggestText = option.getText().string();
        log.info("suggestText is " + suggestText);
      }
    }
}
```

搜索响应结果的解析综合示例

前面我们分别介绍了搜索响应结果 SearchResponse 中包含的各种信息的解析，下面通过一段代码综合展示 SearchResponse 的解析。代码以同步方式执行搜索请求，共分为三层，分别是 Controller 层、Service 层和 ServiceImpl 实现层。

在 Controller 层中新建 SearchController 类，代码如下所示：

```
package com.niudong.esdemo.controller;
import org.springframework.beans.factory.annotation.Autowired;
import org.springframework.web.bind.annotation.RequestMapping;
import org.springframework.web.bind.annotation.RestController;
import com.niudong.esdemo.service.SearchService;
@RestController
@RequestMapping("/springboot/es/search")
public class SearchController {
  @Autowired
  private SearchService searchService;

  //以同步方式执行 SearchRequest
  @RequestMapping("/sr")
  public String executeSearchRequest() {
    searchService.executeSearchRequest();

    return "Execute SearchRequest success!";
  }

}
```

在 Service 层中，添加新类 SearchService，代码如下所示：

```
package com.niudong.esdemo.service;
public interface SearchService {
  //以同步方式执行 SearchRequest
  public void executeSearchRequest();
}
```

在 ServiceImpl 实现层中，添加新类 SearchServiceImpl，核心代码如下所示：

```java
//以同步方式执行 SearchRequest
public void executeSearchRequest() {
  SearchRequest searchRequest = buildSearchRequest("content", "货币");
  // 运行
  try {
    SearchResponse searchResponse = restClient.search(searchRequest,
        RequestOptions.DEFAULT);
    log.info(searchResponse.toString());
    // 解析 SearchResponse
    processSearchResponse(searchResponse);
  } catch (Exception e) {
    e.printStackTrace();
  } finally {
    // 关闭Elasticsearch的连接
    closeEs();
  }
}
// 解析 SearchResponse
private void processSearchResponse(SearchResponse searchResponse) {
  if (searchResponse == null) {
    return;
  }
  // 获取 HTTP 状态代码
  RestStatus status = searchResponse.status();
  // 获取请求执行时间
  TimeValue took = searchResponse.getTook();
  // 获取请求是否提前终止
  Boolean terminatedEarly = searchResponse.isTerminatedEarly();
  // 获取请求是否超时
  boolean timedOut = searchResponse.isTimedOut();
  log.info("status is " + status + ";took is " + took + ";terminatedEarly
      is " + terminatedEarly
      + ";timedOut is " + timedOut);
  // 查看搜索影响的分片总数
  int totalShards = searchResponse.getTotalShards();
  // 执行搜索成功的分片的统计信息
  int successfulShards = searchResponse.getSuccessfulShards();
  // 执行搜索失败的分片的统计信息
  int failedShards = searchResponse.getFailedShards();
  log.info("totalShards is " + totalShards + ";successfulShards is " +
      successfulShards
      + ";failedShards is " + failedShards);
```

```java
for (ShardSearchFailure failure : searchResponse.getShardFailures()) {
  log.info("fail is " + failure.toString());
}
// 获取响应中包含的搜索结果
SearchHits hits = searchResponse.getHits();
// SearchHits 提供了相关结果的全部信息，如点击总数或最高分数
TotalHits totalHits = hits.getTotalHits();
// 点击总数
long numHits = totalHits.value;
// 最高分数
float maxScore = hits.getMaxScore();
log.info("numHits is " + numHits + ";maxScore is " + maxScore);
// 嵌套在 SearchHits 中的是可以迭代的单个搜索结果
SearchHit[] searchHits = hits.getHits();
for (SearchHit hit : searchHits) {
  // SearchHit 提供了对基本信息的访问，如索引、文档 ID 和每次搜索的得分
  String index = hit.getIndex();
  String id = hit.getId();
  float score = hit.getScore();
  log.info("docId is " + id + ";docIndex is " + index + ";docScore is " +
      score);
  // 以 JSON 字符串形式返回文档源
  String sourceAsString = hit.getSourceAsString();
  // 以键值对形式返回文档源
  Map<String, Object> sourceAsMap = hit.getSourceAsMap();
  String documentTitle = (String) sourceAsMap.get("title");
  List<Object> users = (List<Object>) sourceAsMap.get("user");
  Map<String, Object> innerObject = (Map<String, Object>) sourceAsMap.
      get("innerObject");
  log.info(
      "sourceAsString is " + sourceAsString + ";sourceAsMap size is " +
          sourceAsMap.size());
  // 高亮显示
  Map<String, HighlightField> highlightFields = hit.getHighlightFields();
  HighlightField highlight = highlightFields.get("content");
  // 获取包含高亮显示字段内容的一个或多个片段
  Text[] fragments = highlight.fragments();
  String fragmentString = fragments[0].string();
  log.info("fragmentString is " + fragmentString);
}
// 聚合搜索
Aggregations aggregations = searchResponse.getAggregations();
if (aggregations == null) {
  return;
```

```java
    }
    // 按 content 聚合
    Terms byCompanyAggregation = aggregations.get("by_content");
    // 获取以 Elastic 为关键词的 Bucket
    Bucket elasticBucket = byCompanyAggregation.getBucketByKey("Elastic");
    // 获取平均年龄的子聚合
    Avg averageAge = elasticBucket.getAggregations().get("average_age");
    double avg = averageAge.getValue();
    log.info("avg is " + avg);
    // 搜索
    Suggest suggest = searchResponse.getSuggest();
    if (suggest == null) {
      return;
    }
    // 按 content 搜索 Suggest
    TermSuggestion termSuggestion = suggest.getSuggestion("content");
    for (TermSuggestion.Entry entry : termSuggestion.getEntries()) {
      for (TermSuggestion.Entry.Option option : entry) {
        String suggestText = option.getText().string();
        log.info("suggestText is " + suggestText);
      }
    }
    // 在搜索时分析结果
    Map<String, ProfileShardResult> profilingResults = searchResponse.
        getProfileResults();
    if (profilingResults == null) {
      return;
    }
    for (Map.Entry<String, ProfileShardResult> profilingResult :
        profilingResults.entrySet()) {
      String key = profilingResult.getKey();
      ProfileShardResult profileShardResult = profilingResult.getValue();
      log.info("key is " + key + ";profileShardResult is " +
          profileShardResult.toString());
    }
  }
```

随后编译工程，在工程根目录下输入如下命令：

```
mvn clean package
```

通过如下命令启动工程服务：

```
java -jar ./target/esdemo-0.0.1-SNAPSHOT.jar
```

在工程服务启动后,在浏览器中调用如下接口查看搜索响应结果的解析情况:

```
http://localhost:8080/springboot/es/search/sr
```

此时,在服务器中输出如下所示内容:

```
2019-08-03 14:19:51.511  INFO 74120 --- [nio-8080-exec-1] c.n.e.service.impl.SearchServiceImpl     : {"took":4,"timed_ou
t":false,"_shards":{"total":2,"successful":2,"skipped":0,"failed":0},"hits":{"total":{"value":0,"relation":"eq"},"max_sc
ore":null,"hits":[]},"suggest":{"suggest_user":[{"text":"炭","offset":0,"length":1,"options":[]},{"text":"币","offset":1
,"length":1,"options":[]}]}}
2019-08-03 14:19:51.513  INFO 74120 --- [nio-8080-exec-1] c.n.e.service.impl.SearchServiceImpl     : status is OK;took i
s 4ms;terminatedEarly is null;timedOut is false
2019-08-03 14:19:51.526  INFO 74120 --- [nio-8080-exec-1] c.n.e.service.impl.SearchServiceImpl     : totalShards is 2;su
ccessfulShards is 2;failedShards is 0
2019-08-03 14:19:51.527  INFO 74120 --- [nio-8080-exec-1] c.n.e.service.impl.SearchServiceImpl     : numHits is 0;maxSco
re is NaN
```

如果在浏览器中显示如下内容,则证明接口调用成功:

```
Execute SearchRequest success!
```

7.2 滚动搜索

滚动搜索 API,即 Search Scroll API,可通过搜索请求,获取大量搜索结果。滚动搜索有点类似于数据库中的分页查询。

1. 构建滚动搜索请求

为了使用滚动搜索,需要按顺序执行以下步骤。

步骤 1:初始化滚动搜索的上下文信息。

在执行滚动搜索 API 时,滚动搜索会话的初始化必须带有滚动搜索参数的搜索请求,即 SearchRequest。

在执行该搜索请求时,Elasticsearch 会检测到滚动搜索参数的存在,并在相应的时间间隔内保持搜索上下文活动。

带有滚动搜索参数的搜索请求的构建代码如下所示,代码在 ServiceImpl 实现层的 ScrollSearchServiceImpl 类中:

```java
// 构建 SearchRequest
public void buildAndExecuteScrollSearchRequest (String indexName, int size)
{
    // 设置索引名称
    SearchRequest searchRequest = new SearchRequest(indexName);
    SearchSourceBuilder searchSourceBuilder = new SearchSourceBuilder();
    searchSourceBuilder.query(QueryBuilders.matchQuery("title",
        "Elasticsearch"));
```

```java
    // 创建 SearchRequest 及相应的 SearchSourceBuilder。还可以选择设置大小以控制一次检
// 索多少结果
    searchSourceBuilder.size(size);
    searchRequest.source(searchSourceBuilder);

    // 设置滚动间隔
    searchRequest.scroll(TimeValue.timeValueMinutes(1L));
    try {
      SearchResponse searchResponse = restClient.search(searchRequest,
RequestOptions.DEFAULT);

      // 读取返回的滚动 ID，该 ID 指向保持活动状态的搜索上下文，并在后续搜索滚动调用中被需要
      String scrollId = searchResponse.getScrollId();
      // 检索第一批搜索结果
      SearchHits hits = searchResponse.getHits();

    } catch (Exception e) {
      e.printStackTrace();
    } finally {
      // 关闭 Elasticsearch 连接
      closeEs();
    }
  }
```

步骤 2：检索所有相关文档。

首先在 SearchScrollRequest 中设置上文提及的滚动标识符和新的滚动间隔；其次在设置好 SearchScrollRequest 后，将其传送给 searchScroll 方法。

在请求发出后，Elasticsearch 服务器会返回另一批带有新的滚动标识符的结果。依次类推，用户需要在新的 SearchScrollRequest 中设置前文提及的滚动标识符和新的滚动间隔，以便获取下一批次的结果。

这个过程会在一个循环中重复执行，直到不再有任何结果返回。这意味着滚动搜索已经完成，所有匹配的文档都已被检索。

该部分的代码如下所示，代码添加在前文提及的 **buildAndExecuteScrollSearchRequest** 方法中：

```java
//设置滚动标识符
    SearchScrollRequest scrollRequest = new SearchScrollRequest(scrollId);
    scrollRequest.scroll(TimeValue.timeValueSeconds(30));
    SearchResponse searchScrollResponse =
```

```
        restClient.scroll(scrollRequest, RequestOptions.DEFAULT);

    // 读取新的滚动ID，该ID指向保持活动状态的搜索上下文，并在后续搜索滚动调用中被需要
    scrollId = searchScrollResponse.getScrollId();
    // 检索另一批搜索结果
    hits = searchScrollResponse.getHits();
```

此时，buildAndExecuteScrollSearchRequest 方法中的全部代码如下所示：

```
// 构建 SearchRequest
public void buildAndExecuteScrollSearchRequest(String indexName, int size) {
    // 索引名称
    SearchRequest searchRequest = new SearchRequest(indexName);
    SearchSourceBuilder searchSourceBuilder = new SearchSourceBuilder();
    searchSourceBuilder.query(QueryBuilders.matchQuery("title",
        "Elasticsearch"));
    // 创建 SearchRequest 及相应的 SearchSourceBuilder。还可以选择设置大小，以控制一次
    // 检索多少结果
    searchSourceBuilder.size(size);
    searchRequest.source(searchSourceBuilder);
    // 设置滚动间隔
    searchRequest.scroll(TimeValue.timeValueMinutes(1L));
    try {
        SearchResponse searchResponse = restClient.search(searchRequest,
            RequestOptions.DEFAULT);
        //读取返回的滚动ID，该ID指向保持活动状态的搜索上下文，并在后续搜索滚动调用中被需要
        String scrollId = searchResponse.getScrollId();
        // 检索第一批搜索结果
        SearchHits hits = searchResponse.getHits();
        while (hits != null && hits.getHits().length != 0) {
            //设置滚动标识符
            SearchScrollRequest scrollRequest = new SearchScrollRequest(scrollId);
            scrollRequest.scroll(TimeValue.timeValueSeconds(30));
            SearchResponse searchScrollResponse =
                restClient.scroll(scrollRequest, RequestOptions.DEFAULT);

            // 读取新的滚动ID，该ID指向保持活动状态的搜索上下文，并在后续搜索滚动调用中被需要
            scrollId = searchScrollResponse.getScrollId();
            // 检索另一批搜索结果
            hits = searchScrollResponse.getHits();

            log.info("scrollId is " + scrollId);
            log.info(
```

```
            "total hits is " + hits.getTotalHits().value + ";now hits is " +
              hits.getHits().length);
      }
    } catch (Exception e) {
      e.printStackTrace();
    } finally {
      // 关闭Elasticsearch连接
      closeEs();
    }
  }
```

SearchScrollRequest 的可选参数

在 SearchScrollRequest 中，除滚动标识符外，还提供了可选参数供用户进行配置。SearchScrollRequest 提供的主要可选参数是滚动搜索的过期时间，代码如下所示：

```
// 设置滚动搜索的过期时间
scrollRequest.scroll(TimeValue.timeValueSeconds(60L));
scrollRequest.scroll("60s");
```

在实际开发中，如果读者没有为 SearchScrollRequest 设置滚动标识符，则一旦初始滚动时间过期（即初始搜索请求中设置的滚动时间过期），则滚动搜索的上下文也会过期。

2．清除滚动搜索的上下文

在滚动搜索请求执行后，用户可以使用 Clear Scroll API 删除最后一个滚动标识符，以释放滚动搜索的上下文。当滚动搜索超时时间到期时，这个过程也会自动发生。一般在滚动搜索会话后，需立即清除滚动搜索的上下文。

在执行清除滚动搜索上下文的请求之前，需要构建清除滚动搜索请求，即 ClearScrollRequest。ClearScrollRequest 需要把滚动标识符作为参数输入，构建 ClearScrollRequest 的代码如下所示：

```
//构建ClearScrollRequest
public void buildClearScrollRequest(String scrollId){
  ClearScrollRequest request = new ClearScrollRequest();
  //添加单个滚动标识符
  request.addScrollId(scrollId);
}
```

在构建 ClearScrollRequest 时，不仅可以配置单个滚动标识符，还可以配置多个滚动标识符。我们继续在上述方法中添加如下代码：

```
//构建ClearScrollRequest
public void buildClearScrollRequest(String scrollId){
  ClearScrollRequest request = new ClearScrollRequest();
```

```
//添加单个滚动标识符
request.addScrollId(scrollId);

//添加多个滚动标识符
List<String> scrollIds = new ArrayList<>();
scrollIds.add(scrollId);
request.setScrollIds(scrollIds);
}
```

3. 执行清除滚动搜索请求

在 ClearScrollRequest 构建后，即可执行清除滚动搜索请求了。与文档索引请求类似，清除滚动搜索请求也有同步和异步两种执行方式。

同步方式

当以同步方式执行清除滚动搜索请求时，客户端会等待 Elasticsearch 服务器返回的查询结果 ClearScrollResponse。在收到 ClearScrollResponse 后，客户端会继续执行相关的逻辑代码。以同步方式执行的代码如下所示：

```
//以同步方式执行清除滚动搜索请求
public void executeClearScrollRequest(String scrollId) {
  ClearScrollRequest request = new ClearScrollRequest();
  // 添加单个滚动标识符
  request.addScrollId(scrollId);
  try {
    ClearScrollResponse response = restClient.clearScroll(request,
        RequestOptions.DEFAULT);
  } catch (Exception e) {
    e.printStackTrace();
  } finally {
    // 关闭Elasticsearch连接
    closeEs();
  }
}
```

异步方式

当以异步方式执行清除滚动搜索请求时，高级客户端不必同步等待请求结果的返回，可以直接向接口调用方返回异步接口执行成功的结果。

为了处理异步返回的响应信息或处理在请求执行过程中引发的异常信息，用户需要指定监听器。

在异步请求处理后，如果请求执行成功，则调用 ActionListener 类中的 onResponse 方法进行相关逻辑的处理；如果请求执行失败，则调用 ActionListener 类中的 onFailure 方法进行相关逻辑的处理。

以异步方式执行的代码如下所示：

```java
//以异步方式执行清除滚动搜索请求
public void executeClearScrollRequestAsync(String scrollId) {
  ClearScrollRequest request = new ClearScrollRequest();
  // 添加单个滚动标识符
  request.addScrollId(scrollId);
  // 添加监听器
  ActionListener<ClearScrollResponse> listener = new ActionListener
      <ClearScrollResponse>() {
    @Override
    public void onResponse(ClearScrollResponse clearScrollResponse) {
    }
    @Override
    public void onFailure(Exception e) {
    }
  };

  try {
    restClient.clearScrollAsync(request, RequestOptions.DEFAULT, listener);
  } catch (Exception e) {
    e.printStackTrace();
  } finally {
    // 关闭Elasticsearch连接
    closeEs();
  }
}
```

当然，在异步请求执行过程中可能会出现异常，异常的处理与同步方式执行情况相同。

4．解析清除滚动搜索请求的响应结果

不论同步方式，还是异步方式，客户端均需要对请求的返回结果进行处理和解析。ClearScrollResponse 中含有已发布的滚动搜索上下文的信息。

解析清除滚动搜索请求的核心代码如下所示：

```java
// 如果请求成功，则会返回true的结果
boolean success = response.isSucceeded();
// 返回已释放的搜索上下文数
int released = response.getNumFreed();
```

完整代码如下所示：

```java
//以同步方式执行清除滚动搜索请求
 public void executeClearScrollRequest(String scrollId) {
    ClearScrollRequest request = new ClearScrollRequest();
    // 添加单个滚动标识符
    request.addScrollId(scrollId);
    try {
        ClearScrollResponse response = restClient.clearScroll(request,
            RequestOptions.DEFAULT);
        // 如果请求成功，则会返回true的结果
        boolean success = response.isSucceeded();
        // 返回已释放的搜索上下文数
        int released = response.getNumFreed();
        log.info("success is " + success + ";released is  " + released);
    } catch (Exception e) {
        e.printStackTrace();
    } finally {
        // 关闭Elasticsearch连接
        closeEs();
    }
}
```

5．执行滚动搜索请求

在 SearchScrollRequest 构建后，即可执行滚动搜索请求。与文档索引请求类似，滚动搜索请求也有同步和异步两种执行方式。

同步方式

当以同步方式执行滚动搜索请求时，客户端会等待 Elasticsearch 服务器返回的查询结果 SearchResponse。在收到 SearchResponse 后，客户端会继续执行相关的逻辑代码。以同步方式执行的代码如下所示：

```java
SearchResponse searchResponse = client.scroll(scrollRequest, RequestOptions.DEFAULT);
```

以同步方式执行的全部代码详见上文中的 buildAndExecuteScrollSearchRequest 方法。

异步方式

当以异步方式执行滚动搜索请求时，高级客户端不必同步等待请求结果的返回，可以直接向接口调用方返回异步接口执行成功的结果。

为了处理异步返回的响应信息或处理在请求执行过程中引发的异常信息，用户需要指定监听器。以异步方式执行的代码如下所示：

```
client.scrollAsync(scrollRequest, RequestOptions.DEFAULT, scrollListener);
```

其中，scrollListener 是监听器。

在异步请求处理后，如果请求执行成功，则调用 ActionListener 类中的 onResponse 方法进行相关逻辑的处理；如果请求执行失败，则调用 ActionListener 类中的 onFailure 方法进行相关逻辑的处理。

以异步方式执行的代码如下所示：

```java
// 构建 SearchRequest
public void buildAndExecuteScrollSearchRequestAsync(String indexName, int
    size) {
  // 索引名称
  SearchRequest searchRequest = new SearchRequest(indexName);
  SearchSourceBuilder searchSourceBuilder = new SearchSourceBuilder();
  searchSourceBuilder.query(QueryBuilders.matchQuery("title",
      "Elasticsearch"));
  // 创建 SearchRequest 及相应的 SearchSourceBuilder。还可以选择设置大小，以控制一次
  // 检索多少结果
  searchSourceBuilder.size(size);
  searchRequest.source(searchSourceBuilder);
  // 设置滚动间隔
  searchRequest.scroll(TimeValue.timeValueMinutes(1L));
  try {
    SearchResponse searchResponse = restClient.search(searchRequest,
        RequestOptions.DEFAULT);
    // 读取返回的滚动 ID，该 ID 指向保持活动状态的搜索上下文，并在后续搜索滚动调用中被需要
    String scrollId = searchResponse.getScrollId();
    // 检索第一批搜索结果
    SearchHits hits = searchResponse.getHits();
    // 配置监听器
    ActionListener<SearchResponse> scrollListener = new ActionListener
        <SearchResponse>() {
      @Override
      public void onResponse(SearchResponse searchResponse) {
        // 读取新的滚动 ID，该 ID 指向保持活动状态的搜索上下文，并在后续搜索滚动调用中被需要
        String scrollId = searchResponse.getScrollId();
        // 检索另一批搜索结果
        SearchHits hits = searchResponse.getHits();
        log.info("scrollId is " + scrollId);
        log.info("total hits is " + hits.getTotalHits().value + ";now hits
            is "
          + hits.getHits().length);
      }
```

```
            @Override
            public void onFailure(Exception e) {
            }
        };
        while (hits != null && hits.getHits().length != 0) {
            // 设置滚动标识符
            SearchScrollRequest scrollRequest = new SearchScrollRequest(scrollId);
            scrollRequest.scroll(TimeValue.timeValueSeconds(30));
            // 异步执行
            restClient.scrollAsync(scrollRequest, RequestOptions.DEFAULT,
                scrollListener);
        }
    } catch (Exception e) {
        e.printStackTrace();
    } finally {
        // 关闭Elasticsearch连接
        closeEs();
    }
}
```

当然，在异步请求执行过程中可能会出现异常，异常的处理与同步方式执行情况相同。

6. 解析滚动搜索请求的响应结果

滚动搜索请求与搜索 API 有相同的响应结果，即 SearchResponse 对象。

解析滚动搜索请求结果的代码共分为三层，分别是 Controller 层、Service 层和 ServiceImpl 实现层。

在 Controller 层中新增 ScrollSearchController 类。ScrollSearchController 类用于存放有关滚动搜索的请求调用。ScrollSearchController 类中的代码如下所示：

```
package com.niudong.esdemo.controller;
import org.elasticsearch.common.Strings;
import org.springframework.beans.factory.annotation.Autowired;
import org.springframework.web.bind.annotation.RequestMapping;
import org.springframework.web.bind.annotation.RestController;
import com.niudong.esdemo.service.ScrollSearchService;
@RestController
@RequestMapping("/springboot/es/scrollsearch")
public class ScrollSearchController {
    @Autowired
    private ScrollSearchService scrollSearchService;
    // 以同步方式执行 SearchScrollRequest
    @RequestMapping("/sr")
```

```
    public String executeSearchRequest(String indexName, int size) {
      // 参数校验
      if (Strings.isNullOrEmpty(indexName) || size <= 0) {
        return "Parameters are wrong!";
      }
      scrollSearchService.buildAndExecuteScrollSearchRequest(indexName, size);
      return "Execute SearchScrollRequest success!";
    }
}
```

在 Service 层中新增 ScrollSearchService 接口类,该类用于描述滚动搜索的 Service 层接口。ScrollSearchService 类中的代码如下所示:

```
package com.niudong.esdemo.service;
public interface ScrollSearchService {
  // 构建 SearchRequest
  public void buildAndExecuteScrollSearchRequest(String indexName, int size);
}
```

在 ServiceImpl 实现层中新增 ScrollSearchServiceImpl 接口实现类,该类用于描述滚动搜索的 ServiceImpl 层接口的具体实现。ScrollSearchServiceImpl 类中的部分代码如下所示:

```
package com.niudong.esdemo.service.impl;
import java.util.ArrayList;
import java.util.List;
import javax.annotation.PostConstruct;
import org.apache.commons.logging.Log;
import org.apache.commons.logging.LogFactory;
import org.apache.http.HttpHost;
import org.elasticsearch.action.ActionListener;
import org.elasticsearch.action.search.ClearScrollRequest;
import org.elasticsearch.action.search.ClearScrollResponse;
import org.elasticsearch.action.search.SearchRequest;
import org.elasticsearch.action.search.SearchResponse;
import org.elasticsearch.action.search.SearchScrollRequest;
import org.elasticsearch.client.RequestOptions;
import org.elasticsearch.client.RestClient;
import org.elasticsearch.client.RestHighLevelClient;
import org.elasticsearch.common.unit.TimeValue;
import org.elasticsearch.index.query.QueryBuilders;
import org.elasticsearch.search.SearchHits;
import org.elasticsearch.search.builder.SearchSourceBuilder;
import org.springframework.stereotype.Service;
import com.niudong.esdemo.service.ScrollSearchService;
```

```java
import ch.qos.logback.classic.Logger;
@Service
public class ScrollSearchServiceImpl implements ScrollSearchService {
    private static Log log = LogFactory.getLog(ScrollSearchServiceImpl.class);
    private RestHighLevelClient restClient;
    // 初始化连接
    @PostConstruct
    public void initEs() {
        restClient = new RestHighLevelClient(RestClient.builder(new HttpHost
            ("localhost", 9200, "http"),
            new HttpHost("localhost", 9201, "http")));
        log.info("ElasticSearch init in service.");
    }
    // 关闭连接
    public void closeEs() {
        try {
            restClient.close();
        } catch (Exception e) {
            e.printStackTrace();
        }
    }
    // 构建SearchRequest
    public void buildAndExecuteScrollSearchRequest(String indexName, int size)
    {
        // 索引名称
        SearchRequest searchRequest = new SearchRequest(indexName);
        SearchSourceBuilder searchSourceBuilder = new SearchSourceBuilder();
        searchSourceBuilder.query(QueryBuilders.matchQuery("content", "美联储"));
        // 创建SearchRequest及相应的SearchSourceBuilder。还可以选择设置大小,以控制一次
        // 检索多少个结果
        searchSourceBuilder.size(size);
        searchRequest.source(searchSourceBuilder);
        // 设置滚动间隔
        searchRequest.scroll(TimeValue.timeValueMinutes(1L));
        try {
            SearchResponse searchResponse = restClient.search(searchRequest,
                RequestOptions.DEFAULT);
            // 读取返回的滚动ID,该ID指向保持活动状态的搜索上下文,并在后续搜索滚动调用中被需要
            String scrollId = searchResponse.getScrollId();
            // 检索第一批搜索结果
            SearchHits hits = searchResponse.getHits();
            while (hits != null && hits.getHits().length != 0) {
                // 设置滚动标识符
                SearchScrollRequest scrollRequest = new SearchScrollRequest(scrollId);
```

```java
      scrollRequest.scroll(TimeValue.timeValueSeconds(30));
      SearchResponse searchScrollResponse =
          restClient.scroll(scrollRequest, RequestOptions.DEFAULT);
      //读取新的滚动ID,该ID指向保持活动状态的搜索上下文,并在后续搜索滚动调用中被需要
      scrollId = searchScrollResponse.getScrollId();
      // 检索另一批搜索结果
      hits = searchScrollResponse.getHits();
      log.info("scrollId is " + scrollId);
      log.info(
          "total hits is " + hits.getTotalHits().value + ";now hits is " +
              hits.getHits().length);
    }

    //清除滚动搜索上下文的信息
    executeClearScrollRequest(scrollId);
  } catch (Exception e) {
    e.printStackTrace();
  } finally {
    // 关闭Elasticsearch的连接
    closeEs();
  }
}
// 以同步方式执行清除滚动搜索上下文的请求
  public void executeClearScrollRequest(String scrollId) {
    ClearScrollRequest request = new ClearScrollRequest();
    // 添加单个滚动标识符
    request.addScrollId(scrollId);
    try {
      ClearScrollResponse response = restClient.clearScroll(request,
          RequestOptions. DEFAULT);
      // 如果请求成功,则会返回true的结果
      boolean success = response.isSucceeded();
      // 返回已释放的搜索上下文数
      int released = response.getNumFreed();
      log.info("success is " + success + ";released is  " + released);
    } catch (Exception e) {
      e.printStackTrace();
    } finally {
      // 关闭Elasticsearch连接
      closeEs();
    }
  }
}
```

随后编译工程，在工程根目录中输入如下命令：

```
mvn clean package
```

通过如下命令启动工程服务：

```
java -jar ./target/esdemo-0.0.1-SNAPSHOT.jar
```

在工程服务启动后，在浏览器中调用如下接口查看滚动搜索请求的执行情况：

```
http://localhost:8080//springboot/es/scrollsearch/sr?indexName=ultraman&size=1
```

接口执行后，在服务器中输出如下所示内容：

```
2019-09-02 15:02:45.295  INFO 31140 --- [nio-8080-exec-9] o.s.web.servlet.DispatcherServlet        : Completed initializ
ation in 13 ms
2019-09-02 15:02:48.017  INFO 31140 --- [nio-8080-exec-9] c.n.e.s.impl.ScrollSearchServiceImpl     : scrollId is DXF1ZXJ
5QW5kRmVOY2gBAAAAAAAAAEWM2dOUmJFTF9US1NvY3BnUm5jMndqQQ==
2019-09-02 15:02:48.017  INFO 31140 --- [nio-8080-exec-9] c.n.e.s.impl.ScrollSearchServiceImpl     : total hits is 3;now
 hits is 1
2019-09-02 15:02:48.026  INFO 31140 --- [nio-8080-exec-9] c.n.e.s.impl.ScrollSearchServiceImpl     : scrollId is DXF1ZXJ
5QW5kRmVOY2gBAAAAAAAAAEWM2dOUmJFTF9US1NvY3BnUm5jMndqQQ==
2019-09-02 15:02:48.027  INFO 31140 --- [nio-8080-exec-9] c.n.e.s.impl.ScrollSearchServiceImpl     : total hits is 3;now
 hits is 1
2019-09-02 15:02:48.040  INFO 31140 --- [nio-8080-exec-9] c.n.e.s.impl.ScrollSearchServiceImpl     : scrollId is DXF1ZXJ
5QW5kRmVOY2gBAAAAAAAAAEWM2dOUmJFTF9US1NvY3BnUm5jMndqQQ==
2019-09-02 15:02:48.041  INFO 31140 --- [nio-8080-exec-9] c.n.e.s.impl.ScrollSearchServiceImpl     : total hits is 3;now
 hits is 0
2019-09-02 15:02:48.064  INFO 31140 --- [nio-8080-exec-9] c.n.e.s.impl.ScrollSearchServiceImpl     : success is true;rel
eased is  1
```

与搜索内容相关的数据共 3 条，每次滚动搜索数量为 1，因此共输出三次滚动搜索的内容，与预期相同。

此时，如果在调用接口的浏览器中显示如下内容，则表明接口调用成功：

```
Execute SearchScrollRequest success!
```

7.3　批量搜索

在 Elasticsearch 中，不仅提供了单次搜索查询接口，还提供了批量搜索查询接口，即批量搜索 API（MultiSearch API），支持在单次 HTTP 请求中并行执行多个搜索请求。

1. 构建批量搜索请求

在执行批量搜索请求前，需要构建批量搜索请求，即 MultiSearchRequest。在初始化 MultiSearchRequest 时，搜索请求为空，因此用户需要把要执行的所有搜索添到 MultiSearchRequest 中。

构建 MultiSearchRequest 的代码如下所示：

```java
// 构建 MultiSearchRequest
public void buildMultiSearchRequest() {
    MultiSearchRequest request = new MultiSearchRequest();
    // 构建搜索请求对象 1
    SearchRequest firstSearchRequest = new SearchRequest();
    SearchSourceBuilder searchSourceBuilder = new SearchSourceBuilder();
    searchSourceBuilder.query(QueryBuilders.matchQuery("user", "niudong1"));
    firstSearchRequest.source(searchSourceBuilder);
    // 将搜索请求对象 1 添加到 MultiSearchRequest 中
    request.add(firstSearchRequest);
    // 构建搜索请求对象 2
    SearchRequest secondSearchRequest = new SearchRequest();
    searchSourceBuilder = new SearchSourceBuilder();
    searchSourceBuilder.query(QueryBuilders.matchQuery("user", "niudong2"));
    secondSearchRequest.source(searchSourceBuilder);
    // 将搜索请求对象 2 添加到 MultiSearchRequest 中
    request.add(secondSearchRequest);
}
```

可选参数

在 MultiSearchRequest 的构建过程中，有一些可选参数供用户进行配置。MultiSearchRequest 中的 SearchRequest 支持搜索的所有可选参数，如配置索引名称等，代码如下所示：

```java
SearchRequest searchRequest = new SearchRequest("ultraman");
```

2．执行批量搜索请求

在 MultiSearchRequest 构建后，即可执行批量搜索请求。与文档索引请求类似，批量搜索请求也有同步和异步两种执行方式。

同步方式

当以同步方式执行批量搜索请求时，客户端会等待 Elasticsearch 服务器返回的查询结果 MultiSearchResponse。在收到 MultiSearchResponse 后，客户端会继续执行相关的逻辑代码。以同步方式执行的代码如下所示：

```java
MultiSearchResponse response = client.msearch(request, RequestOptions.DEFAULT);
```

以同步方式执行的完整代码如下所示：

```java
// 构建 MultiSearchRequest
public MultiSearchRequest buildMultiSearchRequest(String field, String[] keywords) {
    MultiSearchRequest request = new MultiSearchRequest();
```

```java
    // 构建搜索请求对象1
    SearchRequest firstSearchRequest = new SearchRequest();
    SearchSourceBuilder searchSourceBuilder = new SearchSourceBuilder();
    searchSourceBuilder.query(QueryBuilders.matchQuery(field, keywords[0]));
    firstSearchRequest.source(searchSourceBuilder);
    // 将搜索请求对象1添加到 MultiSearchRequest 中
    request.add(firstSearchRequest);
    // 构建搜索请求对象2
    SearchRequest secondSearchRequest = new SearchRequest();
    searchSourceBuilder = new SearchSourceBuilder();
    searchSourceBuilder.query(QueryBuilders.matchQuery(field, keywords[1]));
    secondSearchRequest.source(searchSourceBuilder);
    // 将搜索请求对象2添加到 MultiSearchRequest 中
    request.add(secondSearchRequest);
    return request;
}
//以同步方式执行 MultiSearchRequest
public void executeMultiSearchRequest(String field, String[] keywords) {
    // 构建 MultiSearchRequest
    MultiSearchRequest request = buildMultiSearchRequest(field,keywords);
    try {
        MultiSearchResponse response = restClient.msearch(request, RequestOptions.
            DEFAULT);
    } catch (Exception e) {
        e.printStackTrace();
    } finally {
        // 关闭 Elasticsearch 连接
        closeEs();
    }
}
```

异步方式

当以异步方式执行批量搜索请求时，高级客户端不必同步等待请求结果的返回，可以直接向接口调用方返回异步接口执行成功的结果。

为了处理异步返回的响应信息或处理在请求执行过程中引发的异常信息，用户需要指定监听器。以异步方式执行的核心代码如下所示：

```java
client.msearchAsync(searchRequest, RequestOptions.DEFAULT, listener);
```

其中，listener 为监听器。

在异步请求处理后，如果请求执行成功，则调用 ActionListener 类中的 onResponse 方法进行相关逻辑的处理；如果请求执行失败，则调用 ActionListener 类中的 onFailure 方法进行相关

逻辑的处理。

以异步方式执行的全部代码如下所示：

```java
// 构建 MultiSearchRequest
public MultiSearchRequest buildMultiSearchRequest(String field, String[]
    keywords) {
  MultiSearchRequest request = new MultiSearchRequest();
  // 构建搜索请求对象1
  SearchRequest firstSearchRequest = new SearchRequest();
  SearchSourceBuilder searchSourceBuilder = new SearchSourceBuilder();
  searchSourceBuilder.query(QueryBuilders.matchQuery(field, keywords[0]));
  firstSearchRequest.source(searchSourceBuilder);
  // 将搜索请求对象1添加到 MultiSearchRequest 中
  request.add(firstSearchRequest);
  // 构建搜索请求对象2
  SearchRequest secondSearchRequest = new SearchRequest();
  searchSourceBuilder = new SearchSourceBuilder();
  searchSourceBuilder.query(QueryBuilders.matchQuery(field, keywords[1]));
  secondSearchRequest.source(searchSourceBuilder);
  // 将搜索请求对象2添加到 MultiSearchRequest 中
  request.add(secondSearchRequest);
  return request;
}
// 以异步方式执行 MultiSearchRequest
public void executeMultiSearchRequestAsync(String field, String[] keywords)
{
  // 构建 MultiSearchRequest
  MultiSearchRequest request = buildMultiSearchRequest(field, keywords);
  // 构建监听器
  ActionListener<MultiSearchResponse> listener = new ActionListener
      <MultiSearchResponse>() {
    @Override
    public void onResponse(MultiSearchResponse response) {
    }
    @Override
    public void onFailure(Exception e) {
    }
  };
  // 以异步方式执行
  try {
    restClient.msearchAsync(request, RequestOptions.DEFAULT, listener);
  } catch (Exception e) {
    e.printStackTrace();
```

```
    } finally {
      // 关闭Elasticsearch连接
      closeEs();
    }
  }
```

当然，在异步请求执行过程中可能会出现异常，异常的处理与同步方式执行情况相同。

3. 解析批量搜索请求的响应结果

不论同步方式，还是异步方式，在批量搜索请求执行后，客户端均需要对返回结果 MultiSearchResponse 进行处理和解析。

MultiSearchResponse 中包含 MultiSearchResponse.item 对象列表，每个 MultiSearchResponse.item 对象对应于 MultiSearchRequest 中的每个搜索请求。

如果批量搜索请求执行失败，则每个 MultiSearchResponse.item 对象的 getFailure 方法中都会包含一个异常信息；反之，如果请求执行成功，则用户可以从每个 MultiSearchResponse.item 对象的 getResponse 方法中获得 SearchResponse，并解析对应的结果信息。

解析 MultiSearchResponse 的代码如下所示，代码共分为三层，分别是 Controller 层、Service 层和 ServiceImpl 实现层。

在 Controller 层中新增 MultiSearchController 类，代码如下所示：

```java
package com.niudong.esdemo.controller;
import java.util.List;
import org.elasticsearch.common.Strings;
import org.springframework.beans.factory.annotation.Autowired;
import org.springframework.web.bind.annotation.RequestMapping;
import org.springframework.web.bind.annotation.RestController;
import com.google.common.base.Splitter;
import com.niudong.esdemo.service.MultiSearchService;
@RestController
@RequestMapping("/springboot/es/multisearch")
public class MultiSearchController {
  @Autowired
  private MultiSearchService multiSearchService;
  // 以同步方式执行MultiSearchRequest
  @RequestMapping("/sr")
  public String executeMultiSearchRequest(String field, String keywords) {
    // 参数校验
    if (Strings.isNullOrEmpty(field) || Strings.isNullOrEmpty(keywords)) {
      return "Parameters are wrong!";
    }
```

```
    // 将英文逗号分隔的字符串切分成数组
    List<String> keywordsList = Splitter.on(",").splitToList(keywords);
    multiSearchService.executeMultiSearchRequest(field,
        keywordsList.toArray(new String[keywordsList.size()]));
    return "Execute MultiSearchRequest success!";
  }
}
```

在 Service 层中新增 MultiSearchService 接口类，该类主要用于描述 Service 层的接口方法，代码如下所示：

```
package com.niudong.esdemo.service;
public interface MultiSearchService {
  // 以同步方式执行 MultiSearchRequest
  public void executeMultiSearchRequest(String field, String[] keywords);
}
```

在 ServiceImpl 实现层中新增 MultiSearchServiceImpl 接口实现类，该类主要用于描述 Service 实现层的接口实现方法，代码如下所示：

```
package com.niudong.esdemo.service.impl;
import javax.annotation.PostConstruct;
import org.apache.commons.logging.Log;
import org.apache.commons.logging.LogFactory;
import org.apache.http.HttpHost;
import org.elasticsearch.action.ActionListener;
import org.elasticsearch.action.search.MultiSearchRequest;
import org.elasticsearch.action.search.MultiSearchResponse;
import org.elasticsearch.action.search.MultiSearchResponse.Item;
import org.elasticsearch.action.search.SearchRequest;
import org.elasticsearch.action.search.SearchResponse;
import org.elasticsearch.client.RequestOptions;
import org.elasticsearch.client.RestClient;
import org.elasticsearch.client.RestHighLevelClient;
import org.elasticsearch.index.query.QueryBuilders;
import org.elasticsearch.search.SearchHit;
import org.elasticsearch.search.SearchHits;
import org.elasticsearch.search.builder.SearchSourceBuilder;
import org.springframework.stereotype.Service;
import com.niudong.esdemo.service.MultiSearchService;
@Service
public class MultiSearchServiceImpl implements MultiSearchService {
  private static Log log = LogFactory.getLog(MultiSearchServiceImpl.class);
  private RestHighLevelClient restClient;
```

```java
// 初始化连接
@PostConstruct
public void initEs() {
  restClient = new RestHighLevelClient(RestClient.builder(new HttpHost
      ("localhost", 9200, "http"),
    new HttpHost("localhost", 9201, "http")));
  log.info("ElasticSearch init in service.");
}
// 关闭连接
public void closeEs() {
  try {
    restClient.close();
  } catch (Exception e) {
    e.printStackTrace();
  }
}
// 构建 MultiSearchRequest
public MultiSearchRequest buildMultiSearchRequest(String field, String[]
    keywords) {
  MultiSearchRequest request = new MultiSearchRequest();
  // 构建搜索请求对象 1
  SearchRequest firstSearchRequest = new SearchRequest();
  SearchSourceBuilder searchSourceBuilder = new SearchSourceBuilder();
  searchSourceBuilder.query(QueryBuilders.matchQuery(field, keywords[0]));
  firstSearchRequest.source(searchSourceBuilder);
  // 将搜索请求对象 1 添加到 MultiSearchRequest 中
  request.add(firstSearchRequest);
  // 构建搜索请求对象 2
  SearchRequest secondSearchRequest = new SearchRequest();
  searchSourceBuilder = new SearchSourceBuilder();
  searchSourceBuilder.query(QueryBuilders.matchQuery(field, keywords[1]));
  secondSearchRequest.source(searchSourceBuilder);
  // 将搜索请求对象 2 添加到 MultiSearchRequest 中
  request.add(secondSearchRequest);
  return request;
}
// 以同步方式执行 MultiSearchRequest
public void executeMultiSearchRequest(String field, String[] keywords) {
  // 构建 MultiSearchRequest
  MultiSearchRequest request = buildMultiSearchRequest(field, keywords);
  try {
    MultiSearchResponse response = restClient.msearch(request,
        RequestOptions.DEFAULT);
```

```
      // 解析返回结果MultiSearchResponse
      processMultiSearchResponse(response);
    } catch (Exception e) {
      e.printStackTrace();
    } finally {
      // 关闭Elasticsearch连接
      closeEs();
    }
  }
  // 解析返回结果MultiSearchResponse
  private void processMultiSearchResponse(MultiSearchResponse response) {
    // 获取返回结果集合
    Item[] items = response.getResponses();
    // 判断返回结果集合是否为空
    if (items == null || items.length <= 0) {
      log.info("items is null.");
      return;
    }
    for (Item item : items) {
      Exception exception = item.getFailure();
      if (exception != null) {
        log.info("eception is " + exception.toString());
      }
      SearchResponse searchResponse = item.getResponse();
      SearchHits hits = searchResponse.getHits();
      if (hits.getTotalHits().value <= 0) {
        log.info("hits.getTotalHits().value is 0.");
        return;
      }
      SearchHit[] hitArray = hits.getHits();
      for (int i = 0; i < hitArray.length; i++) {
        SearchHit hit = hitArray[i];
        log.info("id is " + hit.getId() + ";index is " + hit.getIndex() + ";source is "
            + hit.getSourceAsString());
      }
    }
  }
}
```

随后编译工程，在工程根目录下输入如下命令：

```
mvn clean package
```

通过如下命令启动工程服务：

```
java -jar ./target/esdemo-0.0.1-SNAPSHOT.jar
```

在工程服务启动后，在浏览器中调用如下接口查看批量搜索的执行情况：

```
http://localhost:8080/springboot/es/multisearch/sr?field=content&keywords=美联储,空军
```

在服务器中输出如下所示内容：

```
2019-09-02 20:26:24.855  INFO 32668 --- [nio-8080-exec-1] c.n.e.s.impl.MultiSearchServiceImpl      : id is 3;index is ul
traman;source is {"content":"从此前美联储降息历程来看，美联储降息将打开全球各国央行的降息窗口"}
2019-09-02 20:26:24.856  INFO 32668 --- [nio-8080-exec-1] c.n.e.s.impl.MultiSearchServiceImpl      : id is 3;index is ul
traman1;source is {"content":"从此前美联储降息历程来看，美联储降息将打开全球各国央行的降息窗口"}
2019-09-02 20:26:24.857  INFO 32668 --- [nio-8080-exec-1] c.n.e.s.impl.MultiSearchServiceImpl      : id is 2;index is ul
traman;source is {"content":"自6月起，市场对于美联储降息的预期愈发强烈"}
2019-09-02 20:26:24.857  INFO 32668 --- [nio-8080-exec-1] c.n.e.s.impl.MultiSearchServiceImpl      : id is 2;index is ul
traman1;source is {"content":"自6月起，市场对于美联储降息的预期愈发强烈"}
2019-09-02 20:26:24.858  INFO 32668 --- [nio-8080-exec-1] c.n.e.s.impl.MultiSearchServiceImpl      : id is 1;index is ul
traman;source is {"content":"事实上，自今年年初开始，美联储就已传递出货币政策或将转向的迹象"}
2019-09-02 20:26:24.858  INFO 32668 --- [nio-8080-exec-1] c.n.e.s.impl.MultiSearchServiceImpl      : id is 1;index is ul
traman1;source is {"content":"事实上，自今年年初开始，美联储就已传递出货币政策或将转向的迹象"}
```

索引名称 ultraman 和 ultraman1 中仅仅包含了"美联储"关键词的信息，因此共搜索出 6 条数据，与数据的实际情况相同。

此时，如果浏览器页面中的输出内容如下，则表明接口执行成功：

```
Execute MultiSearchRequest success!
```

7.4 跨索引字段搜索

Elasticsearch 不仅提供了在特定索引下的字段搜索接口，还提供了跨索引的字段搜索接口，即 Field Capabilities API。

1. 构建跨索引字段搜索请求

在执行跨索引字段搜索请求前，需要构建跨索引字段搜索请求，即 FieldCapabilitiesRequest。

FieldCapabilitiesRequest 中包含了要搜索的字段列表及一个可选的目标索引名称列表。如果没有提供目标索引名称列表，则默认对所有索引执行相关请求。

需要指出的是，字段列表（即 fields 参数）支持通配符的表示方法。例如，text_*将返回与表达式匹配的所有字段。

构建 FieldCapabilitiesRequest 的代码如下所示：

```
// 构建 FieldCapabilitiesRequest
public FieldCapabilitiesRequest buildFieldCapabilitiesRequest() {
    FieldCapabilitiesRequest request =
```

```
        new FieldCapabilitiesRequest().fields("content").indices("ultraman",
            "ultraman1");
    return request;
}
```

类似地，FieldCapabilitiesRequest 也有可选参数列表供用户进行配置。FieldCapabilitiesRequest 中的主要可选参数是 IndicesOptions，用于解析不可用的索引及展开通配符表达式。

可选参数配置代码如下所示：

```
// 构建 FieldCapabilitiesRequest
public FieldCapabilitiesRequest buildFieldCapabilitiesRequest() {
    FieldCapabilitiesRequest request =
        new  FieldCapabilitiesRequest().fields("content").indices("ultraman",
"ultraman1");

    // 配置可选参数 IndicesOptions：解析不可用的索引及展开通配符表达式
    request.indicesOptions(IndicesOptions.lenientExpandOpen());

    return request;
}
```

2．执行跨索引字段搜索请求

在 FieldCapabilitiesRequest 构建后，即可执行跨索引字段搜索请求。与文档索引请求类似，跨索引字段搜索请求也有同步和异步两种执行方式。

同步方式

当以同步方式执行跨索引字段搜索请求时，客户端会等待 Elasticsearch 服务器返回的查询结果 FieldCapabilitiesResponse。在收到 FieldCapabilitiesResponse 后，客户端会继续执行相关的逻辑代码。以同步方式执行的代码如下所示：

```
// 构建 FieldCapabilitiesRequest
public FieldCapabilitiesRequest buildFieldCapabilitiesRequest(String field,
    String[] indices) {
    FieldCapabilitiesRequest request =
        new FieldCapabilitiesRequest().fields(field).indices(indices[0],
            indices[1]);
    // 配置可选参数 IndicesOptions：解析不可用的索引及展开通配符表达式
    request.indicesOptions(IndicesOptions.lenientExpandOpen());
    return request;
}
//以同步方式执行跨索引字段搜索请求
public void executeFieldCapabilitiesRequest(String field, String[] indices
```

```
  {
  // 构建 FieldCapabilitiesRequest
  FieldCapabilitiesRequest request= buildFieldCapabilitiesRequest(field, indices);
  try {
    FieldCapabilitiesResponse response = restClient.fieldCaps(request,
        RequestOptions.DEFAULT);
  } catch (Exception e) {
    e.printStackTrace();
  } finally {
    // 关闭 Elasticsearch 连接
    closeEs();
  }
}
```

异步方式

当以异步方式执行跨索引字段搜索请求时,高级客户端不必同步等待请求结果的返回,可以直接向接口调用方返回异步接口执行成功的结果。

为了处理异步返回的响应信息或处理在请求执行过程中引发的异常信息,用户需要指定监听器。以异步方式执行的核心代码如下所示:

```
client.fieldCapsAsync(request, RequestOptions.DEFAULT, listener);
```

其中,listener 为监听器。

在异步请求处理后,如果请求执行成功,则调用 ActionListener 类中的 onResponse 方法进行相关逻辑的处理;如果请求执行失败,则调用 ActionListener 类中的 onFailure 方法进行相关逻辑的处理。

以异步方式执行的全部代码如下所示:

```
// 构建 FieldCapabilitiesRequest
public FieldCapabilitiesRequest buildFieldCapabilitiesRequest(String field,
    String[] indices) {
  FieldCapabilitiesRequest request =
      new FieldCapabilitiesRequest().fields(field).indices(indices[0],
          indices[1]);
  // 配置可选参数 indicesOptions: 解析不可用的索引及展开通配符表达式
  request.indicesOptions(IndicesOptions.lenientExpandOpen());
  return request;
}
// 以异步方式执行跨索引字段搜索请求
public void executeFieldCapabilitiesRequestAsync(String field, String[]
    indices) {
  // 构建 FieldCapabilitiesRequest
```

```
        FieldCapabilitiesRequest request = buildFieldCapabilitiesRequest(field,
            indices);
        // 配置监听器
        ActionListener<FieldCapabilitiesResponse> listener =
            new ActionListener<FieldCapabilitiesResponse>() {
                @Override
                public void onResponse(FieldCapabilitiesResponse response) {
                }
                @Override
                public void onFailure(Exception e) {
                }
            };
        // 执行异步请求
        try {
            restClient.fieldCapsAsync(request, RequestOptions.DEFAULT, listener);
        } catch (Exception e) {
            e.printStackTrace();
        } finally {
            // 关闭Elasticsearch连接
            closeEs();
        }
    }
```

当然,在异步请求执行过程中可能会出现异常,异常的处理与同步方式执行情况相同。

3. 解析跨索引字段搜索请求的响应结果

不论同步方式,还是异步方式,在跨索引字段搜索请求执行后,客户端均需要对响应结果 FieldCapabilitiesResponse 进行处理和解析。

FieldCapabilitiesResponse 中包含了每个索引中数据能否被搜索和聚合的信息,还包含了被搜索字段在对应索引中的贡献值。

以同步请求方式为例,FieldCapabilitiesResponse 中的解析代码共分为三层,分别是 Controller 层、Service 层和 ServiceImpl 实现层。

在 Controller 层中新增 FieldCapabilitiesController,用来编写跨索引字段搜索的接口,代码如下所示:

```
package com.niudong.esdemo.controller;
import java.util.List;
import org.elasticsearch.common.Strings;
import org.springframework.beans.factory.annotation.Autowired;
import org.springframework.web.bind.annotation.RequestMapping;
import org.springframework.web.bind.annotation.RestController;
```

```java
import com.google.common.base.Splitter;
import com.niudong.esdemo.service.FieldCapabilitiesService;
@RestController
@RequestMapping("/springboot/es/fieldsearch")
public class FieldCapabilitiesController {
  @Autowired
  private FieldCapabilitiesService fieldCapabilitiesService;

  // 以同步方式执行 MultiSearchRequest
  @RequestMapping("/sr")
  public String executeFieldSearchRequest(String field, String indices) {
    // 参数校验
    if (Strings.isNullOrEmpty(field) || Strings.isNullOrEmpty(indices)) {
      return "Parameters are wrong!";
    }
    // 将英文逗号分隔的字符串切分成数组
    List<String> indicesList = Splitter.on(",").splitToList(indices);
    fieldCapabilitiesService.executeFieldCapabilitiesRequest(field,
        indicesList.toArray(new String[indicesList.size()]));
    return "Execute FieldSearchRequest success!";
  }
}
```

在 Service 层中新增 FieldCapabilitiesService，用来编写跨索引字段搜索的 Service 接口定义，代码如下所示：

```java
package com.niudong.esdemo.service;
public interface FieldCapabilitiesService {
  // 以同步方式执行跨索引字段搜索请求
  public void executeFieldCapabilitiesRequest(String field, String[]
      indices);
}
```

在 ServiceImpl 实现层新增 FieldCapabilitiesServiceImpl 类，用来编写跨索引字段搜索的 Service 接口的具体实现逻辑，代码如下所示：

```java
package com.niudong.esdemo.service.impl;
import java.util.Map;
import java.util.Set;
import javax.annotation.PostConstruct;
import org.apache.commons.logging.Log;
import org.apache.commons.logging.LogFactory;
import org.apache.http.HttpHost;
import org.elasticsearch.action.ActionListener;
```

```java
import org.elasticsearch.action.fieldcaps.FieldCapabilities;
import org.elasticsearch.action.fieldcaps.FieldCapabilitiesRequest;
import org.elasticsearch.action.fieldcaps.FieldCapabilitiesResponse;
import org.elasticsearch.action.support.IndicesOptions;
import org.elasticsearch.client.RequestOptions;
import org.elasticsearch.client.RestClient;
import org.elasticsearch.client.RestHighLevelClient;
import org.springframework.stereotype.Service;
import com.niudong.esdemo.service.FieldCapabilitiesService;
@Service
public class FieldCapabilitiesServiceImpl implements FieldCapabilitiesService {
  private static Log log = LogFactory.getLog(FieldCapabilitiesServiceImpl.
      class);
  private RestHighLevelClient restClient;
  // 初始化连接
  @PostConstruct
  public void initEs() {
    restClient = new RestHighLevelClient(RestClient.builder(new HttpHost
        ("localhost", 9200, "http"),
      new HttpHost("localhost", 9201, "http")));
    log.info("ElasticSearch init in service.");
  }
  // 关闭连接
  public void closeEs() {
    try {
      restClient.close();
    } catch (Exception e) {
      e.printStackTrace();
    }
  }
  // 构建FieldCapabilitiesRequest
  public FieldCapabilitiesRequest buildFieldCapabilitiesRequest(String field,
     String[] indices) {
    FieldCapabilitiesRequest request =
        new FieldCapabilitiesRequest().fields(field).indices(indices[0],
            indices[1]);
    // 配置可选参数indicesOptions:解析不可用的索引及展开通配符表达式
    request.indicesOptions(IndicesOptions.lenientExpandOpen());
    return request;
  }
  // 以同步方式执行跨索引字段搜索请求
  public void executeFieldCapabilitiesRequest(String field, String[] indices)
  {
    // 构建FieldCapabilitiesRequest
```

```java
    FieldCapabilitiesRequest request = buildFieldCapabilitiesRequest(field,
        indices);
    try {
      FieldCapabilitiesResponse response = restClient.fieldCaps(request,
          RequestOptions.DEFAULT);
      // 处理返回结果 FieldCapabilitiesResponse
      processFieldCapabilitiesResponse(response, field, indices);
    } catch (Exception e) {
      e.printStackTrace();
    } finally {
      // 关闭 Elasticsearch 连接
      closeEs();
    }
}
// 处理返回结果 FieldCapabilitiesResponse
private void processFieldCapabilitiesResponse(FieldCapabilitiesResponse
    response, String field,
    String[] indices) {
  // 获取字段中可能含有的类型的映射
  Map<String, FieldCapabilities> fieldResponse = response.getField(field);
  Set<String> set = fieldResponse.keySet();
  // 获取文本字段类型下的数据
  FieldCapabilities textCapabilities = fieldResponse.get("text");
  // 数据能否被搜索到
  boolean isSearchable = textCapabilities.isSearchable();
  log.info("isSearchable is " + isSearchable);
  // 数据能否聚合
  boolean isAggregatable = textCapabilities.isAggregatable();
  log.info("isAggregatable is " + isAggregatable);
  // 获取特定字段类型下的索引
  String[] indicesArray = textCapabilities.indices();
  if (indicesArray != null) {
    log.info("indicesArray is " + indicesArray.length);
  }
  // field 字段不能被搜索到的索引集合
  String[] nonSearchableIndices = textCapabilities.nonSearchableIndices();
  if (nonSearchableIndices != null) {
    log.info("nonSearchableIndices is " + nonSearchableIndices.length);
  }
  // field 字段不能被聚合到的索引集合
  String[] nonAggregatableIndices = textCapabilities.nonAggregatableIndices();
  if (nonAggregatableIndices != null) {
    log.info("nonAggregatableIndices is " + nonAggregatableIndices.length);
```

```
    }
  }
```

随后编译工程,在工程根目录下输入如下命令:

```
mvn clean package
```

通过如下命令启动工程服务:

```
java -jar ./target/esdemo-0.0.1-SNAPSHOT.jar
```

在工程服务启动后,在浏览器中调用如下接口查看跨索引字段搜索的执行情况:

```
http://localhost:8080/springboot/es/fieldsearch/sr?field=content&indices=ultraman,ultraman1
```

接口调用后,在服务器中输出如下所示内容:

```
2019-09-03 11:52:35.378  INFO 29064 --- [nio-8080-exec-1] c.n.e.s.i.FieldCapabilitiesServiceImpl   : isSearchable is true
2019-09-03 11:52:35.378  INFO 29064 --- [nio-8080-exec-1] c.n.e.s.i.FieldCapabilitiesServiceImpl   : isAggregatable is false
```

在 ultraman 和 ultraman1 两个索引中,搜索字段 content 均能搜索到数据,但不能进行聚合。

此时,如果调用上述接口的浏览器中显示如下内容,则表明接口调用成功:

```
Execute FieldSearchRequest success!
```

7.5 搜索结果的排序评估

Elasticsearch 提供了对搜索结果进行排序评估的接口,即 Ranking Evaluation API。Elasticsearch 提供了 rankeval 方法,对一组搜索请求的结果进行排序评估,以便衡量搜索结果的质量。

首先为搜索请求提供一组手动评级的文档,随后评估批量搜索请求的质量,并计算搜索相关指标,如返回结果的平均倒数排名、精度或折扣累积收益等。

1. 构建排序评估请求

在对搜索结果进行排序评估之前,需要构建排序评估请求,即 RankEvalRequest。在创建 RankEvalRequest 之前,需要创建 RankEvalRequest 的依赖对象 RankEvalSpec。RankEvalSpec 用于描述评估规则,用户需要定义 RankEvalRequest 的计算指标及每个搜索请求的分级文档列表。

此外,在创建排序评估请求时,需要将目标索引名称和 RankEvalSpec 作为参数。

创建排序评估对象 RankEvalRequest 的代码如下所示：

```
// 构建 RankEvalRequest
public RankEvalRequest buildRankEvalRequest(String index, String
    documentId, String field, String content) {
  EvaluationMetric metric = new PrecisionAtK();
  List<RatedDocument> ratedDocs = new ArrayList<>();
  // 添加按索引名称、ID 和分级指定的分级文档
  ratedDocs.add(new RatedDocument(index, documentId, 1));
  SearchSourceBuilder searchQuery = new SearchSourceBuilder();
  // 创建要评估的搜索查询
  searchQuery.query(QueryBuilders.matchQuery(field, content));
  // 将前三部分合并为 RatedRequest
  RatedRequest ratedRequest = new RatedRequest("content_query", ratedDocs,
      searchQuery);
  List<RatedRequest> ratedRequests = Arrays.asList(ratedRequest);
  // 创建排序评估规范
  RankEvalSpec specification = new RankEvalSpec(ratedRequests, metric);
  // 创建排序评估请求
  RankEvalRequest request = new RankEvalRequest(specification, new String[]
      {index});
  return request;
}
```

2. 执行排序评估请求

在 RankEvalRequest 构建后，即可执行排序评估请求。与文档索引请求类似，排序评估请求也有同步和异步两种执行方式。

同步方式

当以同步方式执行排序评估请求时，客户端会等待 Elasticsearch 服务器返回的查询结果 RankEvalResponse。在收到 RankEvalResponse 后，客户端会继续执行相关的逻辑代码。以同步方式执行的核心代码如下所示：

```
RankEvalResponse response = client.rankEval(request, RequestOptions.DEFAULT);
```

以同步方式执行的全部代码如下所示：

```
// 构建 RankEvalRequest
 public RankEvalRequest buildRankEvalRequest(String index, String
    documentId, String field,
    String content) {
  EvaluationMetric metric = new PrecisionAtK();
  List<RatedDocument> ratedDocs = new ArrayList<>();
```

```
    // 添加按索引名称、ID和分级指定的分级文档
    ratedDocs.add(new RatedDocument(index, documentId, 1));
    SearchSourceBuilder searchQuery = new SearchSourceBuilder();
    // 创建要评估的搜索查询
    searchQuery.query(QueryBuilders.matchQuery(field, content));
    // 将前三部分合并为 RatedRequest
    RatedRequest ratedRequest = new RatedRequest("content_query", ratedDocs,
        searchQuery);
    List<RatedRequest> ratedRequests = Arrays.asList(ratedRequest);
    // 创建排序评估规范
    RankEvalSpec specification = new RankEvalSpec(ratedRequests, metric);
    // 创建排序评估请求
    RankEvalRequest request = new RankEvalRequest(specification, new String[]
        {index});
    return request;
}
// 以同步方式执行 RankEvalRequest
public void executeRankEvalRequest(String index, String documentId, String
    field,
    String content) {
    // 构建 RankEvalRequest
    RankEvalRequest request = buildRankEvalRequest(index, documentId, field,
        content);
    try {
        RankEvalResponse response = restClient.rankEval(request, RequestOptions.
            DEFAULT);
    } catch (Exception e) {
        e.printStackTrace();
    } finally {
        // 关闭Elasticsearch连接
        closeEs();
    }
}
```

异步方式

当以异步方式执行排序评估请求时,高级客户端不必同步等待请求结果的返回, 可以直接向接口调用方返回异步接口执行成功的结果。

为了处理异步返回的响应信息或处理在请求执行过程中引发的异常信息,用户需要指定监听器。以异步方式执行的核心代码如下所示:

```
client.rankEvalAsync(request, RequestOptions.DEFAULT, listener);
```

其中,listener 为监听器。

在异步请求处理后,如果请求执行成功,则调用 ActionListener 类中的 onResponse 方法进行相关逻辑的处理;如果请求执行失败,则调用 ActionListener 类中的 onFailure 方法进行相关逻辑的处理。

以异步方式执行的代码如下所示:

```
// 构建 RankEvalRequest
 public RankEvalRequest buildRankEvalRequest(String index, String documentId,
     String field,
     String content) {
  EvaluationMetric metric = new PrecisionAtK();
  List<RatedDocument> ratedDocs = new ArrayList<>();
  // 添加按索引名称、ID 和分级指定的分级文档
  ratedDocs.add(new RatedDocument(index, documentId, 1));
  SearchSourceBuilder searchQuery = new SearchSourceBuilder();
  // 创建要评估的搜索查询
  searchQuery.query(QueryBuilders.matchQuery(field, content));
  // 将前三部分合并为 RatedRequest
  RatedRequest ratedRequest = new RatedRequest("content_query", ratedDocs,
      searchQuery);
  List<RatedRequest> ratedRequests = Arrays.asList(ratedRequest);
  // 创建排序评估规范
  RankEvalSpec specification = new RankEvalSpec(ratedRequests, metric);
  // 创建排序评估请求
  RankEvalRequest request = new RankEvalRequest(specification, new String[]
      {index});
  return request;
}
// 以异步方式执行 RankEvalRequest
 public void executeRankEvalRequestAsync(String index, String documentId,
     String field,
     String content) {
  // 构建 RankEvalRequest
  RankEvalRequest request = buildRankEvalRequest(index, documentId, field,
      content);
  // 构建监听器
  ActionListener<RankEvalResponse> listener = new ActionListener
      <RankEvalResponse>() {
   @Override
   public void onResponse(RankEvalResponse response) {
   }
   @Override
   public void onFailure(Exception e) {
```

```
            }
        };
        try {
            restClient.rankEvalAsync(request, RequestOptions.DEFAULT, listener);
        } catch (Exception e) {
            e.printStackTrace();
        } finally {
            // 关闭Elasticsearch连接
            closeEs();
        }
    }
```

当然,在异步请求执行过程中可能会出现异常,异常的处理与同步方式执行情况相同。

3. 解析排序评估请求的响应结果

不论同步方式,还是异步方式,在排序评估请求执行后,客户端均需要对响应结果 RankEvalResponse 进行处理和解析。RankEvalResponse 中不仅包含了总体评估分数和查询每个搜索请求的分数,还包含了有关搜索命中的详细信息及每个结果度量计算的详细信息。

RankEvalResponse 的解析代码共分为三层,以同步方式为例,分别是 Controller 层、Service 层和 ServiceImpl 实现层。

其中,在 Controller 层中新增 RankEvalController 类,用于编写排序评估请求对应接口,代码如下所示:

```
package com.niudong.esdemo.controller;
import org.elasticsearch.common.Strings;
import org.springframework.beans.factory.annotation.Autowired;
import org.springframework.web.bind.annotation.RequestMapping;
import org.springframework.web.bind.annotation.RestController;
import com.niudong.esdemo.service.RankEvalService;
@RestController
@RequestMapping("/springboot/es/ranksearch")
public class RankEvalController {
    @Autowired
    private RankEvalService rankEvalService;

    // 以同步方式执行RankEvalResponse
    @RequestMapping("/sr")
    public String executeRankEvalRequest(String indexName, String document,
        String field, String content) {
        // 参数校验
        if (Strings.isNullOrEmpty(indexName) || Strings.isNullOrEmpty(document) ||
            Strings.isNullOrEmpty(field) || Strings.isNullOrEmpty(content)) {
```

```
        return "Parameters are wrong!";
    }
    rankEvalService.executeRankEvalRequest(indexName, document, field, content);
    return "Execute RankEvalRequest success!";
    }
}
```

在 Service 层新增 RankEvalService 接口类,用于编写 Service 层中的业务接口定义,代码如下所示:

```
package com.niudong.esdemo.service;
public interface RankEvalService {
    // 以同步方式执行 RankEvalRequest
    public void executeRankEvalRequest(String index, String documentId, String
        field, String content);
}
```

在 ServiceImpl 层中新增 RankEvalServiceImpl 接口实现类,用于编写 Service 层中业务接口的逻辑实现,代码如下所示:

```
package com.niudong.esdemo.service.impl;
import java.util.ArrayList;
import java.util.Arrays;
import java.util.List;
import java.util.Map;
import javax.annotation.PostConstruct;
import org.apache.commons.logging.Log;
import org.apache.commons.logging.LogFactory;
import org.apache.http.HttpHost;
import org.elasticsearch.action.ActionListener;
import org.elasticsearch.client.RequestOptions;
import org.elasticsearch.client.RestClient;
import org.elasticsearch.client.RestHighLevelClient;
import org.elasticsearch.index.query.QueryBuilders;
import org.elasticsearch.index.rankeval.EvalQueryQuality;
import org.elasticsearch.index.rankeval.EvaluationMetric;
import org.elasticsearch.index.rankeval.MetricDetail;
import org.elasticsearch.index.rankeval.PrecisionAtK;
import org.elasticsearch.index.rankeval.RankEvalRequest;
import org.elasticsearch.index.rankeval.RankEvalResponse;
import org.elasticsearch.index.rankeval.RankEvalSpec;
import org.elasticsearch.index.rankeval.RatedDocument;
```

```java
import org.elasticsearch.index.rankeval.RatedRequest;
import org.elasticsearch.index.rankeval.RatedSearchHit;
import org.elasticsearch.search.builder.SearchSourceBuilder;
import org.springframework.stereotype.Service;
import com.niudong.esdemo.service.RankEvalService;
@Service
public class RankEvalServiceImpl implements RankEvalService {
  private static Log log = LogFactory.getLog(RankEvalServiceImpl.class);
  private RestHighLevelClient restClient;
  // 初始化连接
  @PostConstruct
  public void initEs() {
    restClient = new RestHighLevelClient(RestClient.builder(new HttpHost
        ("localhost", 9200, "http"),
      new HttpHost("localhost", 9201, "http")));
    log.info("ElasticSearch init in service.");
  }
  // 关闭连接
  public void closeEs() {
    try {
      restClient.close();
    } catch (Exception e) {
      e.printStackTrace();
    }
  }

  // 构建 RankEvalRequest
  public RankEvalRequest buildRankEvalRequest(String index, String documentId,
      String field,
      String content) {
    EvaluationMetric metric = new PrecisionAtK();
    List<RatedDocument> ratedDocs = new ArrayList<>();
    // 添加按索引名称、ID和分级指定的分级文档
    ratedDocs.add(new RatedDocument(index, documentId, 1));
    SearchSourceBuilder searchQuery = new SearchSourceBuilder();
    // 创建要评估的搜索查询
    searchQuery.query(QueryBuilders.matchQuery(field, content));
    // 将前三个部分合并为 RatedRequest
    RatedRequest ratedRequest = new RatedRequest("content_query", ratedDocs,
        searchQuery);
    List<RatedRequest> ratedRequests = Arrays.asList(ratedRequest);
    // 创建排序评估规范
    RankEvalSpec specification = new RankEvalSpec(ratedRequests, metric);
    // 创建排序评估请求
```

```java
    RankEvalRequest request = new RankEvalRequest(specification, new String[]
        {index});
    return request;
}
// 以同步方式执行 RankEvalRequest
public void executeRankEvalRequest(String index, String documentId, String
    field,
    String content) {
    // 构建 RankEvalRequest
    RankEvalRequest request = buildRankEvalRequest(index, documentId, field,
        content);
    try {
        RankEvalResponse response = restClient.rankEval(request, RequestOptions.
            DEFAULT);
        // 处理 RankEvalResponse
        processRankEvalResponse(response);
    } catch (Exception e) {
        e.printStackTrace();
    } finally {
        // 关闭 Elasticsearch 连接
        closeEs();
    }
}
// 处理 RankEvalResponse
private void processRankEvalResponse(RankEvalResponse response) {
    // 总体评价结果
    double evaluationResult = response.getMetricScore();
    log.info("evaluationResult is " + evaluationResult);
    Map<String, EvalQueryQuality> partialResults = response. getPartialResults();
    // 获取关键词 content_query 对应的评估结果
    EvalQueryQuality evalQuality = partialResults.get("content_query");
    log.info("content_query id is " + evalQuality.getId());
    // 每部分结果的度量分数
    double qualityLevel = evalQuality.metricScore();
    log.info("qualityLevel is " + qualityLevel);
    List<RatedSearchHit> hitsAndRatings = evalQuality.getHitsAndRatings();
    RatedSearchHit ratedSearchHit = hitsAndRatings.get(2);
    // 在分级搜索命中里包含完全成熟的搜索命中 SearchHit
    log.info("SearchHit id is " + ratedSearchHit.getSearchHit().getId());
    // 分级搜索命中还包含一个可选的<integer>分级 Optional<Integer>，如果文档在请求中
    // 未获得分级，则该分级不存在
    log.info("rate's isPresent is " + ratedSearchHit.getRating().isPresent());
    MetricDetail metricDetails = evalQuality.getMetricDetails();
    String metricName = metricDetails.getMetricName();
```

```
        // 度量详细信息，以请求中使用的度量命名
        log.info("metricName is " + metricName);

        PrecisionAtK.Detail detail = (PrecisionAtK.Detail) metricDetails;
        // 在转换到请求中使用的度量之后，度量详细信息提供了对度量计算部分的深入了解
        log.info("detail's relevantRetrieved is " + detail.getRelevantRetrieved
            ());
        log.info("detail's retrieved is " + detail.getRetrieved());
    }
}
```

随后编译工程，在工程根目录下输入如下命令：

```
mvn clean package
```

通过如下命令启动工程服务：

```
java -jar ./target/esdemo-0.0.1-SNAPSHOT.jar
```

在工程服务启动后，在浏览器中调用如下接口查看排序评估的情况：

```
http://localhost:8080/springboot/es/ranksearch/sr?indexName=ultraman&document=1&field=content&content=美联储
```

请求执行后，在服务器中输出如下所示内容：

```
2019-09-03 15:04:38.632  INFO 39048 --- [nio-8080-exec-2] c.n.e.service.impl.RankEvalServiceImpl   : evaluationResult is 0.3333333333333333
2019-09-03 15:04:38.634  INFO 39048 --- [nio-8080-exec-2] c.n.e.service.impl.RankEvalServiceImpl   : content_query id is content_query
2019-09-03 15:04:38.645  INFO 39048 --- [nio-8080-exec-2] c.n.e.service.impl.RankEvalServiceImpl   : qualityLevel is 0.3333333333333333
2019-09-03 15:04:38.647  INFO 39048 --- [nio-8080-exec-2] c.n.e.service.impl.RankEvalServiceImpl   : SearchHit id is 1
2019-09-03 15:04:38.650  INFO 39048 --- [nio-8080-exec-2] c.n.e.service.impl.RankEvalServiceImpl   : rate's isPresent is true
2019-09-03 15:04:38.651  INFO 39048 --- [nio-8080-exec-2] c.n.e.service.impl.RankEvalServiceImpl   : metricName is precision
2019-09-03 15:04:38.662  INFO 39048 --- [nio-8080-exec-2] c.n.e.service.impl.RankEvalServiceImpl   : detail's relevantRetrieved is 1
2019-09-03 15:04:38.662  INFO 39048 --- [nio-8080-exec-2] c.n.e.service.impl.RankEvalServiceImpl   : detail's retrieved is 3
```

此时，如果在调用接口的浏览器中输出如下内容，则表明接口调用成功：

```
Execute RankEvalRequest success!
```

7.6 搜索结果解释

除排序评估接口外，Elasticsearch 还提供了搜索结果解释 API，即 Explain API。Explain API 用于为查询请求和相关的文档计算解释性的分数。无论文档是否匹配这个查询请求，Elasticsearch 服务器都可以给用户提供一些有用的反馈。

1. 构建搜索结果解释请求

在发起搜索结果解释请求前，需要先构建搜索结果解释请求，即 ExplainRequest。ExplainRequest 有两个必选参数，即索引名称和文档 ID，同时需要通过 QueryBuilder 来构建查询表达式。

构建搜索结果解释请求的代码如下所示：

```
// 构建 ExplainRequest
  public ExplainRequest buildExplainRequest(String indexName, String
      document, String field, String content) {
   ExplainRequest request = new ExplainRequest(indexName, document);
   request.query(QueryBuilders.termQuery(field, content));

   return request;
}
```

类似地，ExplainRequest 的构建也有可选参数供用户进行配置。ExplainRequest 的可选参数主要有路由、搜索首选项、存储字段的控制、是否搜索文档源等。

配制 ExplainRequest 的可选参数的代码如下所示：

```
// 构建 ExplainRequest
  public ExplainRequest buildExplainRequest(String indexName, String
document, String field, String content) {
    ExplainRequest request = new ExplainRequest(indexName, document);
    request.query(QueryBuilders.termQuery(field, content));
    // 设置路由
    request.routing("routing");
    // 使用首选参数，例如执行搜索以首选本地碎片。默认值是在分片之间随机进行的
    request.preference("_local");
    // 设置为"真"，以检索解释的文档源。还可以通过使用"包含源代码"和"排除源代码"来检索
    // 部分文档
    request.fetchSourceContext(new FetchSourceContext(true, new String[]
        {field}, null));
    // 允许控制一部分的存储字段（要求在映射中单独存储该字段），并将其返回作为说明文档
    request.storedFields(new String[] {field});

    return request;
  }
```

2. 执行搜索结果解释请求

在 ExplainRequest 构建后，即可执行搜索结果解释请求。与文档索引请求类似，搜索结果解释请求也有同步和异步两种执行方式。

同步方式

当以同步方式执行搜索结果解释请求时,客户端会等待 Elasticsearch 服务器返回的查询结果 ExplainResponse。在收到 ExplainResponse 后,客户端会继续执行相关的逻辑代码。以同步方式执行的核心代码如下所示:

```java
ExplainResponse response = client.explain(request, RequestOptions.DEFAULT);
```

以同步方式执行的完整代码如下所示:

```java
// 构建ExplainRequest
public ExplainRequest buildExplainRequest(String indexName, String
    document, String field, String content) {
  ExplainRequest request = new ExplainRequest(indexName, document);
  request.query(QueryBuilders.termQuery(field, content));

  return request;
}
// 以同步方式执行ExplainRequest
public void executeExplainRequest(String indexName, String document,
    String field,
    String content) {
  // 构建ExplainRequest
  ExplainRequest request = buildExplainRequest(indexName, document, field,
      content);
  // 执行请求,接收返回结果
  try {
    ExplainResponse response = restClient.explain(request, RequestOptions.
        DEFAULT);
  } catch (Exception e) {
    e.printStackTrace();
  } finally {
    // 关闭Elasticsearch的连接
    closeEs();
  }
}
```

异步方式

当以异步方式执行搜索结果解释请求时,高级客户端不必同步等待请求结果的返回,可以直接向接口调用方返回异步接口执行成功的结果。

为了处理异步返回的响应信息或处理在请求执行过程中引发的异常信息,用户需要指定监听器。以异步方式执行的核心代码如下所示:

```
client.explainAsync(request, RequestOptions.DEFAULT, listener);
```

其中，listener 是监听器。

在异步请求处理后，如果请求执行成功，则调用 ActionListener 类中的 onResponse 方法进行相关逻辑的处理；如果请求执行失败，则调用 ActionListener 类中的 onFailure 方法进行相关逻辑的处理。

以异步方式执行的完整代码如下所示：

```java
// 构建 ExplainRequest
public ExplainRequest buildExplainRequest(String indexName, String
    document, String field, String content) {
  ExplainRequest request = new ExplainRequest(indexName, document);
  request.query(QueryBuilders.termQuery(field, content));

  return request;
}
// 以异步方式执行 ExplainRequest
public void executeExplainRequestAsync(String indexName, String document,
    String field,
    String content) {
  // 构建 ExplainRequest
  ExplainRequest request = buildExplainRequest(indexName, document, field,
      content);
  // 构建监听器
  ActionListener<ExplainResponse> listener = new ActionListener <ExplainResponse>
      () {
    @Override
    public void onResponse(ExplainResponse explainResponse) {
    }
    @Override
    public void onFailure(Exception e) {
    }
  };

  // 执行请求，接收返回结果
  try {
    restClient.explainAsync(request, RequestOptions.DEFAULT, listener);
  } catch (Exception e) {
    e.printStackTrace();
  } finally {
    // 关闭 Elasticsearch 连接
    closeEs();
  }
}
```

}
```

当然，在异步请求执行过程中可能会出现异常，异常的处理与同步方式执行情况相同。

**3. 解析搜索结果解释请求的响应结果**

不论同步方式，还是异步方式，在搜索合理解释请求执行后，客户端均需要对请求的响应结果 ExplainResponse 进行处理和解析。以同步方式为例，代码共分为三层，分别是 Controller 层、Service 层和 ServiceImpl 实现层。

其中，在 Controller 层中添加新类 ExplainController，用来编写与搜索结果解释请求相关的接口，代码如下所示：

```java
package com.niudong.esdemo.controller;
import org.elasticsearch.common.Strings;
import org.springframework.beans.factory.annotation.Autowired;
import org.springframework.web.bind.annotation.RequestMapping;
import org.springframework.web.bind.annotation.RestController;
import com.niudong.esdemo.service.ExplainService;
/**
 *
 */
@RestController
@RequestMapping("/springboot/es/explainsearch")
public class ExplainController {
  @Autowired
  private ExplainService explainService;
  // 以同步方式执行 ExplainRequest
  @RequestMapping("/sr")
  public String executeExplainRequest(String indexName, String document,
      String field,
      String content) {
    // 参数校验
    if (Strings.isNullOrEmpty(indexName) || Strings.isNullOrEmpty(document)
        || Strings.isNullOrEmpty(field) || Strings.isNullOrEmpty(content)) {
      return "Parameters are wrong!";
    }
    explainService.executeExplainRequest(indexName, document, field, content);
    return "Execute ExplainRequest success!";
  }
}
```

在 Service 层中添加新类 ExplainService，用来编写 Service 层中的搜索结果解释方法声明，代码如下所示：
```

```
package com.niudong.esdemo.service;
public interface ExplainService {
  // 以同步方式执行 ExplainRequest
  public void executeExplainRequest(String indexName, String document,
      String field,
      String content);
}
```

在 ServiceImpl 实现层中新增 ExplainServiceImpl 类,用来编写 Service 层中的搜索结果解释方法的具体逻辑实现,代码如下所示:

```
package com.niudong.esdemo.service.impl;
import java.util.Map;
import javax.annotation.PostConstruct;
import org.apache.commons.logging.Log;
import org.apache.commons.logging.LogFactory;
import org.apache.http.HttpHost;
import org.apache.lucene.search.Explanation;
import org.elasticsearch.action.ActionListener;
import org.elasticsearch.action.explain.ExplainRequest;
import org.elasticsearch.action.explain.ExplainResponse;
import org.elasticsearch.client.RequestOptions;
import org.elasticsearch.client.RestClient;
import org.elasticsearch.client.RestHighLevelClient;
import org.elasticsearch.common.document.DocumentField;
import org.elasticsearch.index.get.GetResult;
import org.elasticsearch.index.query.QueryBuilders;
import org.elasticsearch.search.fetch.subphase.FetchSourceContext;
import org.springframework.stereotype.Service;
import com.niudong.esdemo.service.ExplainService;
@Service
public class ExplainServiceImpl implements ExplainService {
  private static Log log = LogFactory.getLog(ExplainServiceImpl.class);
  private RestHighLevelClient restClient;
  // 初始化连接
  @PostConstruct
  public void initEs() {
    restClient = new RestHighLevelClient(RestClient.builder(new HttpHost
        ("localhost", 9200, "http"),
        new HttpHost("localhost", 9201, "http")));
    log.info("ElasticSearch init in service.");
  }
  // 关闭连接
  public void closeEs() {
```

```java
    try {
      restClient.close();
    } catch (Exception e) {
      e.printStackTrace();
    }
  }
  // 构建ExplainRequest
  public ExplainRequest buildExplainRequest(String indexName, String
      document, String field,
    String content) {
    ExplainRequest request = new ExplainRequest(indexName, document);
    request.query(QueryBuilders.termQuery(field, content));
    return request;
  }
  // 以同步方式执行ExplainRequest
  public void executeExplainRequest(String indexName, String document,
      String field,
    String content) {
    // 构建ExplainRequest
    ExplainRequest request = buildExplainRequest(indexName, document, field,
        content);
    // 执行请求，接收返回结果
    try {
      ExplainResponse response = restClient.explain(request,
          RequestOptions.DEFAULT);
      // 解析ExplainResponse
      processExplainResponse(response);
    } catch (Exception e) {
      e.printStackTrace();
    } finally {
      // 关闭Elasticsearch连接
      closeEs();
    }
  }
  // 解析ExplainResponse
  private void processExplainResponse(ExplainResponse response) {
    // 解释文档的索引名称
    String index = response.getIndex();
    // 解释文档的ID
    String id = response.getId();
    // 查看解释的文档是否存在
    boolean exists = response.isExists();
    log.info("index is " + index + ";id is " + id + ";exists is " + exists);
    // 解释的文档与提供的查询之间是否匹配（匹配是从后台的Lucene解释中检索的，如果
```

```java
// Lucene 解释建模匹配，则返回 true，否则返回 false）
boolean match = response.isMatch();
// 查看是否存在此请求的 Lucene 解释
boolean hasExplanation = response.hasExplanation();
log.info("match is " + match + ";hasExplanation is " + hasExplanation);
// 获取 Lucene 解释对象（如果存在）
Explanation explanation = response.getExplanation();
if (explanation != null) {
  log.info("explanation is " + explanation.toString());
}
// 如果检索到源或存储字段，则获取 getresult 对象
GetResult getResult = response.getGetResult();
if (getResult == null) {
  return;
}
// getresult 内部包含两个映射，用于存储提取的源字段和存储的字段
// 以 Map 形式检索源
Map<String, Object> source = getResult.getSource();
if (source == null) {
  return;
}
for (String str : source.keySet()) {
  log.info("str key is " + str);
}
// 以映射形式检索指定的存储字段
Map<String, DocumentField> fields = getResult.getFields();
if (fields == null) {
  return;
}
for (String str : fields.keySet()) {
  log.info("field str key is " + str);
}
}
}
```

随后编译工程，在工程根目录下输入如下命令：

```
mvn clean package
```

通过如下命令启动工程服务：

```
java -jar ./target/esdemo-0.0.1-SNAPSHOT.jar
```

在工程服务启动后，在浏览器中调用如下接口查看搜索结果解释请求的执行情况：

```
http://localhost:8080/springboot/es/explainsearch/sr?indexName=ultraman&docu
ment=1&field=content&content=美联储
```

请求执行后，在服务器中输出如下所示内容：

```
2019-09-04 14:58:34.216  INFO 48744 --- [nio-8080-exec-8] c.n.e.service.impl.ExplainServiceImpl    : index is ultraman;i
d is 1;exists is true
2019-09-04 14:58:34.216  INFO 48744 --- [nio-8080-exec-8] c.n.e.service.impl.ExplainServiceImpl    : match is false;hasE
xplanation is true
2019-09-04 14:58:34.219  INFO 48744 --- [nio-8080-exec-8] c.n.e.service.impl.ExplainServiceImpl    : explanation is 0.0
 = no matching term
```

此时，如果在执行请求的浏览器中输出如下内容，则表明接口调用成功：

```
Execute ExplainRequest success!
```

7.7 统计

除前文提及的排序评估接口、搜索解释接口外，Elasticsearch 还提供了统计 API，即 Count API。统计接口用于执行查询请求，并返回与请求匹配的统计结果。

1．构建统计请求

在执行统计请求前，需要先构建统计请求，即 CountRequest。CountRequest 的用法与 SearchRequest 类似，且二者都是基于 SearchSourceBuilder 实例的。

构建 CountRequest 的代码如下所示：

```
// 构建 CountRequest
public CountRequest buildCountRequest() {
// 创建 CountRequest。如果没有参数，则对所有索引运行
CountRequest countRequest = new CountRequest();

// 大多数搜索参数都需要添加到 SearchSourceBuilder 中
SearchSourceBuilder searchSourceBuilder = new SearchSourceBuilder();

// 向 SearchSourceBuilder 中添加"全部匹配"查询
searchSourceBuilder.query(QueryBuilders.matchAllQuery());

// 将 SearchSourceBuilder 添加到 CountRequest 中
countRequest.source(searchSourceBuilder);
return countRequest;
}
```

类似地，CountRequest 还提供了可选参数列表供用户进行配置。CountRequest 的参数列表主要有索引配置项、路由配置项、IndicesOptions 控制项、首选参数配置项等。

在 CountRequest 中配置可选参数的代码如下所示：

```
// 构建 CountRequest
public CountRequest buildCountRequest(String indexName,String routeName) {
    // 将请求限制为特定名称的索引
    CountRequest countRequest = new CountRequest(indexName).
        // 设置路由参数
        routing(routeName)
        // 设置 IndiceOptions，控制如何解析不可用索引及如何展开通配符表达式
        .indicesOptions(IndicesOptions.lenientExpandOpen())
        // 使用首选参数，例如执行搜索以首选本地分片。默认值是在分片之间随机选择的
        .preference("_local");

    return countRequest;
}
```

正如前文所提及的，在构建 CountRequest 时使用了 SearchSourceBuilder 实例。在统计接口的调用中，大多数控制搜索行为的选项都可以在 SearchSourceBuilder 实例中进行设置。设置常见选项的代码如下所示：

```
// 构建 CountRequest
public CountRequest buildCountRequest(String indexName, String routeName,
    String field,
    String content) {
  // 将请求限制为特定名称的索引
  CountRequest countRequest = new CountRequest(indexName).
      // 设置路由参数
      routing(routeName)
      // 设置 IndicesOptions，控制如何解析不可用索引及如何展开通配符表达式
      .indicesOptions(IndicesOptions.lenientExpandOpen())
      // 使用首选参数，例如执行搜索以首选本地分片。默认值是在分片之间随机选择的
      .preference("_local");
  // 使用默认选项创建 SearchSourceBuilder
  SearchSourceBuilder sourceBuilder = new SearchSourceBuilder();
  // 设置查询可以是任意类型的 QueryBuilder
  sourceBuilder.query(QueryBuilders.termQuery(field, content));
  // 将 SearchSourceBuilder 添加到 CountRequest 中
  countRequest.source(sourceBuilder);
  return countRequest;
}
```

2. 执行统计请求

在 CountRequest 构建后，即可执行统计请求。与文档索引请求类似，统计请求也有同步

和异步两种执行方式。

同步方式

当以同步方式执行统计请求时,客户端会等待 Elasticsearch 服务器返回的查询结果 CountResponse。在收到 CountResponse 后,客户端会继续执行相关的逻辑代码。以同步方式执行的核心代码如下所示:

```
CountResponse countResponse = client
    .count(countRequest, RequestOptions.DEFAULT);
```

以同步方式执行的全部代码如下所示:

```
// 构建 CountRequest
  public CountRequest buildCountRequest() {
    // 创建 CountRequest。如果没有参数,则对所有索引运行
    CountRequest countRequest = new CountRequest();
    // 大多数搜索参数都需要添加到 SearchSourceBuilder 中
    SearchSourceBuilder searchSourceBuilder = new SearchSourceBuilder();
    // 向 SearchSourceBuilder 中添加 "全部匹配" 查询
    searchSourceBuilder.query(QueryBuilders.matchAllQuery());
    // 把 SearchSourceBuilder 添加到 CountRequest 中
    countRequest.source(searchSourceBuilder);
    return countRequest;
  }
// 以同步方式执行 CountRequest
  public void executeCountRequest() {
    CountRequest countRequest = buildCountRequest();
    try {
      CountResponse countResponse = restClient.count(countRequest, RequestOptions.DEFAULT);
    } catch (Exception e) {
      e.printStackTrace();
    } finally {
      // 关闭 Elasticsearch 连接
      closeEs();
    }
  }
```

异步方式

当以异步方式执行统计请求时,高级客户端不必同步等待请求结果的返回,可以直接向接口调用方返回异步接口执行成功的结果。

为了处理异步返回的响应信息或处理在请求执行过程中引发的异常信息,用户需要指定监

听器。以异步方式执行的核心代码如下所示：

```
client.countAsync(countRequest, RequestOptions.DEFAULT, listener);
```

其中，listener 为监听器对象。

在异步请求处理后，如果请求执行成功，则调用 ActionListener 类中的 onResponse 方法进行相关逻辑的处理；如果请求执行失败，则调用 ActionListener 类中的 onFailure 方法进行相关逻辑的处理。

以异步方式执行的全部代码如下所示：

```
// 以异步方式执行 CountRequest
public void executeCountRequestAsync() {
  // 构建 CountRequest
  CountRequest countRequest = buildCountRequest();
  // 构建监听器
  ActionListener<CountResponse> listener = new ActionListener
      <CountResponse>() {
   @Override
   public void onResponse(CountResponse countResponse) {
   }
   @Override
   public void onFailure(Exception e) {
   }
  };
  try {
    restClient.countAsync(countRequest, RequestOptions.DEFAULT, listener);
  } catch (Exception e) {
    e.printStackTrace();
  } finally {
    // 关闭 Elasticsearch 的连接
    closeEs();
  }
}
```

当然，在异步请求执行过程中可能会出现异常，异常的处理与同步方式执行情况相同。

3．解析统计请求的响应结果

不论同步方式，还是异步方式，在统计请求执行后，客户端均需要对请求的响应结果 CountResponse 进行处理和解析。

CountResponse 提供了统计请求对应的结果命中总数和统计执行本身的详细信息，如 HTTP 状态代码、请求是否提前终止等。

CountResponse 还提供了与统计请求对应的分片总数、成功执行与失败执行的分片的统计信息，以及有关分片级别执行的信息等。用户可以通过遍历 ShardSearchFailures 数组来处理可能的失败信息。

以同步方式执行为例，CountResponse 的解析代码如下所示。代码共分为三层，分别是 Controller 层、Service 层和 ServiceImpl 实现层。

其中，在 Controller 层中新增 CountController 类，用来编写与统计相关的接口，代码如下所示：

```java
package com.niudong.esdemo.controller;
import org.springframework.beans.factory.annotation.Autowired;
import org.springframework.web.bind.annotation.RequestMapping;
import org.springframework.web.bind.annotation.RestController;
import com.niudong.esdemo.service.CountService;
@RestController
@RequestMapping("/springboot/es/countsearch")
public class CountController {
  @Autowired
  private CountService countService;
  //以同步方式执行 CountRequest
  @RequestMapping("/sr")
  public String executeCount() {
    countService.executeCountRequest();
    return "Execute CountRequest success!";
  }
}
```

在 Service 层中新增 CountService 类，用来编写 Service 层中接口方法的声明，代码如下所示：

```java
package com.niudong.esdemo.service;
public interface CountService {
  // 以同步方式执行 CountRequest
  public void executeCountRequest();
}
```

在 ServiceImpl 实现层中新增 CountServiceImpl 类，用来编写具体的业务逻辑实现，代码如下所示：

```java
package com.niudong.esdemo.service.impl;
import javax.annotation.PostConstruct;
import org.apache.commons.logging.Log;
```

```java
import org.apache.commons.logging.LogFactory;
import org.apache.http.HttpHost;
import org.elasticsearch.action.ActionListener;
import org.elasticsearch.action.search.ShardSearchFailure;
import org.elasticsearch.action.support.IndicesOptions;
import org.elasticsearch.client.RequestOptions;
import org.elasticsearch.client.RestClient;
import org.elasticsearch.client.RestHighLevelClient;
import org.elasticsearch.client.core.CountRequest;
import org.elasticsearch.client.core.CountResponse;
import org.elasticsearch.index.query.QueryBuilders;
import org.elasticsearch.rest.RestStatus;
import org.elasticsearch.search.builder.SearchSourceBuilder;
import org.springframework.stereotype.Service;
import com.niudong.esdemo.service.CountService;
@Service
public class CountServiceImpl implements CountService {
    private static Log log = LogFactory.getLog(CountServiceImpl.class);
    private RestHighLevelClient restClient;
    // 初始化连接
    @PostConstruct
    public void initEs() {
        restClient = new RestHighLevelClient(RestClient.builder(new HttpHost
            ("localhost", 9200, "http"),
            new HttpHost("localhost", 9201, "http")));
        log.info("ElasticSearch init in service.");
    }
    // 关闭连接
    public void closeEs() {
        try {
            restClient.close();
        } catch (Exception e) {
            e.printStackTrace();
        }
    }
    // 构建CountRequest
    public CountRequest buildCountRequest() {
        // 创建CountRequest。如果没有参数,则将针对所有索引运行
        CountRequest countRequest = new CountRequest();
        // 大多数搜索参数都需要添加到SearchSourceBuilder中
        SearchSourceBuilder searchSourceBuilder = new SearchSourceBuilder();
        // 向SearchSourceBuilder中添加"全部匹配"查询
        searchSourceBuilder.query(QueryBuilders.matchAllQuery());
```

```java
    // 将 SearchSourceBuilder 添加到 CountRequest 中
    countRequest.source(searchSourceBuilder);
    return countRequest;
}
// 以同步方式执行 CountRequest
public void executeCountRequest() {
    CountRequest countRequest = buildCountRequest();
    try {
        CountResponse countResponse = restClient.count(countRequest, RequestOptions.
            DEFAULT);
        // 解析 CountResponse
        processCountResponse(countResponse);
    } catch (Exception e) {
        e.printStackTrace();
    } finally {
        // 关闭 Elasticsearch 连接
        closeEs();
    }
}
// 解析 CountResponse
private void processCountResponse(CountResponse countResponse) {
    // 统计请求对应的结果命中总数
    long count = countResponse.getCount();
    // HTTP 状态代码
    RestStatus status = countResponse.status();
    // 请求是否提前终止
    Boolean terminatedEarly = countResponse.isTerminatedEarly();
    log.info("count is " + count + ";status is " + status.getStatus() +
        ";terminatedEarly is "
        + terminatedEarly);
    // 与统计请求对应的分片总数
    int totalShards = countResponse.getTotalShards();
    // 执行统计请求跳过的分片数量
    int skippedShards = countResponse.getSkippedShards();
    // 执行统计请求成功的分片数量
    int successfulShards = countResponse.getSuccessfulShards();
    // 执行统计请求失败的分片数量
    int failedShards = countResponse.getFailedShards();
    log.info("totalShards is " + totalShards + ";skippedShards is " +
        skippedShards
        + ";successfulShards is " + successfulShards + ";failedShards is " +
        failedShards);
    // 通过遍历 ShardSearchFailures 数组来处理可能的失败信息
```

```
    if (countResponse.getShardFailures() == null) {
      return;
    }
    for (ShardSearchFailure failure : countResponse.getShardFailures()) {
      log.info("fail index is " + failure.index());
    }
  }
}
```

随后编译工程,在工程根目录下输入如下命令:

```
mvn clean package
```

通过如下命令启动工程服务:

```
java -jar ./target/esdemo-0.0.1-SNAPSHOT.jar
```

在工程服务启动后,在浏览器中调用如下接口查看统计请求的情况:

```
http://localhost:8080/springboot/es/countsearch/sr
```

接口执行后,在服务器中输出如下所示内容:

```
2019-09-04 20:13:02.119  INFO 54400 --- [nio-8080-exec-1] c.n.e.service.impl.CountServiceImpl      : count is 6;status is 200;terminatedEarly is null
2019-09-04 20:13:02.119  INFO 54400 --- [nio-8080-exec-1] c.n.e.service.impl.CountServiceImpl      : totalShards is 2;skippedShards is 0;successfulShards is 2;failedShards is 0
```

目前,在 Elasticsearch 中存在 2 个索引,索引名称分别是 ultraman 和 ultraman1。每个索引各有 3 条数据,因此输出的 6 个数据与预期相符;而返回码 200 表明接口执行成功。

此时,如果在执行请求的浏览器中显示如下内容,则表明接口执行成功:

```
Execute CountRequest success!
```

7.8 搜索过程解析

下面介绍在 Elasticsearch 内部是如何搜索文档的。

7.8.1 对已知文档的搜索

如果被搜索的文档(不论是单个文档,还是批量文档)能够从主分片或任意一个副本分片中被检索到,则与索引文档过程相同,对已知文档的搜索也会用到路由算法,Elasticsearch 中的路由算法如下所示:

```
shard = hash(routing) % number_of_primary_shards
```

下面以图 7-3 为例，展示在主分片上搜索一个文档的必要步骤。

（1）客户端给主节点 1 发送文档的 Get 请求，此时主节点 1 就成为协同节点。主节点使用路由算法算出文档所在的主分片；随后协同节点将请求转发给主分片所在的节点 2，当然，也可以基于轮询算法转发给副本分片。

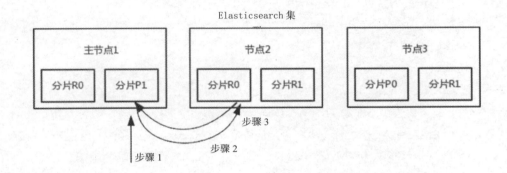

图 7-3　已知文档的搜索

（2）如图 7-3 所示，主节点 1 根据文档的 ID 确定文档属于分片 R0。分片 R0 对应的副本分片在三个节点上都有。此时，主节点 1 转发请求到节点 2。

（3）节点 2 在本地分片进行搜索，并将目标文档信息作为结果返给主节点 1。

对于读请求，为了在各节点间负载均衡，请求节点一般会为每个请求选择不同的分片，一般采用轮询算法循环在所有副本分片中进行请求。

7.8.2　对未知文档的搜索

除对已知文档的搜索外，大部分请求实际上是不知道查询条件会命中哪些文档的。这些被查询条件命中的文档可能位于 Elasticsearch 集群中的任意位置上。因此，搜索请求的执行不得不去询问每个索引中的每一个分片。

在 Elasticsearch 中，搜索过程分为查询阶段（Query Phase）和获取阶段（Fetch Phase）。

在查询阶段，查询请求会广播到索引中的每一个主分片和备份中，每一个分片都会在本地执行检索，并在本地各建立一个优先级队列（Priority Queue）。该优先级队列是一份根据文档相关度指标进行排序的列表，列表的长度由 from 和 size 两个分页参数决定。

查询阶段可以再细分成 3 个小的子阶段：

（1）客户端发送一个检索请求给某节点 A，此时节点 A 会创建一个空的优先级队列，并配置好分页参数 from 与 size。

（2）节点 A 将搜索请求发送给该索引中的每一个分片，每个分片在本地执行检索，并将结果添加到本地优先级队列中。

（3）每个分片返回本地优先级序列中所记录的 ID 与 sort 值，并发送给节点 A。节点 A 将这些值合并到自己的本地的优先级队列中，并做出全局的排序。

在获取阶段，主要是基于上一阶段找到所要搜索文档数据的具体位置，将文档数据内容取回并返回给客户端。

在 Elasticsearch 中，默认的搜索类型就是上面介绍的 Query then Fetch。上述描述运作方式就是 Query then fetch。Query then Fetch 有可能会出现打分偏离的情形，幸好，Elasticsearch 还提供了一个称为"DFS Query then Fetch"的搜索方式，它和 Query then Fetch 基本相同，但是它会执行一个预查询来计算整体文档的 frequency。其处理过程如下所示：

（1）预查询每个分片，询问 Term 和 Document Frequency 等信息。

（2）发送查询请求到每个分片。

（3）找到各个分片中所有匹配的文档，并使用全局的 Term/Document Frequency 信息进行打分。在执行过程中依然需要对结果构建一个优先队列，如排序等。

（4）返回关于结果的元数据到请求节点。需要指出的是，此时实际文档还没有发送到请求节点，发送的只是分数。

（5）请求节点将来自所有分片的分数合并起来，并在请求节点上进行排序，文档被按照查询要求进行选择。最终，实际文档从它们各自所在的独立的分片上被检索出来，结果被返回给读者。

7.8.3　对词条的搜索

前面介绍了 Elasticsearch 是如何对已知文档和未知文档进行搜索的，那么具体到一个分片，Elasticsearch 是如何按照词条进行搜索的呢？

前文中已提及，Elasticsearch 分别为每个文档中的字段建立了一个倒排索引。倒排索引示意图如图 2-6 所示。

当要搜索中文词 1 时，我们通过倒排索引可以获悉，与待搜索中文词 1 的文档是中文网页 1、中文网页 2 和中文网页 3。

当词条数量较少时，我们可以顺序遍历词条获取结果，但如果词条有成千上万个呢？

Elasticsearch 为了能快速找到某个词条,它对所有的词条都进行了排序,随后使用二分法查找词条,其查找效率为 log(N)。这个过程就像查字典一样,因此排序词条的集合也称为 Term Dictionary。

为了提高查询性能,Elasticsearch 直接通过内存查找词条,而非从磁盘中读取。但当词条太多时,显然 Term Dictionary 也会很大,此时全部放在内存有些不现实,于是引入了 Term Index。

Term Index 就像字典中的索引页,其中的内容如字母 A 开头的有哪些词条,这些词条分别在哪页。通过 Term Index,Elasticsearch 可以快速地定位到 Term Dictionary 的某个 OffSet(位置偏移),然后从这个位置再往后顺序查找。

前面提及了单个词条的搜索方法,而在实际应用中,更常见的往往是多个词条拼接成的"联合查询",那么 Elasticsearch 是如何基于倒排索引实现快速查询的呢?

核心思想是利用跳表快速做"与"运算,还有一种方法是利用 BitSet(位图)按位"与"运算。

先来看利用跳表快速做"与"运算的方式,首先介绍跳表,如图 7-4 所示。

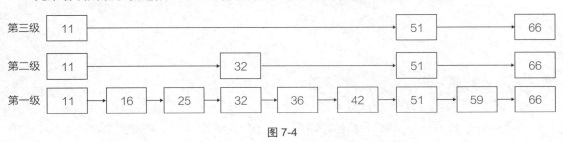

图 7-4

跳表由多级链表组成,上一级是下一级元素的子集,因此上一级往往数据较少。在查找时,数据从上级向下级逐级查找。如查找数据 36,从第三级中找到元素 11,从第二级中到数据 32,从第三级中的数据 32 之后查找到 36,总共用了 3 次查找。

那么跳表是如何用在多词条索引中的呢?

通过倒排索引,每个词条都会有一个命中的文档 ID 列表(如果能命中的话),此时,我们可以找到这些文档 ID 列表中最短的那一个。接下来,用最短的文档 ID 列表中的 ID 逐个在其他文档 ID 列表中进行查找,都能找到的 ID 即为多个词条的交集结果。

再来看 BitSet 按位"与"运算的方式。

BitSet 中的每一位只有两个可能取值,即 0 或 1。如果数据存在,则对应的标记置为 1,否则置为 0。

因此，将词条的文档 ID 列表转化为位图后，将多个词条对应的位图取"与"运算，即可得到交集结果。

7.9 知识点关联

在数据存储领域，游标与滚动搜索的思想一脉相承。在数据库领域，游标是一个十分重要的方法。

与使用滚动搜索类似，在使用游标时一般也有 4 个步骤：

（1）声明游标，并把游标与数据库查询结果集联系起来。

（2）打开游标。

（3）使用游标操作数据，每次都需要指定下一次游标的 NEXT 值，NEXT 为默认的游标提取选项。

（4）关闭游标。

游标和 C 语言中的文件句柄类似，一旦打开文件，该文件句柄就可代表该文件。同理，游标亦然。

7.10 小结

本章主要介绍了搜索 API 的使用，主要涉及搜索、滚动搜索、批量搜索、跨索引字段搜索、搜索结果的排序评估、搜索结果解释和统计等接口。

第 8 章
索引实战

爱著目录
略述鸿烈

本章介绍索引相关 API 的使用。在介绍这些 API 的使用之前，我们先简单介绍什么是索引。

存储数据的行为就叫作索引。在 Elasticsearch 中，文档会归属于一种类型，这些类型会存在于索引中。

Elasticsearch 集群和数据库中核心概念的对应关系如表 8-1 所示。

表 8-1

Elasticsearch 集群	关系数据库
索引	数据库
类型	表
文档	行数据
字段	列数据

需要指出的是，"索引"一词在 Elasticsearch 中有着不同的属性和含义。

（1）索引作为名词时。一个索引就好比传统关系数据库中的数据库，它是存储相关文档的地方。

（2）索引作为动词时。"索引文档"指的是将一个文档存储到索引（这里是名词）里，以便该文档可以被检索或者查询。这类似于关系数据库 SQL 语句中的 INSERT 关键字，不同的是，如果文档已经存在，则新的文档将覆盖旧的文档。

（3）指代倒排索引。在默认情况下，文档中的所有字段都会被索引，即字段都有一个倒排索引，也就是说，所有字段都可被搜索。

高级客户端支持以下索引相关的 API：

（1）Analyze API。

（2）Create Index API。

（3）Delete Index API。

（4）Indices Exists API。

（5）Open Index API。

（6）Close Index API。

（7）Shrink Index API。

（8）Split Index API。

（9）Refresh API。

（10）Flush API。

（11）Flush Synced API。

（12）Clear Cache API。

（13）Force Merge API。

（14）Rollover Index API。

（15）Get Index API。

（16）Index Aliases API。

（17）Exists Alias API。

（18）Get Alias API。

这些索引相关的 API 可以分为三类，索引的增删改查、索引的刷新及合并/拆分、索引别名的使用。下面一一介绍。

8.1 字段索引分析

前文提及，在 Elasticsearch 中，当文档被索引到索引文件时会进行分词操作，在用户查询时，也会基于分词进行检索。

但有的时候，明明被索引的文档包含了下一步要搜索的关键词，但为什么搜索不出来呢？由于索引过程对用户来说是黑盒操作，因此用户十分好奇文档索引后的分词结果是什么？

对于这些场景,我们都可以使用分析接口,即 Analyze API,来分析字段是如何建立索引的。

1. 构建分析请求

在执行分析请求之前,需要构建分析请求,即 AnalyzeRequest。AnalyzeRequest 中一般包含要分析的文本,并指定如何执行分析。

构建 AnalyzeRequest 的代码添加在 IndexServiceImpl 类中,如下所示:

```
    // 构建AnalyzeRequest
    public void buildAnalyzeRequest() {
    AnalyzeRequest request = new AnalyzeRequest();
    // 要包含的文本。多个字符串被视为多值字段
    request.text("中国天眼系统首次探测到宇宙深处的神秘射电信号");
    // 内置分析器
request.analyzer("standard");
}
```

如果是对英文字符串进行分析,则代码如下所示,详见 buildAnalyzeRequest 方法:

```
    // 要包含的文本。多个字符串被视为多值字段
    request.text("Some text to analyze", "Some more text to analyze");
    // 内置分析器
    request.analyzer("english");
```

上述对中、英文字符串的分析采用了内置分析器,Elasticsearch 还支持用户自定义分析器。自定义分析器的代码如下所示,详见 buildAnalyzeRequest 方法:

```
        // 自定义分析器1
        request.text("<b>Some text to analyze</b>");
        //配置字符筛选器
        request.addCharFilter("html_strip");
        //配置标记器
        request.tokenizer("standard");
        //添加内置标记筛选器
        request.addTokenFilter("lowercase");

        //自定义分析器2
        Map < String, Object > stopFilter = new HashMap<>();
        stopFilter.put("type", "stop");
        //自定义令牌筛选器tokenfilter的配置
        stopFilter.put("stopwords", new String[] {
           "to"
        });
```

```
    //添加自定义标记筛选器
    request.addTokenFilter(stopFilter);
```

类似地，AnalyzeRequest 也有可选参数供用户进行配置，其可选参数主要有 explain 字段和属性配置，代码如下所示，详见 buildAnalyzeRequest 方法：

```
// 可选参数
    // 将explain设置为true，为响应添加更多详细信息
    request.explain(true);
    // 设置属性，允许只返回用户感兴趣的令牌属性
    request.attributes("keyword", "type");
```

2. 执行分析请求

在 AnalyzeRequest 构建后，即可执行分析请求。与文档索引请求类似，分析请求也有同步和异步两种执行方式。

同步方式

当以同步方式执行分析请求时，客户端会等待 Elasticsearch 服务器返回的查询结果 AnalyzeResponse。在收到 AnalyzeResponse 后，客户端会继续执行相关的逻辑代码。以同步方式执行的核心代码如下所示：

```
AnalyzeResponse response = client.indices().analyze(request, RequestOptions.DEFAULT);
```

以同步方式执行的全部代码添加在 IndexServiceImpl 类中，如下所示：

```
    // 构建AnalyzeRequest
 public AnalyzeRequest buildAnalyzeRequest(String text) {
    AnalyzeRequest request = new AnalyzeRequest();
    // 要包含的文本。多个字符串被视为多值字段
    request.text(text);
    // 内置分析器
    request.analyzer("standard");
    return request;
 }
 // 以同步方式执行AnalyzeRequest
 public void executeAnalyzeRequest(String text) {
    // 构建AnalyzeRequest
    AnalyzeRequest request = buildAnalyzeRequest(text);
    try {
       AnalyzeResponse response = restClient.indices().analyze(request, RequestOptions.DEFAULT);
    } catch (Exception e) {
```

```
      e.printStackTrace();
    } finally {
      // 关闭 Elasticsearch 的连接
      closeEs();
    }
  }
```

异步方式

当以异步方式执行 AnalyzeRequest 时,高级客户端不必同步等待请求结果的返回,可以直接向接口调用方返回异步接口执行成功的结果。

为了处理异步返回的响应信息或处理在请求执行过程中引发的异常信息,用户需要指定监听器。以异步方式执行的核心代码如下所示:

```
client.indices().analyzeAsync(request, RequestOptions.DEFAULT, listener);
```

其中,listener 为监听器。

在异步请求处理后,如果请求执行成功,则调用 ActionListener 类中的 onResponse 方法进行相关逻辑的处理;如果请求执行失败,则调用 ActionListener 类中的 onFailure 方法进行相关逻辑的处理。

以异步方式执行的全部代码添加在 IndexServiceImpl 类中,如下所示:

```
  // 以异步方式执行 AnalyzeRequest
  public void executeAnalyzeRequestAsync(String text) {
    // 构建 AnalyzeRequest
    AnalyzeRequest request = buildAnalyzeRequest(text);
    // 构建监听器
    ActionListener<AnalyzeResponse> listener = new ActionListener
        <AnalyzeResponse>() {
      @Override
      public void onResponse(AnalyzeResponse analyzeTokens) {
      }
      @Override
      public void onFailure(Exception e) {
      }
    };

    try {
      restClient.indices().analyzeAsync(request, RequestOptions.DEFAULT,
          listener);
    } catch (Exception e) {
      e.printStackTrace();
```

```
    } finally {
      // 关闭Elasticsearch连接
      closeEs();
    }
}
```

当然，在异步请求执行过程中可能会出现异常，异常的处理与同步方式执行情况相同。

3. 解析分析请求的响应结果

不论同步方式，还是异步方式，客户端均需要对 AnalyzeResponse 进行处理和解析。以同步方式为例，AnalyzeResponse 的解析代码共分为三层，分别是 Controller 层、Service 层和 ServiceImpl 实现层。

在 Controller 层中新增 IndexController 类，用来编写索引操作相关的接口，代码如下所示：

```java
package com.niudong.esdemo.controller;
import org.elasticsearch.common.Strings;
import org.springframework.beans.factory.annotation.Autowired;
import org.springframework.web.bind.annotation.RequestMapping;
import org.springframework.web.bind.annotation.RestController;
import com.niudong.esdemo.service.IndexService;
@RestController
@RequestMapping("/springboot/es/indexsearch")
public class IndexController {
  @Autowired
  private IndexService indexService;
  // 以同步方式执行 IndexRequest
  @RequestMapping("/sr")
  public String executeIndex(String text) {
    // 参数校验
    if (Strings.isNullOrEmpty(text)) {
      return " Parameters are wrong!";
    }
    indexService.executeAnalyzeRequest(text);
    return "Execute IndexRequest success!";
  }
}
```

在 Service 层中新增 IndexService 类，用来编写 Service 层中索引操作相关方法的声明，代码如下所示：

```java
package com.niudong.esdemo.service;
  public interface IndexService {
  // 以同步方式执行 AnalyzeRequest
```

```
    public void executeAnalyzeRequest(String text);
}
```

在 ServiceImpl 实现层中新增 IndexServiceImpl 类，用来编写 Service 层中索引操作相关方法的具体实现，代码如下所示：

```
package com.niudong.esdemo.service.impl;
import java.util.HashMap;
import java.util.List;
import java.util.Map;
import javax.annotation.PostConstruct;
import org.apache.commons.logging.Log;
import org.apache.commons.logging.LogFactory;
import org.apache.http.HttpHost;
import org.elasticsearch.action.ActionListener;
import org.elasticsearch.action.admin.indices.analyze.AnalyzeRequest;
import org.elasticsearch.action.admin.indices.analyze.AnalyzeResponse;
import org.elasticsearch.action.admin.indices.analyze.DetailAnalyzeResponse;
import org.elasticsearch.client.RequestOptions;
import org.elasticsearch.client.RestClient;
import org.elasticsearch.client.RestHighLevelClient;
import org.springframework.stereotype.Service;
import com.niudong.esdemo.service.IndexService;
@Service
public class IndexServiceImpl implements IndexService {
  private static Log log = LogFactory.getLog(IndexServiceImpl.class);
  private RestHighLevelClient restClient;
  // 初始化连接
  @PostConstruct
  public void initEs() {
    restClient = new RestHighLevelClient(RestClient.builder(new HttpHost
        ("localhost", 9200, "http"),
      new HttpHost("localhost", 9201, "http")));
    log.info("ElasticSearch init in service.");
  }
  // 关闭连接
  public void closeEs() {
    try {
      restClient.close();
    } catch (Exception e) {
      e.printStackTrace();
    }
  }
  // 构建 AnalyzeRequest
```

```java
public AnalyzeRequest buildAnalyzeRequest(String text) {
    AnalyzeRequest request = new AnalyzeRequest();
    // 要包含的文本。多个字符串被视为多值字段
    request.text(text);
    // 内置分析器
    request.analyzer("standard");
    return request;
}
// 以同步方式执行 AnalyzeRequest
public void executeAnalyzeRequest(String text) {
    // 构建 AnalyzeRequest
    AnalyzeRequest request = buildAnalyzeRequest(text);
    try {
        AnalyzeResponse response = restClient.indices().analyze(request,
            RequestOptions.DEFAULT);
        // 解析 AnalyzeResponse
        processAnalyzeResponse(response);
    } catch (Exception e) {
        e.printStackTrace();
    } finally {
        // 关闭 Elasticsearch 连接
        closeEs();
    }
}
// 解析 AnalyzeResponse
private void processAnalyzeResponse(AnalyzeResponse response) {
    // AnalyzeToken 保存了有关分析生成的单个令牌的信息
    List<AnalyzeResponse.AnalyzeToken> tokens = response.getTokens();
    if (tokens == null) {
        return;
    }
    for (AnalyzeResponse.AnalyzeToken token : tokens) {
        log.info(token.getTerm() + " start offset is " + token.getStartOffset()
                + ";end offset is "
            + token.getEndOffset() + ";position is" + token.getPosition());
    }
    // 如果把 explain 设置为 true，则通过 detail 方法返回信息
    // DetailAnalyzeResponse 包含有关分析链中不同子步骤生成的令牌的更详细的信息
    DetailAnalyzeResponse detail = response.detail();
    if (detail == null) {
        return;
    }
    log.info("detail is " + detail.toString());
}
```

}

随后编译工程，在工程根目录下输入如下命令：

```
mvn clean package
```

通过如下命令启动工程服务：

```
java -jar ./target/esdemo-0.0.1-SNAPSHOT.jar
```

在工程服务启动后，在浏览器中调用如下接口查看字段索引分析的情况：

http://localhost:8080/springboot/es/indexsearch/sr?text=中国天眼系统首次探测到宇宙深处的神秘射电信号

此时，在服务器中输出内容如下所示：

```
2019-09-05 20:17:26.790  INFO 54068 --- [nio-8080-exec-1] c.n.e.service.impl.IndexServiceImpl      : 中 start offset is 0;end offset is 1;position is0
2019-09-05 20:17:26.790  INFO 54068 --- [nio-8080-exec-1] c.n.e.service.impl.IndexServiceImpl      : 国 start offset is 1;end offset is 2;position is1
2019-09-05 20:17:26.803  INFO 54068 --- [nio-8080-exec-1] c.n.e.service.impl.IndexServiceImpl      : 天 start offset is 2;end offset is 3;position is2
2019-09-05 20:17:26.806  INFO 54068 --- [nio-8080-exec-1] c.n.e.service.impl.IndexServiceImpl      : 眼 start offset is 3;end offset is 4;position is3
2019-09-05 20:17:26.807  INFO 54068 --- [nio-8080-exec-1] c.n.e.service.impl.IndexServiceImpl      : 系 start offset is 4;end offset is 5;position is4
2019-09-05 20:17:26.828  INFO 54068 --- [nio-8080-exec-1] c.n.e.service.impl.IndexServiceImpl      : 统 start offset is 5;end offset is 6;position is5
2019-09-05 20:17:26.829  INFO 54068 --- [nio-8080-exec-1] c.n.e.service.impl.IndexServiceImpl      : 首 start offset is 6;end offset is 7;position is6
2019-09-05 20:17:26.831  INFO 54068 --- [nio-8080-exec-1] c.n.e.service.impl.IndexServiceImpl      : 次 start offset is 7;end offset is 8;position is7
2019-09-05 20:17:26.833  INFO 54068 --- [nio-8080-exec-1] c.n.e.service.impl.IndexServiceImpl      : 探 start offset is 8;end offset is 9;position is8
2019-09-05 20:17:26.834  INFO 54068 --- [nio-8080-exec-1] c.n.e.service.impl.IndexServiceImpl      : 测 start offset is 9;end offset is 10;position is9
2019-09-05 20:17:26.840  INFO 54068 --- [nio-8080-exec-1] c.n.e.service.impl.IndexServiceImpl      : 到 start offset is 10;end offset is 11;position is10
2019-09-05 20:17:26.879  INFO 54068 --- [nio-8080-exec-1] c.n.e.service.impl.IndexServiceImpl      : 宇 start offset is 11;end offset is 12;position is11
2019-09-05 20:17:26.884  INFO 54068 --- [nio-8080-exec-1] c.n.e.service.impl.IndexServiceImpl      : 宙 start offset is 12;end offset is 13;position is12
2019-09-05 20:17:26.884  INFO 54068 --- [nio-8080-exec-1] c.n.e.service.impl.IndexServiceImpl      : 深 start offset is 13;end offset is 14;position is13
2019-09-05 20:17:26.885  INFO 54068 --- [nio-8080-exec-1] c.n.e.service.impl.IndexServiceImpl      : 处 start offset is 14;end offset is 15;position is14
2019-09-05 20:17:26.885  INFO 54068 --- [nio-8080-exec-1] c.n.e.service.impl.IndexServiceImpl      : 的 start offset is 15;end offset is 16;position is15
2019-09-05 20:17:26.886  INFO 54068 --- [nio-8080-exec-1] c.n.e.service.impl.IndexServiceImpl      : 神 start offset is 16;end offset is 17;position is16
2019-09-05 20:17:26.925  INFO 54068 --- [nio-8080-exec-1] c.n.e.service.impl.IndexServiceImpl      : 秘 start offset is 17;end offset is 18;position is17
2019-09-05 20:17:26.930  INFO 54068 --- [nio-8080-exec-1] c.n.e.service.impl.IndexServiceImpl      : 射 start offset is 18;end offset is 19;position is18
2019-09-05 20:17:26.931  INFO 54068 --- [nio-8080-exec-1] c.n.e.service.impl.IndexServiceImpl      : 电 start offset is 19;end offset is 20;position is19
2019-09-05 20:17:26.932  INFO 54068 --- [nio-8080-exec-1] c.n.e.service.impl.IndexServiceImpl      : 信 start offset is 20;end offset is 21;position is20
2019-09-05 20:17:26.961  INFO 54068 --- [nio-8080-exec-1] c.n.e.service.impl.IndexServiceImpl      : 号 start offset is 21;end offset is 22;position is21
```

8.2 创建索引

通过创建索引接口，创建用户所需的索引。

1. 构建创建索引请求

在执行创建索引请求前,需要构建索引请求,即 CreateIndexRequest。在构建 CreateIndexRequest 时,主要参数有索引名称,以及与其关联的特定设置,如分片数量和副本数量。代码添加在 IndexServiceImpl 类中,如下所示:

```java
// 创建索引
public void buildIndexRequest(String index, int shardsNumber, int
    replicasNumber) {
  CreateIndexRequest request = new CreateIndexRequest(index);
  // 配置分片数量和副本数量
  request.settings(Settings.builder().put("index.number_of_shards",
      shardsNumber)
    .put("index.number_of_replicas", replicasNumber));
}
```

在构建 CreateIndexRequest 时,还可以使用文档类型的 Map 映射来创建索引。Elasticsearch 提供了字符串、Map 映射、XContentBuilder 对象等不同的方式提供映射源,在上述 buildIndexRequest 方法中新增如下代码:

```java
  // 配置映射源
    // 以字符串方式提供映射源
    request.mapping("{\n" + "  \"properties\": {\n" + "    \"message\": {\n"
        + "      \"type\": \"text\"\n" + "    }\n" + "  }\n" + "}",
            XContentType.JSON);
    // 以 Map 方式提供映射源
    Map<String, Object> message = new HashMap<>();
    message.put("type", "text");
    Map<String, Object> properties = new HashMap<>();
    properties.put("message", message);
    Map<String, Object> mapping = new HashMap<>();
    mapping.put("properties", properties);
    request.mapping(mapping);
    // 以 XContentBuilder 方式提供映射源
    try {
      XContentBuilder builder = XContentFactory.jsonBuilder();
      builder.startObject();
      {
        builder.startObject("properties");
        {
          builder.startObject("message");
          {
            builder.field("type", "text");
```

```
            }
            builder.endObject();
          }
          builder.endObject();
        }
        builder.endObject();
        request.mapping(builder);
    } catch (Exception e) {
        e.printStackTrace();
    } finally {
```

在创建索引的过程中，还可以为索引设置别名。设置别名的方式有两种，一种是在创建索引时设置，另一种是通过提供整个索引源来设置。两种设置索引别名的代码添加在上述 buildIndexRequest 方法中，如下所示：

```
// 设置别名:在索引创建时设置
request.alias(new Alias(index + "_alias").filter(QueryBuilders.termQuery
            ("user", "niudong")));
// 设置别名:通过提供整个索引源
request.source("{\n" +
    "    \"settings\" : {\n" +
    "        \"number_of_shards\" : 1,\n" +
    "        \"number_of_replicas\" : 0\n" +
    "    },\n" +
    "    \"mappings\" : {\n" +
    "        \"properties\" : {\n" +
    "            \"message\" : { \"type\" : \"text\" }\n" +
    "        }\n" +
    "    },\n" +
    "    \"aliases\" : {\n" +
    "        \"niudong_alias\" : {}\n" +
    "    }\n" +
    "}", XContentType.JSON);
```

在构建 CreateIndexRequest 时，Elasticsearch 还提供了可选参数供用户进行配置。可选参数主要有等待所有节点确认创建索引的超时时间、从节点连接到主节点的超时时间、在请求响应返回前活动状态的分片数量和拷贝数量。相关代码添加在 buildIndexRequest 方法中，如下所示：

```
// 可选参数配置
    // 等待所有节点确认创建索引的超时时间
    request.setTimeout(TimeValue.timeValueMinutes(2));
```

```
    // 从节点连接到主节点的超时时间
    request.setMasterTimeout(TimeValue.timeValueMinutes(1));

    // 在请求响应返回前活动状态的分片数量
    request.waitForActiveShards(ActiveShardCount.from(2));
    // 在请求响应返回前活动状态的拷贝数量
    request.waitForActiveShards(ActiveShardCount.DEFAULT);
```

2. 执行创建索引请求

在 CreateIndexRequest 构建后，即可执行创建索引请求。与文档索引请求类似，创建索引请求也有同步和异步两种执行方式。

同步方式

当以同步方式执行创建索引请求时，客户端会等待 Elasticsearch 服务器返回的查询结果 CreateIndexResponse。在收到 CreateIndexResponse 后，客户端会继续执行相关的逻辑代码。以同步方式执行的代码添加在 IndexServiceImpl 类中，如下所示：

```
    // 创建索引请求
    public CreateIndexRequest buildIndexRequest(String index) {
      CreateIndexRequest request = new CreateIndexRequest(index);
      // 配置默认分片数量和副本数量
      request.settings(
          Settings.builder().put("index.number_of_shards",
              3).put("index.number_of_replicas", 2));
      return request;
    }
    // 以同步方式执行创建索引请求
    public void executeIndexRequest(String index) {
      // 创建索引请求
      CreateIndexRequest request = buildIndexRequest(index);
      try {
        CreateIndexResponse createIndexResponse =
            restClient.indices().create(request, RequestOptions.DEFAULT);
      } catch (Exception e) {
        e.printStackTrace();
      } finally {
        // 关闭Elasticsearch连接
        closeEs();
      }
    }
```

异步方式

当以异步方式执行创建索引请求时,高级客户端不必同步等待请求结果的返回,可以直接向接口调用方返回异步接口执行成功的结果。

为了处理异步返回的响应信息或处理在请求执行过程中引发的异常信息,用户需要指定监听器。以异步方式执行的核心代码如下所示:

```
client.indices().createAsync(request, RequestOptions.DEFAULT, listener);
```

其中,listener 是监听器。

在异步请求处理后,如果请求执行成功,则调用 ActionListener 类中的 onResponse 方法进行相关逻辑的处理;如果请求执行失败,则调用 ActionListener 类中的 onFailure 方法进行相关逻辑的处理。

以异步方式执行的全部代码添加在 IndexServiceImpl 类中,如下所示:

```
// 以异步方式执行创建索引请求
public void executeIndexRequestAsync(String index) {
  // 创建索引请求
  CreateIndexRequest request = buildIndexRequest(index);
  // 创建监听器
  ActionListener<CreateIndexResponse> listener = new ActionListener
      <CreateIndexResponse>() {
    @Override
    public void onResponse(CreateIndexResponse createIndexResponse) {
    }
    @Override
    public void onFailure(Exception e) {
    }
  };
  // 异步执行
  try {
    restClient.indices().createAsync(request, RequestOptions.DEFAULT,
        listener);
  } catch (Exception e) {
    e.printStackTrace();
  } finally {
    // 关闭Elasticsearch连接
    closeEs();
  }
}
```

当然,在异步请求执行过程中可能会出现异常,异常的处理与同步方式执行情况相同。

3. 解析创建索引请求的响应结果

不论同步方式，还是异步方式，在创建索引请求执行后，客户端均需要对响应结果 CreateIndexResponse 进行处理和解析。CreateIndexResponse 包含有关已执行操作的信息，解析 CreateIndexResponse 的代码共分为三层，分别是 Controller 层、Service 层和 ServiceImpl 实现层。

在 Controller 层的 IndexController 类中新增如下代码：

```
//以同步方式执行 CreateIndexRequest
@RequestMapping("/create/sr")
public String executeCreateIndexRequest(String indexName) {
  // 参数校验
  if (Strings.isNullOrEmpty(indexName)) {
    return "Parameters are wrong!";
  }
  indexService.executeIndexRequest(indexName);
  return "Execute CreateIndexRequest success!";
}
```

在 Service 层的 IndexService 类中新增如下代码：

```
// 以同步方式执行创建索引请求
public void executeIndexRequest(String index);
```

在 ServiceImpl 实现层的 IndexServiceImpl 类中新增如下代码：

```
// 以同步方式执行创建索引请求
public void executeIndexRequest(String index) {
  // 创建索引请求
  CreateIndexRequest request = buildIndexRequest(index);
  try {
    CreateIndexResponse createIndexResponse =
        restClient.indices().create(request, RequestOptions.DEFAULT);
    // 解析 CreateIndexResponse
    processCreateIndexResponse(createIndexResponse);
  } catch (Exception e) {
    e.printStackTrace();
  } finally {
    // 关闭 Elasticsearch 连接
    closeEs();
  }
}
// 解析 CreateIndexResponse
private void processCreateIndexResponse(CreateIndexResponse
```

```
createIndexResponse) {
    // 所有节点是否已确认请求
    boolean acknowledged = createIndexResponse.isAcknowledged();
    // 是否在超时前为索引中的每个分片启动了所需数量的分片副本
    boolean shardsAcknowledged = createIndexResponse.isShardsAcknowledged();
    log.info("acknowledged is " + acknowledged + ";shardsAcknowledged is " +
        shardsAcknowledged);
}
```

随后编译工程，在工程根目录下输入如下命令：

```
mvn clean package
```

通过如下命令启动工程服务：

```
java -jar ./target/esdemo-0.0.1-SNAPSHOT.jar
```

在工程服务启动后，在浏览器中调用如下接口查看索引创建情况：

```
http://localhost:8080/springboot/es/indexsearch/create/sr?indexName=ultraman2
```

请求执行后，如果在服务器中输出如下所示内容，则表明课件索引已经创建成功：

```
2019-09-06 10:29:05.758  INFO 78164 --- [nio-8080-exec-1] c.n.e.service.impl.IndexServiceImpl      : acknowledged is true;shardsAcknowledged is true
```

8.3 获取索引

在索引创建后，用户可以通过获取索引请求查看索引的创建情况。

1. 构建获取索引请求

在获取索引请求前，需要构建获取索引请求，即 GetIndexRequest。GetIndexRequest 需要一个或多个索引参数，代码添加在 IndexServiceImpl 类中，如下所示：

```
// 构建获取索引请求
public GetIndexRequest buildGetIndexRequest(String index) {
    GetIndexRequest request = new GetIndexRequest(index);
    return request;
}
```

GetIndexRequest 提供了可选参数列表供用户进行配置，可选参数主要有 IndicesOptions 和 includeDefaults。IndicesOptions 用于控制解析不可用索引及展开通配符表达式。includeDefaults 的值如果设置为 true，则对于未在索引上显式设置的内容，将返回默认值。代码添加在 IndexServiceImpl 类中的 buildGetIndexRequest 方法内，如下所示：

```
// 如果设置为 true，则对于未在索引上显式设置的内容，将返回默认值
    request.includeDefaults(true);
    // 控制解析不可用索引及展开通配符表达式
    request.indicesOptions(IndicesOptions.lenientExpandOpen());
```

2．执行获取索引请求

在 GetIndexRequest 构建后，即可执行获取索引请求。与创建索引请求类似，获取索引请求也有同步和异步两种执行方式。

同步方式

当以同步方式执行获取索引请求时，客户端会等待 Elasticsearch 服务器返回的查询结果 GetIndexResponse。在收到 GetIndexResponse 后，客户端会继续执行相关的逻辑代码。以同步方式执行的代码添加在 IndexServiceImpl 类中，如下所示：

```
    // 以同步方式执行 GetIndexRequest
    public void excuteGetIndexRequest(String index) {
    GetIndexRequest request = buildGetIndexRequest(index);

    try {
      GetIndexResponse getIndexResponse = restClient.indices().get(request,
          RequestOptions.DEFAULT);
    } catch (Exception e) {
      e.printStackTrace();
    } finally {
      // 关闭 Elasticsearch 连接
      closeEs();
    }
  }
```

异步方式

当以异步方式执行获取索引请求时，高级客户端不必同步等待请求结果的返回，可以直接向接口调用方返回异步接口执行成功的结果。

为了处理异步返回的响应信息或处理在请求执行过程中引发的异常信息，用户需要指定监听器。以异步方式执行的核心代码如下所示：

```
client.indices().getAsync(request, RequestOptions.DEFAULT, listener);
```

其中，listener 为监听器。

在异步请求处理后，如果请求执行成功，则调用 ActionListener 类中的 onResponse 方法进

行相关逻辑的处理；如果请求执行失败，则调用 ActionListener 类中的 onFailure 方法进行相关逻辑的处理。

以异步方式执行的全部代码添加在 IndexServiceImpl 类中，如下所示：

```java
// 以异步方式执行 GetIndexRequest
public void excuteGetIndexRequestAsync(String index) {
  GetIndexRequest request = buildGetIndexRequest(index);
  // 构建监听器
  ActionListener<GetIndexResponse> listener = new ActionListener
      <GetIndexResponse>() {
    @Override
    public void onResponse(GetIndexResponse getIndexResponse) {
    }
    @Override
    public void onFailure(Exception e) {
    }
  };
  try {
    restClient.indices().getAsync(request, RequestOptions.DEFAULT, listener);
  } catch (Exception e) {
    e.printStackTrace();
  } finally {
    // 关闭 Elasticsearch 连接
    closeEs();
  }
}
```

当然，在异步请求执行过程中可能会出现异常，异常的处理与同步方式执行情况相同。

3. 解析获取索引请求的响应结果

不论同步方式，还是异步方式，在获取索引请求执行后，客户端均需要对响应结果 GetIndexResponse 进行处理和解析。GetIndexResponse 中包含了有关已执行操作的信息。解析代码共分为三层，分别是 Controller 层、Service 层和 ServiceImpl 实现层。

其中，在 Controller 层的 IndexController 类中新增如下代码：

```java
// 以同步方式执行 GetIndexRequest
@RequestMapping("/get/sr")
public String executeGetIndexRequest(String indexName) {
  // 参数校验
  if (Strings.isNullOrEmpty(indexName)) {
    return "Parameters are wrong!";
  }
```

```
  indexService.excuteGetIndexRequest(indexName);
  return "Execute GetIndexRequest success!";
}
```

在 Service 层的 IndexService 类中新增如下代码：

```
// 以同步方式执行 GetIndexRequest
public void excuteGetIndexRequest(String index);
```

在 ServiceImpl 实现层的 IndexServiceImpl 类中新增如下代码：

```
// 以同步方式执行 GetIndexRequest
public void excuteGetIndexRequest(String index) {
  GetIndexRequest request = buildGetIndexRequest(index);
  try {
    GetIndexResponse getIndexResponse = restClient.indices().get(request,
        RequestOptions.DEFAULT);
    // 解析 GetIndexResponse
    processGetIndexResponse(getIndexResponse, index);
  } catch (Exception e) {
    e.printStackTrace();
  } finally {
    // 关闭 Elasticsearch 连接
    closeEs();
  }
}
// 解析 GetIndexResponse
private void processGetIndexResponse(GetIndexResponse getIndexResponse,
    String index) {
  // 检索不同类型的映射到索引的映射元数据 MappingMetadata
  MappingMetaData indexMappings = getIndexResponse.getMappings().get(index);
  if (indexMappings == null) {
    return;
  }
  // 检索文档类型和文档属性的映射
  Map<String, Object> indexTypeMappings = indexMappings.getSourceAsMap();
  for (String str : indexTypeMappings.keySet()) {
    log.info("key is " + str);
  }
  // 获取索引的别名列表
  List<AliasMetaData> indexAliases = getIndexResponse.getAliases().get
      (index);
  if (indexAliases == null) {
    return;
  }
```

```java
log.info("indexAliases is " + indexAliases.size());
// 获取为索引设置字符串 index.number_shards 的值。该设置是默认设置的一部分
//（includeDefault 为 true），如果未显式指定设置，则将检索默认设置
String numberOfShardsString = getIndexResponse.getSetting(index, "index.
    number_of_shards");
// 检索索引的所有设置
Settings indexSettings = getIndexResponse.getSettings().get(index);
// 设置对象提供了更多的灵活性。在这里，它被用来提取作为整数的碎片的设置 index.number
Integer numberOfShards = indexSettings.getAsInt("index.number_of_shards",
    null);
// 获取默认设置 index.refresh_interval（includeDefault 默认设置为 true，如果
// includeDefault 设置为 false，则 getIndexResponse.defaultSettings()将返回空映射
TimeValue time = getIndexResponse.getDefaultSettings().get(index).
    getAsTime("index.refresh_interval", null);
log.info("numberOfShardsString is " + numberOfShardsString
    +";indexSettings is "
    + indexSettings.toString() + ";numberOfShards is " + numberOfShards.
    intValue() + ";time is "

    + time.getMillis());
}
```

随后编译工程，在工程根目录下输入如下命令：

```
mvn clean package
```

通过如下命令启动工程服务：

```
java -jar ./target/esdemo-0.0.1-SNAPSHOT.jar
```

在工程服务启动后，在浏览器中调用如下接口查看索引获取情况：

```
http://localhost:8080/springboot/es/indexsearch/get/sr?indexName=ultraman2
```

请求执行后，在服务器中输出如下所示内容：

```
2019-09-06 11:09:41.832  INFO 30192 --- [nio-8080-exec-1] c.n.e.service.impl.IndexServiceImpl      : indexAliases is 0
2019-09-06 11:09:41.835  INFO 30192 --- [nio-8080-exec-1] c.n.e.service.impl.IndexServiceImpl      : numberOfShardsStrin
g is 3;indexSettings is {"index.creation_date":"1567736944366","index.number_of_replicas":"2","index.number_of_shards":"
3","index.provided_name":"ultraman2","index.uuid":"3MJyBU9yQ-mOxyjN1NQDCQ","index.version.created":"7020099"};numberOfSh
ards is 3;time is 1000
```

从中可以看出，我们在创建索引时没有配置别名，因此别名列表长度为 0；分片数量与配置相同，其他参数为默认值。

8.4 删除索引

在 Elasticsearch 中，索引既可以创建，也可以删除。

1. 构建删除索引请求

在执行删除索引请求前，需要构建删除索引请求，即 DeleteIndexRequest。DeleteIndexRequest 需要以索引名称作为参数，代码添加在 IndexServiceImpl 类中，如下所示：

```java
//构建删除索引请求
public DeleteIndexRequest buildDeleteIndexRequest(String index) {
    DeleteIndexRequest request = new DeleteIndexRequest(index);
    return request;
}
```

DeleteIndexRequest 提供了可选参数供用户进行配置。在 DeleteIndexRequest 中，用户可以配置所有节点删除索引的确认等待超时时间、从节点连接到主节点的超时时间和 IndicesOptions。其中，IndicesOptions 用于控制解析不可用索引及展开通配符表达式。代码添加在 IndexServiceImpl 类的 buildDeleteIndexRequest 方法中，如下所示：

```java
// 配置可选参数
    // 等待所有节点删除索引的确认超时时间
    request.timeout(TimeValue.timeValueMinutes(2));
    // 等待所有节点删除索引的确认的超时时间
    request.timeout("2m");
    // 从节点连接到主节点的超时时间
    request.masterNodeTimeout(TimeValue.timeValueMinutes(1));
    // 从节点连接到主节点的超时时间
    request.masterNodeTimeout("1m");
    // 设置 IndicesOptions，控制解析不可用索引及展开通配符表达式
    request.indicesOptions(IndicesOptions.lenientExpandOpen());
```

2. 执行删除索引请求

在 DeleteIndexRequest 构建后，即可执行删除索引请求。与创建索引请求类似，删除索引请求也有同步和异步两种执行方式。

同步方式

当以同步方式执行删除索引请求时，客户端会等待 Elasticsearch 服务器返回的查询结果 DeleteIndexResponse。在收到 DeleteIndexResponse 后，客户端会继续执行相关的逻辑代码。以同步方式执行的代码添加在 IndexServiceImpl 类中，如下所示：

```java
//以同步方式执行 DeleteIndexRequest
 public void executeDeleteIndexRequest(String index) {
   DeleteIndexRequest request = buildDeleteIndexRequest(index);
   try {
     AcknowledgedResponse deleteIndexResponse =
        restClient.indices().delete(request, RequestOptions.DEFAULT);
   } catch (Exception e) {
     e.printStackTrace();
   } finally {
   // 关闭 Elasticsearch 连接
     closeEs();
   }
 }
```

异步方式

当以异步方式执行删除索引请求时,高级客户端不必同步等待请求结果的返回,可以直接向接口调用方返回异步接口执行成功的结果。

为了处理异步返回的响应信息或处理在请求执行过程中引发的异常信息,用户需要指定监听器。以异步方式执行的核心代码如下所示:

```java
client.indices().deleteAsync(request, RequestOptions.DEFAULT, listener);
```

其中,listener 为监听器。

在异步请求处理后,如果请求执行成功,则调用 ActionListener 类中的 onResponse 方法进行相关逻辑的处理;如果请求执行失败,则调用 ActionListener 类中的 onFailure 方法进行相关逻辑的处理。

以异步方式执行的全部代码添加在 IndexServiceImpl 类中,如下所示:

```java
// 以异步方式执行 DeleteIndexRequest
 public void executeDeleteIndexRequestAsync(String index) {
   DeleteIndexRequest request = buildDeleteIndexRequest(index);
   // 构建监听器
   ActionListener<AcknowledgedResponse> listener = new ActionListener
      <AcknowledgedResponse>() {
     @Override
     public void onResponse(AcknowledgedResponse deleteIndexResponse) {
     }
     @Override
     public void onFailure(Exception e) {
     }
   };
```

```
    try {
      restClient.indices().deleteAsync(request, RequestOptions.DEFAULT,
          listener);
    } catch (Exception e) {
      e.printStackTrace();
    } finally {
      // 关闭 Elasticsearch 连接
      closeEs();
    }
  }
```

当然，在异步请求执行过程中可能会出现异常，异常的处理与同步方式执行情况相同。

3. 解析删除索引请求的响应结果

不论同步方式，还是异步方式，在删除索引请求执行后，客户端均需要对响应结果 AcknowledgedResponse 进行解析和处理。代码添加在 IndexServiceImpl 类中，共分为三层，分别是 Controller 层、Service 层和 ServiceImpl 实现层。

在 Controller 层的 IndexController 类中新增如下代码：

```
//以同步方式执行 DeleteIndexRequest
@RequestMapping("/delete/sr")
public String executeDeleteIndexRequest(String indexName) {
  // 参数校验
  if (Strings.isNullOrEmpty(indexName)) {
    return "Parameters are wrong!";
  }
  indexService.executeDeleteIndexRequest(indexName);
  return "Execute DeleteIndexRequest success!";
}
```

在 Service 层的 IndexService 类中新增如下代码：

```
// 以同步方式执行 DeleteIndexRequest
public void executeDeleteIndexRequest(String index);
```

在 ServiceImpl 实现层的 IndexServiceImpl 类中新增如下代码：

```
// 以同步方式执行 DeleteIndexRequest
public void executeDeleteIndexRequest(String index) {
  DeleteIndexRequest request = buildDeleteIndexRequest(index);
  try {
    AcknowledgedResponse deleteIndexResponse =
        restClient.indices().delete(request, RequestOptions.DEFAULT);
    // 解析 AcknowledgedResponse
```

```
        processAcknowledgedResponse(deleteIndexResponse);
    } catch (Exception e) {
        e.printStackTrace();
    } finally {
        // 关闭 Elasticsearch 连接
        closeEs();
    }
}
// 解析 AcknowledgedResponse
private void processAcknowledgedResponse(AcknowledgedResponse
    deleteIndexResponse) {
    // 所有节点是否已确认请求
    boolean acknowledged = deleteIndexResponse.isAcknowledged();
    log.info("acknowledged is " + acknowledged);
}
```

随后编译工程，在工程根目录下输入如下命令：

```
mvn clean package
```

通过如下命令启动工程服务：

```
java -jar ./target/esdemo-0.0.1-SNAPSHOT.jar
```

在工程服务启动后，在浏览器中调用如下接口删除索引情况：

```
http://localhost:8080/springboot/es/indexsearch/delete/sr?indexName=ultraman3
```

请求执行成功后，在服务器控制台输出如下所示内容：

```
2019-09-06 11:35:43.309  INFO 77484 --- [io-8080-exec-10] c.n.e.service.impl.IndexServiceImpl      : acknowledged is true
```

8.5 索引存在验证

想要查看目标索引是否已经创建，用户需要使用索引存在验证接口，即 Exists API。

1. 构建索引存在验证请求

在高级客户端中，用户称需要基于 GetIndexRequest 才能发起索引存在验证请求。GetIndexRequest 的必选参数是索引名，代码添加在 IndexServiceImpl 类中，如下所示：

```
// 构建索引存在验证请求
public GetIndexRequest buildExistsIndexRequest(String index) {
```

```
        GetIndexRequest request = new GetIndexRequest(index);
        return request;
    }
```

在索引存在验证请求中，GetIndexRequest 的可选参数有是否从主节点返回本地信息或检索状态、是否回归到适合人类的格式、是否返回每个索引的所有默认设置和 IndicesOptions。其中，IndicesOptions 用于解析不可用的索引及展开通配符表达式，代码添加在 IndexServiceImpl 类中，如下所示：

```
// 从主节点返回本地信息或检索状态
    request.local(false);
    // 回归到适合人类的格式
    request.humanReadable(true);
    // 是否返回每个索引的所有默认设置
    request.includeDefaults(false);
    // 控制如何解析不可用索引及如何展开通配符表达式
    request.indicesOptions(IndicesOptions.lenientExpandOpen());
```

2. 执行索引存在验证请求

在 GetIndexRequest 构建完成后，即可执行索引存在验证请求。与文档索引请求类似，索引存在验证请求也有同步和异步两种执行方式。

同步方式

当以同步方式执行索引存在验证请求时，客户端会等待 Elasticsearch 服务器返回的布尔型查询结果。以同步方式执行的代码添加在 IndexServiceImpl 类中，如下所示：

```
    // 以同步方式执行索引存在验证请求
    public void executeExistsIndexRequest(String index) {
        GetIndexRequest request = buildExistsIndexRequest(index);
        try {
            boolean exists = restClient.indices().exists(request, RequestOptions.
                DEFAULT);
            log.info("exists is " + exists);
        } catch (Exception e) {
            e.printStackTrace();
        } finally {
            // 关闭Elasticsearch连接
            closeEs();
        }
    }
```

异步方式

当以异步方式执行索引存在验证请求时,高级客户端不必同步等待请求结果的返回,可以直接向接口调用方返回异步接口执行成功的结果。

为了处理异步返回的响应信息或处理在请求执行过程中引发的异常信息,用户需要指定监听器。以异步方式执行的核心代码如下所示:

```
client.indices().existsAsync(request, RequestOptions.DEFAULT, listener);
```

其中,listener 为监听器。

在异步请求处理完成后,如果请求执行成功,则调用 ActionListener 类中的 onResponse 方法进行相关逻辑的处理;如果请求执行失败,则调用 ActionListener 类中的 onFailure 方法进行相关逻辑的处理。

以异步方式执行的全部代码添加在 IndexServiceImpl 类中,如下所示:

```
// 以异步方式执行索引存在验证请求
public void executeExistsIndexRequestAsync(String index) {
  GetIndexRequest request = buildExistsIndexRequest(index);
  // 构建监听器
  ActionListener<Boolean> listener = new ActionListener<Boolean>() {
    @Override
    public void onResponse(Boolean exists) {
    }
    @Override
    public void onFailure(Exception e) {
    }
  };
  try {
    restClient.indices().existsAsync(request, RequestOptions.DEFAULT,
      listener);
  } catch (Exception e) {
    e.printStackTrace();
  } finally {
    // 关闭 Elasticsearch 连接
    closeEs();
  }
}
```

当然,在异步请求执行过程中可能会出现异常,异常的处理与同步方式执行情况相同。

3. 解析索引存在验证请求的响应结果

不论同步方式，还是异步方式，在索引存在验证请求执行后，客户端均会得到索引是否存在的结果。代码添加在 IndexServiceImpl 类中，共分为三层，分别是 Controller 层、Service 层和 ServiceImpl 实现层。

在 Controller 层的 IndexController 类中新增如下代码：

```java
// 以同步方式执行 ExistsIndexRequest
@RequestMapping("/exists/sr")
public String executeExistsIndexRequest(String indexName) {
  // 参数校验
  if (Strings.isNullOrEmpty(indexName)) {
    return "Parameters are wrong!";
  }
  indexService.executeExistsIndexRequest(indexName);
  return "Execute ExistsIndexRequest success!";
}
```

在 Service 层的 IndexService 类中新增如下代码：

```java
//以同步方式执行索引存在验证请求
public void executeExistsIndexRequest(String index);
```

在 ServiceImpl 实现层的 IndexServiceImpl 类中的代码和"同步方式"部分一致，不再赘述。

随后编译工程，在工程根目录下输入如下命令：

```
mvn clean package
```

通过如下命令启动工程服务：

```
java -jar ./target/esdemo-0.0.1-SNAPSHOT.jar
```

在工程服务启动后，在浏览器中调用如下接口查看索引 ultraman 的存在情况：

```
http://localhost:8080/springboot/es/indexsearch/exists/sr?indexName=ultraman
```

请求执行后，在服务器中输出如下所示内容，与预期相符：

```
2019-09-06 14:05:14.685  INFO 78932 --- [nio-8080-exec-4] c.n.e.service.impl.IndexServiceImpl      : exists is true
```

8.6 打开索引

在索引创建后,即可通过打开索引接口打开索引。

1. 构建打开索引请求

在执行打开索引请求前,需要构建打开索引请求,即 OpenIndexRequest。OpenIndexRequest 需要以索引名称作为参数,代码添加在 IndexServiceImpl 类中,如下所示:

```
// 构建 OpenIndexRequest
public OpenIndexRequest buildOpenIndexRequest(String index) {
  OpenIndexRequest request = new OpenIndexRequest(index);
  return request;
}
```

在构建打开索引请求时,OpenIndexRequest 提供了可选参数供用户进行配置。OpenIndexRequest 的可选参数有所有节点确认索引打开的超时时间、从节点连接到主节点的超时时间、请求返回响应前活跃的分片数量和 IndicesOptions。IndicesOptions 用于控制解析不可用索引及展开通配符表达式,代码添加在 IndexServiceImpl 类中的 buildOpenIndexRequest()方法内,如下所示:

```
// 配置可选参数
  // 所有节点确认索引打开的超时时间
  request.timeout(TimeValue.timeValueMinutes(2));
  request.timeout("2m");
  // 从节点连接到主节点的超时时间
  request.masterNodeTimeout(TimeValue.timeValueMinutes(1));
  request.masterNodeTimeout("1m");
  // 请求返回响应前活跃的分片数量
  request.waitForActiveShards(2);
  request.waitForActiveShards(ActiveShardCount.DEFAULT);
  // 设置 IndicesOptions
  request.indicesOptions(IndicesOptions.strictExpandOpen());
```

2. 执行打开索引请求

在 OpenIndexRequest 构建后,即可执行打开索引请求。与创建索引请求类似,打开索引请求也有同步和异步两种执行方式。

同步方式

当以同步方式执行打开索引请求时,客户端会等待 Elasticsearch 服务器返回的查询结果 OpenIndexResponse。在收到 OpenIndexResponse 后,客户端会继续执行相关的逻辑代码。以同步方式执行的代码添加在 IndexServiceImpl 类中，如下所示：

```
// 以同步方式执行 OpenIndexRequest
public void executeOpenIndexRequest(String index) {
  OpenIndexRequest request = buildOpenIndexRequest(index);
  try {
    OpenIndexResponse openIndexResponse =
        restClient.indices().open(request, RequestOptions.DEFAULT);
  } catch (Exception e) {
    e.printStackTrace();
  } finally {
    // 关闭 Elasticsearch 连接
    closeEs();
  }
}
```

异步方式

当以异步方式执行打开索引请求时，高级客户端不必同步等待请求结果的返回，可以直接向接口调用方返回异步接口执行成功的结果。

为了处理异步返回的响应信息或处理在请求执行过程中引发的异常信息，用户需要指定监听器。以异步方式执行的核心代码如下所示：

```
client.indices().openAsync(request, RequestOptions.DEFAULT, listener);
```

其中，listener 为监听器。

在异步请求处理后，如果请求执行成功，则调用 ActionListener 类中的 onResponse 方法进行相关逻辑的处理；如果请求执行失败，则调用 ActionListener 类中的 onFailure 方法进行相关逻辑的处理。

以异步方式执行的全部代码添加在 IndexServiceImpl 类中，如下所示：

```
// 以异步方式执行 OpenIndexRequest
public void executeOpenIndexRequestAsync(String index) {
  OpenIndexRequest request = buildOpenIndexRequest(index);
  // 构建监听器
  ActionListener<OpenIndexResponse> listener = new ActionListener
      <OpenIndexResponse>() {
    @Override
    public void onResponse(OpenIndexResponse openIndexResponse) {
    }
    @Override
    public void onFailure(Exception e) {
```

```
    };
    try {
        restClient.indices().openAsync(request, RequestOptions.DEFAULT, listener);
    } catch (Exception e) {
        e.printStackTrace();
    } finally {
        // 关闭Elasticsearch连接
        closeEs();
    }
}
```

当然，在异步请求执行过程中可能会出现异常，异常的处理与同步方式执行情况相同。

3．解析打开索引请求的响应结果

不论同步方式，还是异步方式，在打开索引请求执行后，客户端均需要对响应结果 OpenIndexResponse 进行解析和处理。代码添加在 IndexServiceImpl 类中，共分为三层，分别是 Controller 层、Service 层和 ServiceImpl 实现层。

在 Controller 层的 IndexController 类中新增如下代码：

```
// 以同步方式执行OpenIndexRequest
@RequestMapping("/open/sr")
public String executeOpenIndexRequest(String indexName) {
    // 参数校验
    if (Strings.isNullOrEmpty(indexName)) {
        return "Parameters are wrong!";
    }
    indexService.executeOpenIndexRequest(indexName);
    return "Execute OpenIndexRequest success!";
}
```

在 Service 层的 IndexService 类中新增如下代码：

```
// 以同步方式执行OpenIndexRequest
public void executeOpenIndexRequest(String index);
```

在 ServiceImpl 实现层的 IndexServiceImpl 类中新增如下代码：

```
// 以同步方式执行OpenIndexRequest
public void executeOpenIndexRequest(String index) {
    OpenIndexRequest request = buildOpenIndexRequest(index);
    try {
        OpenIndexResponse openIndexResponse =
            restClient.indices().open(request, RequestOptions.DEFAULT);
        // 解析OpenIndexResponse
```

```
            processOpenIndexResponse(openIndexResponse);
        } catch (Exception e) {
            e.printStackTrace();
        } finally {
            // 关闭Elasticsearch连接
            closeEs();
        }
    }
    // 解析 OpenIndexResponse
    private void processOpenIndexResponse(OpenIndexResponse openIndexResponse)
    {
        // 所有节点是否已确认请求
        boolean acknowledged = openIndexResponse.isAcknowledged();
        // 是否在超时前为索引中的每个分片启动了所需数量的分片副本
        boolean shardsAcked = openIndexResponse.isShardsAcknowledged();
        log.info("acknowledged is " + acknowledged + ";shardsAcked is " +
            shardsAcked);
    }
```

随后编译工程，在工程根目录下输入如下命令：

```
mvn clean package
```

通过如下命令启动工程服务：

```
java -jar ./target/esdemo-0.0.1-SNAPSHOT.jar
```

在工程服务启动后，在浏览器中调用如下接口查看索引 ultraman 的打开情况：

```
http://localhost:8080/springboot/es/indexsearch/open/sr?indexName=ultraman
```

在请求执行后，服务器输出如下所示内容：

```
2019-09-06 14:19:45.202  INFO 76468 --- [nio-8080-exec-1] c.n.e.service.impl.IndexServiceImpl      : acknowledged is tru
e;shardsAcked is true
```

8.7 关闭索引

用完索引后，还需要关闭索引，这时就需要使用关闭索引接口。

1. 构建关闭索引请求

在执行关闭索引请求前，需要构建关闭索引请求，即 CloseIndexRequest。CloseIndexRequest 必须以索引名称作为参数，代码添加在 IndexServiceImpl 类中，如下所示：

```java
// 构建CloseIndexRequest
public CloseIndexRequest buildCloseIndexRequest(String index) {
  CloseIndexRequest request = new CloseIndexRequest(index);
  return request;
}
```

在构建关闭索引请求时，CloseIndexRequest 提供了可选参数列表供用户进行配置。CloseIndexRequest 支持的可选参数有所有节点确认索引关闭的超时时间、从节点连接到主节点的超时时间和 IndicesOptions。IndicesOptions 用于控制解析不可用索引及展开通配符表达式。代码添加在 IndexServiceImpl 类中的 buildCloseIndexRequest 方法内，如下所示：

```java
// 配置可选参数
  // 所有节点确认索引关闭的超时时间
  request.timeout(TimeValue.timeValueMinutes(2));
  request.timeout("2m");
  // 从节点连接到主节点的超时时间
  request.masterNodeTimeout(TimeValue.timeValueMinutes(1));
  request.masterNodeTimeout("1m");
  // 用于控制解析不可用索引及展开通配符表达式
  request.indicesOptions(IndicesOptions.lenientExpandOpen());
```

2. 执行关闭索引请求

在 CloseIndexRequest 构建后，即可执行关闭索引请求。与创建索引请求类似，关闭索引请求也有同步和异步两种执行方式。

同步方式

当以同步方式执行关闭索引请求时，客户端会等待 Elasticsearch 服务器返回的查询结果 AcknowledgedResponse。在收到 AcknowledgedResponse 后，客户端会继续执行相关的逻辑代码。以同步方式执行的代码添加在 IndexServiceImpl 类中，如下所示：

```java
// 以同步方式执行关闭索引请求
public void executeCloseIndexRequest(String index) {
  CloseIndexRequest request = buildCloseIndexRequest(index);
  try {
    AcknowledgedResponse closeIndexResponse =
      restClient.indices().close(request, RequestOptions.DEFAULT);

    // 所有节点是否已确认请求
    boolean acknowledged = closeIndexResponse.isAcknowledged();
    log.info(index + " acknowledged is " + acknowledged);
  } catch (Exception e) {
    e.printStackTrace();
```

```
    } finally {
        // 关闭Elasticsearch连接
        closeEs();
    }
}
```

异步方式

当以异步方式执行关闭索引请求时,高级客户端不必同步等待请求结果的返回,可以直接向接口调用方返回异步接口执行成功的结果。

为了处理异步返回的响应信息或处理在请求执行过程中引发的异常信息,用户需要指定监听器。以异步方式执行的核心代码如下所示:

```
client.indices().closeAsync(request, RequestOptions.DEFAULT, listener);
```

其中,listener为监听器。

在异步请求处理后,如果请求执行成功,则调用ActionListener类中的onResponse方法进行相关逻辑的处理;如果请求执行失败,则调用ActionListener类中的onFailure方法进行相关逻辑的处理。

以异步方式执行的全部代码添加在IndexServiceImpl类中,如下所示:

```
// 以异步方式执行关闭索引请求
public void executeCloseIndexRequestAsync(String index) {
    CloseIndexRequest request = buildCloseIndexRequest(index);
    // 构建监听器
    ActionListener<AcknowledgedResponse> listener = new ActionListener
        <AcknowledgedResponse>() {
        @Override
        public void onResponse(AcknowledgedResponse closeIndexResponse) {
        }
        @Override
        public void onFailure(Exception e) {
        }
    };
    try {
        restClient.indices().closeAsync(request, RequestOptions.DEFAULT, listener);
    } catch (Exception e) {
        e.printStackTrace();
    } finally {
        // 关闭Elasticsearch连接
        closeEs();
    }
}
```

当然，在异步请求执行过程中可能会出现异常，异常的处理与同步方式执行情况相同。

3. 解析关闭索引请求的响应结果

以同步方式执行的代码如下所示，AcknowledgedResponse 的 isAcknowledged 方法会返回异步布尔值，表示索引是否已经关闭。

代码添加在 IndexServiceImpl 类中，共分为三层，分别是 Controller 层、Service 层和 ServiceImpl 实现层。

在 Controller 层的 IndexController 类中新增如下代码：

```java
// 以同步方式执行CloseIndexRequest
@RequestMapping("/close/sr")
public String executeCloseIndexRequest(String indexName) {
  // 参数校验
  if (Strings.isNullOrEmpty(indexName)) {
    return "Parameters are wrong!";
  }
  indexService.executeCloseIndexRequest(indexName);
  return "Execute CloseIndexRequest success!";
}
```

在 Service 层的 IndexService 类中新增如下代码：

```java
// 以同步方式执行CloseIndexRequest
public void executeCloseIndexRequest(String index);
```

在 ServiceImpl 实现层的 IndexServiceImpl 类中的代码与"同步方式"中的代码一致，不再赘述。

随后编译工程，在工程根目录下输入如下命令：

```
mvn clean package
```

通过如下命令启动工程服务：

```
java -jar ./target/esdemo-0.0.1-SNAPSHOT.jar
```

在工程服务启动后，在浏览器中调用如下接口查看索引 ultraman 的关闭情况：

```
http://localhost:8080/springboot/es/indexsearch/close/sr?indexName=ultraman
```

请求执行后，在服务器中输出如下所示内容：

```
2019-09-06 14:41:57.947  INFO 82424 --- [nio-8080-exec-1] c.n.e.service.impl.IndexServiceImpl      : ultraman acknowledged is true
```

8.8 缩小索引

如果要缩小索引,则需要使用缩小索引接口,即 Shrink API。

1. 构建缩小索引大小请求

在执行调整索引大小请求前,需要构建调整索引大小请求,即 ResizeRequest。ResizeRequest 的必选参数是两个字符串参数,分别是源索引名称和目标索引名称。代码添加在 IndexServiceImpl 类中,如下所示:

```java
// 构建ResizeRequest
public ResizeRequest buildResizeRequest(String sourceIndex, String
    targetIndex) {
    ResizeRequest request = new ResizeRequest(targetIndex, sourceIndex);
    return request;
}
```

在执行调整索引大小请求时,ResizeRequest 提供了可选参数列表供用户进行配置。ResizeRequest 支持的可选参数有所有节点确认索引打开的超时时间、从节点连接到主节点的超时时间、请求返回前需要等待的活跃状态的分片数量、在缩小索引上的目标索引中的分片数、删除从源索引复制的分配要求、与目标索引关联的别名等。代码添加在 IndexServiceImpl 类中的 buildResizeRequest 方法内,如下所示:

```java
// 配置可选参数
    // 所有节点确认索引打开的超时时间
    request.timeout(TimeValue.timeValueMinutes(2));
    request.timeout("2m");
    // 从节点连接到主节点的超时时间
    request.masterNodeTimeout(TimeValue.timeValueMinutes(1));
    request.masterNodeTimeout("1m");
    // 请求返回前需要等待的活跃状态的分片数量
    request.setWaitForActiveShards(2);
    request.setWaitForActiveShards(ActiveShardCount.DEFAULT);
    //在缩小索引上的目标索引中的分片数、删除从源索引复制的分配要求
    request.getTargetIndexRequest().settings(Settings.builder().put
("index.number_of_shards",2).putNull("index.routing.allocation.require._
name"));
    // 与目标索引关联的别名
    request.getTargetIndexRequest().alias(new Alias(targetIndex + "_alias"));
```

2. 执行缩小索引请求

在构建 ResizeRequest 之后,即可执行缩小索引请求。与创建索引请求类似,缩小索引请求也有同步和异步两种执行方式。

同步方式

当以同步方式执行缩小索引请求时,客户端会等待 Elasticsearch 服务器返回的查询结果 ResizeResponse。在收到 ResizeResponse 后,客户端会继续执行相关的逻辑代码。以同步方式执行的代码添加在 IndexServiceImpl 类中,如下所示:

```
//以同步方式执行ResizeRequest
public void executeResizeRequest(String sourceIndex, String targetIndex) {
  ResizeRequest request = buildResizeRequest(sourceIndex, targetIndex);
  try {
    ResizeResponse resizeResponse = restClient.indices().shrink(request,
        RequestOptions.DEFAULT);
    // 解析 ResizeResponse
    processResizeResponse(resizeResponse);
  } catch (Exception e) {
    e.printStackTrace();
  } finally {
    // 关闭Elasticsearch连接
    closeEs();
  }
}
```

异步方式

当以异步方式执行缩小索引请求时,高级客户端不必同步等待请求结果的返回,可以直接向接口调用方返回异步接口执行成功的结果。

为了处理异步返回的响应信息或处理在请求执行过程中引发的异常信息,用户需要指定监听器。以异步方式执行的核心代码如下所示:

```
client.indices().shrinkAsync(request, RequestOptions.DEFAULT, listener);
```

其中,listener 为监听器。

在异步请求处理后,如果请求执行成功,则调用 ActionListener 类中的 onResponse 方法进行相关逻辑的处理;如果请求执行失败,则调用 ActionListener 类中的 onFailure 方法进行相关逻辑的处理。

以异步方式执行的全部代码如下所示:

```
// 以异步方式执行ResizeRequest
```

```java
    public void executeResizeRequestAsync(String sourceIndex, String
        targetIndex) {
    ResizeRequest request = buildResizeRequest(sourceIndex, targetIndex);
    // 构建监听器
    ActionListener<ResizeResponse> listener = new ActionListener
        <ResizeResponse>() {
        @Override
        public void onResponse(ResizeResponse resizeResponse) {
        }
        @Override
        public void onFailure(Exception e) {
        }
    };

    try {
        restClient.indices().shrinkAsync(request, RequestOptions.DEFAULT, listener);
    } catch (Exception e) {
        e.printStackTrace();
    } finally {
        // 关闭Elasticsearch连接
        closeEs();
    }
}
```

当然，在异步请求执行过程中可能会出现异常，异常的处理与同步方式执行情况相同。

3. 解析缩小索引请求的响应结果

不论同步方式，还是异步方式，在缩小索引请求执行后，客户端均需要对响应结果 ResizeResponse 进行处理和解析。

代码添加在 IndexServiceImpl 类中，共分为三层，分别是 Controller 层、Service 层和 ServiceImpl 实现层。

在 Controller 层的 IndexController 类中新增如下代码：

```java
    // 以同步方式执行ResizeRequest
    @RequestMapping("/resize/sr")
    public String executeResizeRequest(String sourceIndexName, String
        targetIndexName) {
        // 参数校验
        if (Strings.isNullOrEmpty(sourceIndexName) || Strings.isNullOrEmpty
            (targetIndexName)) {
            return "Parameters are wrong!";
        }
```

```
        indexService.executeResizeRequest(sourceIndexName, targetIndexName);
        return "Execute ResizeRequest success!";
    }
```

在 Service 层的 IndexService 类中新增如下代码:

```
// 以同步方式执行 ResizeRequest
    public void executeResizeRequest(String sourceIndex, String targetIndex);
```

在 ServiceImpl 实现层的 IndexServiceImpl 类中新增如下代码:

```
// 解析 ResizeResponse
    private void processResizeResponse(ResizeResponse resizeResponse) {
        // 所有节点是否已确认请求
        boolean acknowledged = resizeResponse.isAcknowledged();
        // 是否在超时前为索引中的每个分片启动了所需数量的分片副本
        boolean shardsAcked = resizeResponse.isShardsAcknowledged();
        log.info("acknowledged is " + acknowledged + ";shardsAcked is " +
            shardsAcked);
    }
```

随后编译工程,在工程根目录下输入如下命令:

```
mvn clean package
```

通过如下命令启动工程服务:

```
java -jar ./target/esdemo-0.0.1-SNAPSHOT.jar
```

在工程服务启动后,在浏览器中调用如下接口查看索引 ultraman6 缩小的情况:

```
http://localhost:8080/springboot/es/indexsearch/resize/sr?sourceIndexName=ultraman6&targetIndexName=ultraman7
```

请求执行后,在服务器中输出如下所示内容,可见缩小索引请求执行成功:

```
2019-09-06 15:22:06.474  INFO 83236 --- [nio-8080-exec-2] c.n.e.service.impl.IndexServiceImpl      : acknowledged is true;shardsAcked is true
```

8.9 拆分索引

如果要拆分索引,则需要调用拆分索引接口,即 Split API。

1. 构建拆分索引请求

在拆分索引请求前同样需要构建 ResizeRequest。在 8.8 节中,曾构建 ResizeRequest 代码,两个请求接口的区别在于，在拆分索引请求中，需要将"调整大小"的类型设置为"拆分"。代码添加在 IndexServiceImpl 类中，如下所示：

```
// 构建 ResizeRequest
public ResizeRequest buildSplitRequest(String sourceIndex, String
    targetIndex) {
  ResizeRequest request = new ResizeRequest(targetIndex, sourceIndex);
  //把"调整大小"的类型设置为"拆分"
  request.setResizeType(ResizeType.SPLIT);
  return request;
}
```

在拆分索引请求构建过程中，ResizeRequest 可配置的可选参数列表与缩小索引请求构建过程中的相同，不再赘述。

2. 执行拆分索引请求

在 ResizeRequest 构建后，即可执行拆分索引请求。与创建索引请求类似，拆分索引请求也有同步和异步两种执行方式。

同步方式

当以同步方式执行拆分索引请求时，客户端会等待 Elasticsearch 服务器返回的查询结果 ResizeResponse。在收到 ResizeResponse 后，客户端继续执行相关的逻辑代码。以同步方式执行的核心代码添加在 IndexServiceImpl 类中，如下所示：

```
// 以同步方式执行 ResizeRequest
public void executeSplitRequest(String sourceIndex, String targetIndex) {
  ResizeRequest request = buildSplitRequest(sourceIndex, targetIndex);
  try {
    ResizeResponse resizeResponse = restClient.indices().split(request,
        RequestOptions.DEFAULT);

    // 解析 ResizeResponse
    processResizeResponse(resizeResponse);
  } catch (Exception e) {
    e.printStackTrace();
  } finally {
    // 关闭 Elasticsearch 连接
    closeEs();
  }
}
```

异步方式

当以异步方式执行拆分索引请求时,高级客户端不必同步等待请求结果的返回,可以直接向接口调用方返回异步接口执行成功的结果。

为了处理异步返回的响应信息或处理在请求执行过程中引发的异常信息,用户需要指定监听器。以异步方式执行的核心代码如下所示:

```
client.indices().splitAsync(request, RequestOptions.DEFAULT,listener);
```

其中,listener 为监听器。

在异步请求处理后,如果请求执行成功,则调用 ActionListener 类中的 onResponse 方法进行相关逻辑的处理;如果请求执行失败,则调用 ActionListener 类中的 onFailure 方法进行相关逻辑的处理。

以异步方式执行的全部代码添加在 IndexServiceImpl 类中,如下所示:

```
// 以异步方式执行ResizeRequest
public void executeSplitRequestAsync(String sourceIndex, String
    targetIndex) {
  ResizeRequest request = buildSplitRequest(sourceIndex, targetIndex);
  // 构建监听器
  ActionListener<ResizeResponse> listener = new ActionListener
      <ResizeResponse>() {
    @Override
    public void onResponse(ResizeResponse resizeResponse) {
    }
    @Override
    public void onFailure(Exception e) {
    }
  };
  try {
    restClient.indices().splitAsync(request, RequestOptions.DEFAULT,
        listener);
  } catch (Exception e) {
    e.printStackTrace();
  } finally {
    // 关闭Elasticsearch连接
    closeEs();
  }
}
```

当然,在异步请求执行过程中可能会出现异常,异常的处理与同步方式执行情况相同。

3. 解析拆分索引请求的响应结果

不论同步方式，还是异步方式，在拆分索引请求执行后，客户端均需要对 ResizeResponse 进行处理和解析。

代码添加在 IndexServiceImpl 类中，共分为三层，分别是 Controller 层、Service 层和 ServiceImpl 实现层。

在 Controller 层的 IndexController 类中新增如下代码：

```
//以同步方式执行 ResizeRequest
@RequestMapping("/split/sr")
public String executeSplitRequest(String sourceIndexName, String
    targetIndexName) {
  // 参数校验
  if (Strings.isNullOrEmpty(sourceIndexName) || Strings.isNullOrEmpty
      (targetIndexName)) {
    return "Parameters are wrong!";
  }
  indexService.executeSplitRequest(sourceIndexName, targetIndexName);
  return "Execute SplitResizeRequest success!";
}
```

在 Service 层的 IndexService 类中新增如下代码：

```
// 以同步方式执行 ResizeRequest
  public void executeSplitRequest(String sourceIndex, String targetIndex);
```

在 ServiceImpl 实现层 IndexServiceImpl 类中的代码与本节的"同步方式"中的代码相同，不再赘述。

随后编译工程，在工程根目录下输入如下命令：

```
mvn clean package
```

通过如下命令启动工程服务：

```
java -jar ./target/esdemo-0.0.1-SNAPSHOT.jar
```

当工程服务启动后，在浏览器中调用如下接口查看索引 ultraman6 的拆分情况：

```
http://localhost:8080/springboot/es/indexsearch/split/sr?sourceIndexName=ultraman6&targetIndexName=ultraman10
```

请求执行后，在服务器中输出的内容如下所示，可见相关节点均已完成索引 ultraman6 的拆分：

```
2019-09-06 16:17:13.897  INFO 83936 --- [nio-8080-exec-1] c.n.e.service.impl.IndexServiceImpl      : acknowledged is true;shardsAcked is true
```

8.10 刷新索引

刷新索引接口可以应用于单个或多个索引,甚至可以应用于全部索引。

1. 构建刷新索引请求

在执行刷新索引请求前,需要构建刷新索引请求,即 RefreshRequest。构建 RefreshRequest 的方法有多种,可以刷新单个索引、多个索引或全部索引。代码添加在 IndexServiceImpl 类中,如下所示:

```java
// 构建刷新索引请求
public RefreshRequest buildRefreshRequest(String index) {
    // 刷新单个索引
    RefreshRequest request = new RefreshRequest(index);
    // 刷新多个索引
    RefreshRequest requestMultiple = new RefreshRequest(index, index);
    // 刷新全部索引
    RefreshRequest requestAll = new RefreshRequest();
    return request;
}
```

在执行 RefreshRequest 时,RefreshRequest 提供了可选参数供用户进行配置,其中最主要的是 IndicesOptions,用于解析不可用索引及展开通配符表达式。代码添加在 IndexServiceImpl 类中的 buildRefreshRequest 方法内,如下所示:

```java
// 解析不可用索引及展开通配符表达式
request.indicesOptions(IndicesOptions.lenientExpandOpen());
```

2. 执行刷新索引请求

在 RefreshRequest 构建后,即可执行刷新索引请求。与创建索引请求类似,刷新索引请求也有同步和异步两种执行方式。

同步方式

当以同步方式执行刷新索引请求时,客户端会等待 Elasticsearch 服务器返回的查询结果 RefreshResponse。在收到 RefreshResponse 后,客户端会继续执行相关的逻辑代码。以同步方式执行的代码添加在 IndexServiceImpl 类中,如下所示:

```java
// 以同步方式执行 RefreshRequest
```

```java
public void executeRefreshRequest(String index) {
  RefreshRequest request = buildRefreshRequest(index);
  try {
    RefreshResponse refreshResponse =
        restClient.indices().refresh(request, RequestOptions.DEFAULT);
    // 解析 RefreshResponse
    processRefreshResponse(refreshResponse);
  } catch (Exception e) {
    e.printStackTrace();
  } finally {
    // 关闭 Elasticsearch 连接
    closeEs();
  }
}
```

异步方式

当以异步方式执行刷新索引请求时，高级客户端不必同步等待请求结果的返回，可以直接向接口调用方返回异步接口执行成功的结果。

为了处理异步返回的响应信息或处理在请求执行过程中引发的异常信息，用户需要指定监听器。以异步方式执行的核心代码如下所示：

```java
client.indices().refreshAsync(request, RequestOptions.DEFAULT, listener);
```

其中，listener 为监听器。

在异步请求处理后，如果请求执行成功，则调用 ActionListener 类中的 onResponse 方法进行相关逻辑的处理；如果请求执行失败，则调用 ActionListener 类中的 onFailure 方法进行相关逻辑的处理。

以异步方式执行的全部代码如下所示：

```java
// 以异步方式执行 RefreshRequest
public void executeRefreshRequestAsync(String index) {
  RefreshRequest request = buildRefreshRequest(index);
  // 构建监听器
  ActionListener<RefreshResponse> listener = new ActionListener
      <RefreshResponse>() {
    @Override
    public void onResponse(RefreshResponse refreshResponse) {
    }
    @Override
    public void onFailure(Exception e) {
    }
```

```
    };
    try {
        restClient.indices().refreshAsync(request, RequestOptions.DEFAULT, listener);
    } catch (Exception e) {
        e.printStackTrace();
    } finally {
        // 关闭Elasticsearch连接
        closeEs();
    }
}
```

当然，在异步请求执行过程中可能会出现异常，异常的处理与同步方式执行情况相同。

3．解析刷新索引请求的响应结果

不论同步方式，还是异步方式，在刷新索引请求执行后，客户端均需要对 RefreshResponse 进行处理和解析。

代码添加在 IndexServiceImpl 类中，共分为三层，分别是 Controller 层、Service 层和 ServiceImpl 实现层。

在 Controller 层的 IndexController 类中新增如下代码：

```
// 以同步方式执行RefreshRequest
@RequestMapping("/refresh/sr")
public String executeRefreshRequest(String indexName) {
    // 参数校验
    if (Strings.isNullOrEmpty(indexName)) {
        return "Parameters are wrong!";
    }
    indexService.executeRefreshRequest(indexName);
    return "Execute RefreshRequest success!";
}
```

在 Service 层的 IndexService 类中新增如下代码：

```
// 以同步方式执行RefreshRequest
public void executeRefreshRequest(String index);
```

在 ServiceImpl 实现层的 IndexServiceImpl 类中新增如下代码：

```
// 解析RefreshResponse
private void processRefreshResponse(RefreshResponse refreshResponse) {
    // 刷新请求命中的分片总数
    int totalShards = refreshResponse.getTotalShards();
    // 刷新成功的分片数
```

```
        int successfulShards = refreshResponse.getSuccessfulShards();
        // 刷新失败的分片数
        int failedShards = refreshResponse.getFailedShards();
        //在一个或多个分片上刷新失败时的失败列表
        DefaultShardOperationFailedException[]  failures  =  refreshResponse.getShardFailures();
        log.info("totalShards is " + totalShards + ";successfulShards is " + successfulShards
                + ";failedShards is " + failedShards + "; failures is "
                + (failures == null ? 0 : failures.length));
    }
```

随后编译工程，在工程根目录下输入如下命令：

```
mvn clean package
```

通过如下命令启动工程服务：

```
java -jar ./target/esdemo-0.0.1-SNAPSHOT.jar
```

在工程服务启动后，在浏览器中调用如下接口查看索引 ultraman8 的刷新情况：

```
http://localhost:8080/springboot/es/indexsearch/refresh/sr?indexName=ultraman8
```

请求执行后，在服务器中输出如下所示内容，表明刷新请求执行成功：

```
2019-09-06 16:37:50.659  INFO 82996 --- [nio-8080-exec-1] c.n.e.service.impl.IndexServiceImpl      : totalShards is 12;successfulShards is 6;failedShards is 0; failures is 0
```

8.11　Flush 刷新

除刷新索引接口外，Elasticsearch 还提供了 Flush 刷新接口。Flush 刷新接口可以应用于单个或多个索引，甚至可以应用于全部索引。

索引的刷新过程是将数据刷新到索引存储，然后清除内部事务日志释放索引内存。在默认情况下，Elasticsearch 使用内存启发式方式根据需要自动触发刷新操作，以清理内存。

1. 构建 Flush 刷新请求

在执行 Flush 刷新请求前，需要构建 Flush 刷新请求，即 FlushRequest。构建 FlushRequest 的方法有多种，可以刷新单个索引、多个索引或全部索引。代码添加在 IndexServiceImpl 类中，如下所示：

```
// 构建刷新索引请求对象
public FlushRequest buildFlushRequest(String index) {
  // 刷新单个索引
  FlushRequest request = new FlushRequest (index);
  // 刷新全部索引
  FlushRequest requestMultiple = new FlushRequest (index, index);
  // 刷新所有索引
  FlushRequest requestAll = new FlushRequest ();
  return request;
}
```

在执行 FlushRequest 时，FlushRequest 还提供了可选参数供用户进行配置，其中最主要的是 IndicesOptions，用于解析不可用索引及展开通配符表达式。代码添加在 IndexServiceImpl 类中的 buildFlushRequest 方法内，如下所示：

```
// 控制解析不可用索引及展开通配符表达式
request.indicesOptions(IndicesOptions.lenientExpandOpen());
```

2．执行 Flush 刷新请求

在构建 FlushRequest 之后，即可执行 Flush 刷新请求。与创建索引请求类似，Flush 刷新请求也有同步和异步两种执行方式。

同步方式

当以同步方式执行 Flush 刷新请求时，客户端会等待 Elasticsearch 服务器返回的查询结果 FlushResponse。在收到 FlushResponse 后，客户端会继续执行相关的逻辑代码。以同步方式执行的代码添加在 IndexServiceImpl 类中，如下所示：

```
// 以同步方式执行 FlushRequest
public void executeFlushRequest(String index) {
  FlushRequest request = buildFlushRequest(index);
  try {
    FlushResponse flushResponse = restClient.indices().flush(request,
        RequestOptions.DEFAULT);

    // 解析 FlushResponse
    processFlushResponse(flushResponse);
  } catch (Exception e) {
    e.printStackTrace();
  } finally {
    // 关闭 Elasticsearch 连接
    closeEs();
  }
}
```

异步方式

当以异步方式执行 Flush 刷新索引请求时,高级客户端不必同步等待请求结果的返回,可以直接向接口调用方返回异步接口执行成功的结果。

为了处理异步返回的响应信息或处理在请求执行过程中引发的异常信息,用户需要指定监听器。以异步方式执行的核心代码如下所示:

```
client.indices().flushAsync(request, RequestOptions.DEFAULT, listener);
```

其中,listener 为监听器。

在异步请求处理后,如果请求执行成功,则调用 ActionListener 类中的 onResponse 方法进行相关逻辑的处理;如果请求执行失败,则调用 ActionListener 类中的 onFailure 方法进行相关逻辑的处理。

以异步方式执行的全部代码如下所示:

```
//以异步方式执行FlushRequest
public void executeFlushRequestAsync(String index) {
  FlushRequest request = buildFlushRequest(index);
  // 构建监听器
  ActionListener<FlushResponse> listener = new ActionListener <FlushResponse>() {
    @Override
    public void onResponse(FlushResponse refreshResponse) {
    }
    @Override
    public void onFailure(Exception e) {
    }
  };
  try {
    restClient.indices().flushAsync(request, RequestOptions.DEFAULT, listener);
  } catch (Exception e) {
    e.printStackTrace();
  } finally {
    // 关闭Elasticsearch连接
    closeEs();
  }
}
```

当然,在异步请求执行过程中可能会出现异常,异常的处理与同步方式执行情况相同。

3. 解析 Flush 刷新请求的响应结果

不论同步方式,还是异步方式,在 Flush 刷新索引请求执行后,客户端均需要对

FlushResponse 进行处理和解析。

代码添加在 IndexServiceImpl 类中，共分为三层，分别是 Controller 层、Service 层和 ServiceImpl 实现层。

在 Controller 层的 IndexController 类中新增如下代码：

```java
//以同步方式执行 FlushRequest
@RequestMapping("/flush/sr")
public String executeFlushRequest(String indexName) {
  // 参数校验
  if (Strings.isNullOrEmpty(indexName)) {
    return "Parameters are wrong!";
  }

  indexService.executeFlushRequest(indexName);
  return "Execute FlushRequest success!";
}
```

在 Service 层的 IndexService 类中新增如下代码：

```java
// 以同步方式执行 FlushRequest
public void executeFlushRequest(String index);
```

在 ServiceImpl 实现层的 IndexServiceImpl 类中新增如下代码：

```java
// 解析 FlushResponse
private void processFlushResponse(FlushResponse flushResponse) {
  // 刷新请求命中的分片总数
  int totalShards = flushResponse.getTotalShards();
  // 刷新成功的分片数
  int successfulShards = flushResponse.getSuccessfulShards();
  // 刷新失败的分片数
  int failedShards = flushResponse.getFailedShards();
  // 在一个或多个分片上刷新失败时的失败列表
  DefaultShardOperationFailedException[] failures = flushResponse.
    getShardFailures();
  log.info("totalShards is " + totalShards + ";successfulShards is " +
    successfulShards
    + ";failedShards is " + failedShards + "; failures is "
    + (failures == null ? 0 : failures.length));
}
```

随后编译工程，在工程根目录下输入如下命令：

```
mvn clean package
```

通过如下命令启动工程服务：

```
java -jar ./target/esdemo-0.0.1-SNAPSHOT.jar
```

在工程服务启动后，在浏览器中调用如下接口查看索引 ultraman8 的 Flush 刷新情况：

```
http://localhost:8080/springboot/es/indexsearch/flush/sr?indexName=ultraman8
```

请求执行后，如果在服务器中输出如下所示内容，则表明 Flush 刷新成功：

```
2019-09-06 17:10:01.470  INFO 85664 --- [nio-8080-exec-1] c.n.e.service.impl.IndexServiceImpl      : totalShards is 12;successfulShards is 6;failedShards is 0; failures is 0
```

8.12 同步 Flush 刷新

Elasticsearch 不仅提供了 Flush 刷新方式，还提供了同步 Flush 刷新方式。与普通 Flush 刷新方式一样，同步 Flush 刷新也可以应用于单个或多个索引，甚至可以应用于全部索引。

什么是同步 Flush 刷新呢？

Elasticsearch 会跟踪每个分片的索引活动，在 5 分钟内未收到任何索引操作的分片会自动标记为非活动状态，这样 Elasticsearch 就可以减少分片资源。

1. 构建同步 Flush 刷新请求

在执行同步 Flush 刷新请求前，需要构建同步 Flush 刷新请求，即 SyncedFlushRequest。与 FlushRequest 类似，构建 SyncedFlushRequest 的方法有多种，可以刷新单个索引、多个索引或全部索引。代码添加在 IndexServiceImpl 类中，如下所示：

```java
// 构建同步 Flush 刷新索引请求
public SyncedFlushRequest buildSyncedFlushRequest(String index) {
    // 刷新单个索引
    SyncedFlushRequest request = new SyncedFlushRequest(index);
    // 刷新多个索引
    SyncedFlushRequest requestMultiple = new SyncedFlushRequest(index, index);
    // 刷新全部索引
    SyncedFlushRequest requestAll = new SyncedFlushRequest();
    return request;
}
```

在执行 SyncedFlushRequest 时，SyncedFlushRequest 还提供了可选参数供用户进行配置，其中主要的是 IndicesOptions，用于解析不可用索引及展开通配符表达式。代码添加在 IndexServiceImpl 类中的 buildSyncedFlushRequest 方法内，如下所示：

```
// 控制解析不可用索引及展开通配符表达式
request.indicesOptions(IndicesOptions.lenientExpandOpen());
```

2. 执行同步 Flush 刷新请求

在 SyncedFlushRequest 构建后，即可执行同步 Flush 刷新请求。与创建索引请求类似，同步 Flush 刷新请求也有同步和异步两种执行方式。

同步方式

当以同步方式执行同步 Flush 刷新请求时，客户端会等待 Elasticsearch 服务器返回的查询结果 SyncedFlushResponse。在收到 SyncedFlushResponse 后，客户端会继续执行相关的逻辑代码。

以同步方式执行的代码添加在 IndexServiceImpl 类中，如下所示：

```
//以同步方式执行 SyncedFlushRequest
public void executeSyncedFlushRequest(String index) {
  SyncedFlushRequest request = buildSyncedFlushRequest(index);
  try {
    SyncedFlushResponse flushResponse = restClient.indices().flushSynced
        (request, RequestOptions.DEFAULT);
    // 解析 SyncedFlushResponse
    processSyncedFlushResponse(flushResponse);
  } catch (Exception e) {
    e.printStackTrace();
  } finally {
    // 关闭 Elasticsearch 连接
    closeEs();
  }
}
```

异步方式

当以异步方式执行同步 Flush 刷新请求时，高级客户端不必同步等待请求结果的返回，可以直接向接口调用方返回异步接口执行成功的结果。

为了处理异步返回的响应信息或处理在请求执行过程中引发的异常信息，用户需要指定监听器。以异步方式执行的核心代码如下所示：

```
client.indices().flushSyncedAsync(request, RequestOptions.DEFAULT, listener);
```

其中，listener 为监听器。

在异步请求处理后，如果请求执行成功，则调用 ActionListener 类中的 onResponse 方法进行相关逻辑的处理；如果请求执行失败，则调用 ActionListener 类中的 onFailure 方法进行相关

逻辑的处理。

以异步方式执行的全部代码如下所示：

```java
// 异步方式执行 SyncedFlushRequest
public void executeSyncedFlushRequestAsync(String index) {
  SyncedFlushRequest request = buildSyncedFlushRequest(index);
  // 构建监听器
  ActionListener<SyncedFlushResponse> listener = new ActionListener
      <SyncedFlushResponse>() {
    @Override
    public void onResponse(SyncedFlushResponse refreshResponse) {
    }
    @Override
    public void onFailure(Exception e) {
    }
  };
  try {
    restClient.indices().flushSyncedAsync(request, RequestOptions.DEFAULT,
        listener);
  } catch (Exception e) {
    e.printStackTrace();
  } finally {
    // 关闭 Elasticsearch 连接
    closeEs();
  }
}
```

当然，在异步请求执行过程中可能会出现异常，异常的处理与同步方式执行情况相同。

3. 解析同步 Flush 刷新请求的响应结果

不论同步方式，还是异步方式，在同步 Flush 刷新索引请求执行后，客户端均需要对 SyncedFlushResponse 进行处理和解析。

代码添加在 IndexServiceImpl 类中，共分为三层，分别是 Controller 层、Service 层和 ServiceImpl 实现层。

在 Controller 层的 IndexController 类中新增如下代码：

```java
// 以同步方式执行 SyncedFlushRequest
@RequestMapping("/syncedflush/sr")
public String executeSyncedFlushRequest(String indexName) {
  // 参数校验
  if (Strings.isNullOrEmpty(indexName)) {
    return "Parameters are wrong!";
```

```
    }
    indexService.executeSyncedFlushRequest(indexName);
    return "Execute SyncedFlushRequest success!";
}
```

在 Service 层的 IndexService 类中新增如下代码：

```
// 以同步方式执行 SyncedFlushRequest
public void executeSyncedFlushRequest(String index);
```

在 ServiceImpl 实现层的 IndexServiceImpl 类中新增如下代码：

```
// 解析 SyncedFlushResponse
private void processSyncedFlushResponse(SyncedFlushResponse flushResponse)
{
    // 刷新请求命中的分片总数
    int totalShards = flushResponse.totalShards();
    // 刷新成功的分片数
    int successfulShards = flushResponse.successfulShards();
    // 刷新失败的分片数
    int failedShards = flushResponse.failedShards();
    log.info("totalShards is " + totalShards + ";successfulShards is " +
        successfulShards
        + ";failedShards is " + failedShards);
}
```

随后编译工程，在工程根目录下输入如下命令：

```
mvn clean package
```

通过如下命令启动工程服务：

```
java -jar ./target/esdemo-0.0.1-SNAPSHOT.jar
```

在工程服务启动后，在浏览器中调用如下接口查看索引 ultraman8 的同步 Flush 刷新情况：

```
http://localhost:8080/springboot/es/indexsearch/syncedflush/sr?indexName=ultraman8
```

请求执行后，在服务器控制台输出如下所示内容，表明同步 Flush 刷新成功。

```
2019-09-06 17:35:12.267  INFO 73596 --- [nio-8080-exec-9] c.n.e.service.impl.IndexServiceImpl      : totalShards is 12;successfulShards is 6;failedShards is 0
```

8.13 清除索引缓存

清除索引缓存接口允许用户清除与单个或多个与索引相关联的所有缓存和特定缓存。

1. 构建清除索引缓存请求

在执行清除索引缓存请求前，需要构建清除索引缓存请求，即 ClearIndicesCacheRequest。与 FlushRequest 类似，构建 ClearIndicesCacheRequest 的方法有多种，可以消除单个索引、多个索引或全部索引。代码添加在 IndexServiceImpl 类中，如下所示：

```java
// 构建清除索引缓存请求
public ClearIndicesCacheRequest buildClearIndicesCacheRequest(String index)
{
  // 清除单个索引
  ClearIndicesCacheRequest request = new ClearIndicesCacheRequest(index);
  // 清除多个索引
  ClearIndicesCacheRequest requestMultiple = new ClearIndicesCacheRequest
  (index, index);
  // 清除全部索引
  ClearIndicesCacheRequest requestAll = new ClearIndicesCacheRequest();
  return request;
}
```

在构建 ClearIndicesCacheRequest 时可以配置可选参数，其中最主要的是 IndicesOptions，用于解析不可用索引及展开通配符表达式。代码添加在 IndexServiceImpl 类中的 buildClearIndicesCacheRequest 方法内，如下所示：

```java
// 用于解析不可用索引及展开通配符表达式
request.indicesOptions(IndicesOptions.lenientExpandOpen());
```

在默认情况下，清除索引缓存 API 会清除所有缓存，但是用户可以通过设置 query、fieldData 或者 request 来显式清除特定高速缓存。代码添加在 IndexServiceImpl 类中的 buildClearIndicesCacheRequest 方法内，如下所示：

```java
// 将查询标志设置为 true
  request.queryCache(true);
  // 将 FieldData 标志设置为 true
  request.fieldDataCache(true);
  // 将请求标志设置为 true
  request.requestCache(true);
  // 设置字段参数
  request.fields("field1", "field2", "field3");
```

2. 执行清除索引缓存请求

在 ClearIndicesCacheRequest 构建后，即可执行清除索引缓存请求。与文档索引请求类似，清除索引缓存请求也有同步和异步两种执行方式。

同步方式

当以同步方式执行清除索引缓存请求时，客户端会等待 Elasticsearch 服务器返回的查询结果 ClearIndicesCacheResponse。在收到 ClearIndicesCacheResponse 后，客户端会继续执行相关的逻辑代码。以同步方式执行的代码添加在 IndexServiceImpl 类中，如下所示：

```
//以同步方式执行 ClearIndicesCacheRequest
public void executeClearIndicesCacheRequest(String index) {
    ClearIndicesCacheRequest request = buildClearIndicesCacheRequest(index);
    try {
        ClearIndicesCacheResponse clearCacheResponse = restClient.indices().
        clearCache(request, RequestOptions.DEFAULT);
        // 解析 ClearIndicesCacheRequest
        processClearIndicesCacheRequest (clearCacheResponse);
    } catch (Exception e) {
        e.printStackTrace();
    } finally {
        // 关闭 Elasticsearch 连接
        closeEs();
    }
}
```

异步方式

当以异步方式执行清除索引缓存请求时，高级客户端不必同步等待请求结果的返回，可以直接向接口调用方返回异步接口执行成功的结果。

为了处理异步返回的响应信息或处理在请求执行过程中引发的异常信息，用户需要指定监听器。以异步方式执行的核心代码如下所示：

```
client.indices().clearCacheAsync(request, RequestOptions.DEFAULT, listener);
```

其中，listener 为监听器。

在异步请求处理后，如果请求执行成功，则调用 ActionListener 类中的 onResponse 方法进行相关逻辑的处理；如果请求执行失败，则调用 ActionListener 类中的 onFailure 方法进行相关逻辑的处理。

以异步方式执行的全部代码如下所示：

```
// 以异步方式执行 ClearIndicesCacheRequest
```

```java
public void executeClearIndicesCacheRequestAsync(String index) {
    ClearIndicesCacheRequest request = buildClearIndicesCacheRequest(index);
    // 构建监听器
    ActionListener<ClearIndicesCacheResponse> listener =
        new ActionListener<ClearIndicesCacheResponse>() {
            @Override
            public void onResponse(ClearIndicesCacheResponse refreshResponse) {
            }
            @Override
            public void onFailure(Exception e) {
            }
        };
    try {
        restClient.indices().clearCacheAsync(request, RequestOptions.DEFAULT,
            listener);
    } catch (Exception e) {
        e.printStackTrace();
    } finally {
        // 关闭Elasticsearch连接
        closeEs();
    }
}
```

当然，在异步请求执行过程中可能会出现异常，异常的处理与同步方式执行情况相同。

3. 解析清除索引缓存请求的响应结果

不论同步方式，还是异步方式，在清除索引缓存请求执行后，客户端均需要对 ClearIndicesCacheResponse 进行处理和解析。

代码添加在 IndexServiceImpl 类中，共分为三层，分别是 Controller 层、Service 层和 ServiceImpl 实现层。

在 Controller 层的 IndexController 类中新增如下代码：

```java
// 以同步方式执行ClearIndicesCacheRequest
@RequestMapping("/clearcache/sr")
public String executeClearIndicesCacheRequest(String indexName) {
    // 参数校验
    if (Strings.isNullOrEmpty(indexName)) {
        return "Parameters are wrong!";
    }
    indexService.executeClearIndicesCacheRequest(indexName);
    return "Execute ClearIndicesCacheRequest success!";
}
```

在 Service 层的 IndexService 类中新增如下代码：

```
// 以同步方式执行 ClearIndicesCacheRequest
public void executeClearIndicesCacheRequest(String index);
```

在 ServiceImpl 实现层的 IndexServiceImpl 类中新增如下代码：

```
// 解析 ClearIndicesCacheRequest
private void processClearIndicesCacheRequest(ClearIndicesCacheResponse
    clearCacheResponse) {
  // 清除索引缓存请求命中的分片总数
  int totalShards = clearCacheResponse.getTotalShards();
  // 清除成功的分片数
  int successfulShards = clearCacheResponse.getSuccessfulShards();
  // 清除失败的分片数
  int failedShards = clearCacheResponse.getFailedShards();
  log.info("totalShards is " + totalShards + ";successfulShards is " +
      successfulShards
    + ";failedShards is " + failedShards);
}
```

随后编译工程，在工程根目录下输入如下命令：

```
mvn clean package
```

通过如下命令启动工程服务：

```
java -jar ./target/esdemo-0.0.1-SNAPSHOT.jar
```

在工程服务启动后，在浏览器中调用如下接口查看索引 ultraman8 的清除缓存后的情况：

```
http://localhost:8080/springboot/es/indexsearch/clearcache/sr?indexName=ultraman8
```

请求执行后，如果在服务器中输出如下所示内容，则表明清除成功：

```
2019-09-06 18:12:25.020  INFO 84892 --- [nio-8080-exec-1] c.n.e.service.impl.IndexServiceImpl      : totalShards is 12;successfulShards is 6;failedShards is 0
```

8.14　强制合并索引

　　Elasticsearch 提供了强制合并索引接口，当用户需要合并单个、多个或全部索引时，该接口会合并依赖于 Lucene 索引在每个分片中保存的分段数。强制合并操作通过合并分段来减少分段数量。如果 HTTP 连接丢失，则请求将在后台继续执行，并且任何新的请求都会被阻塞，直到前面的强制合并完成。

1. 构建强制合并索引请求

在执行强制合并索引请求前，需要构建强制合并索引请求，即 ForceMergeRequest。构建 ForceMergeRequest 的方法有多种，可以强制合并单个索引、多个索引或全部索引。代码添加在 IndexServiceImpl 类中，如下所示：

```
// 构建 ForceMergeRequest
 public ForceMergeRequest buildForceMergeRequest(String index) {
    // 强制合并单个索引
    ForceMergeRequest request = new ForceMergeRequest(index);
    // 强制合并多个索引
    ForceMergeRequest requestMultiple = new ForceMergeRequest(index + "1",
        index + "2");
    // 强制合并全部索引
    ForceMergeRequest requestAll = new ForceMergeRequest();
    return request;
 }
```

在构建 ForceMergeRequest 时，用户可以配置可选参数。可选参数有 flush 标识、唯一删除标识、合并的段数和 IndicesOptions。其中，IndicesOptions 用于解析不可用索引及展开通配符表达式。代码添加在 IndexServiceImpl 类中的 buildForceMergeRequest 方法内，如下所示：

```
// 配置可选参数
    // 设置 IndicesOptions，用于解析不可用索引及展开通配符表达式
    request.indicesOptions(IndicesOptions.lenientExpandOpen());
    // 设置 max_num_segments，以控制合并后的段数
    request.maxNumSegments(1);
    // 将唯一删除标识设置为 true
    request.onlyExpungeDeletes(true);
    // 将 flush 标识设置为 true
    request.flush(true);
```

2. 执行强制合并索引请求

在 ForceMergeRequest 构建后，即可执行强制合并索引请求。与创建索引请求类似，强制合并索引请求也有同步和异步两种执行方式。

同步方式

当以同步方式执行强制合并索引请求时，客户端会等待 Elasticsearch 服务器返回的查询结果 ForceMergeResponse。在收到 ForceMergeResponse 后，客户端会继续执行相关的逻辑代码。

以同步方式执行的代码添加在 IndexServiceImpl 类中，如下所示：

```
// 以同步方式执行 ForceMergeRequest
public void executeForceMergeRequest(String index) {
  ForceMergeRequest request = buildForceMergeRequest(index);
  try {
    ForceMergeResponse forceMergeResponse =
        restClient.indices().forcemerge(request, RequestOptions.DEFAULT);
  } catch (Exception e) {
    e.printStackTrace();
  } finally {
    // 关闭 Elasticsearch 连接
    closeEs();
  }
}
```

异步方式

当以异步方式执行强制合并索引请求时，Java 高级客户端不必同步等待请求结果的返回，可以直接向接口调用方返回异步接口执行成功的结果。

为了处理异步返回的响应信息或处理在请求执行过程中引发的异常信息，用户需要指定监听器。以异步方式执行的核心代码如下所示：

```
client.indices().forcemergeAsync(request, RequestOptions.DEFAULT, listener);
```

其中，listener 为监听器。

在异步请求处理后，如果请求执行成功，则调用 ActionListener 类中的 onResponse 方法进行相关逻辑的处理；如果请求执行失败，则调用 ActionListener 类中的 onFailure 方法进行相关逻辑的处理。

以异步方式执行的全部代码添加在 IndexServiceImpl 类中，如下所示：

```
// 以异步方式执行 ForceMergeRequest
public void executeForceMergeRequestAsync(String index) {
  ForceMergeRequest request = buildForceMergeRequest(index);
  // 构建监听器
  ActionListener<ForceMergeResponse> listener = new ActionListener
      <ForceMergeResponse>() {
    @Override
    public void onResponse(ForceMergeResponse forceMergeResponse) {
    }
    @Override
    public void onFailure(Exception e) {
    }
```

```
    };
    try {
        restClient.indices().forcemergeAsync(request, RequestOptions.DEFAULT,
            listener);
    } catch (Exception e) {
        e.printStackTrace();
    } finally {
        // 关闭 Elasticsearch 连接
        closeEs();
    }
}
```

当然,在异步请求执行过程中可能会出现异常,异常的处理与同步方式执行情况相同。

3. 解析强制合并索引请求的响应结果

不论同步方式,还是异步方式,在强制合并索引请求执行后,客户端均需要对请求的响应结果 ForceMergeResponse 进行处理和解析。

代码共分为三层,分别是 Controller 层、Service 层和 ServiceImpl 实现层。

在 Controller 层的 IndexController 类中新增如下代码:

```
// 以同步方式执行 ForceMergeRequest
@RequestMapping("/merge/sr")
public String executeForceMergeRequest(String indexName) {
    // 参数校验
    if (Strings.isNullOrEmpty(indexName)) {
        return "Parameters are wrong!";
    }
    indexService.executeForceMergeRequest(indexName);
    return "Execute ForceMergeRequest success!";
}
```

在 Service 层的 IndexService 类中新增如下代码:

```
// 以同步方式执行 ForceMergeRequest
public void executeForceMergeRequest(String index);
```

在 ServiceImpl 实现层的 IndexServiceImpl 类中新增如下代码:

```
// 以同步方式执行 ForceMergeRequest
public void executeForceMergeRequest(String index) {
    ForceMergeRequest request = buildForceMergeRequest(index);
    try {
        ForceMergeResponse forceMergeResponse =
            restClient.indices().forcemerge(request, RequestOptions.DEFAULT);
```

```java
    // 解析ForceMergeResponse
    processForceMergeResponse(forceMergeResponse);
} catch (Exception e) {
    e.printStackTrace();
} finally {
    // 关闭Elasticsearch连接
    closeEs();
}
}
// 解析ForceMergeResponse
private void processForceMergeResponse(ForceMergeResponse forceMergeResponse)
{
    // 强制合并索引请求命中的分片总数
    int totalShards = forceMergeResponse.getTotalShards();
    // 强制合并成功的分片数
    int successfulShards = forceMergeResponse.getSuccessfulShards();
    // 强制合并失败的分片数
    int failedShards = forceMergeResponse.getFailedShards();
    // 在一个或多个分片上强制合并失败的失败列表
    DefaultShardOperationFailedException[] failures = forceMergeResponse.
        getShardFailures();
    log.info("totalShards is " + totalShards + ";successfulShards is " +
        successfulShards
        + ";failedShards is " + failedShards + ";failures size is "
        + (failures == null ? 0 : failures.length));
}
```

随后编译工程，在工程根目录下输入如下命令：

```
mvn clean package
```

通过如下命令启动工程服务：

```
java -jar ./target/esdemo-0.0.1-SNAPSHOT.jar
```

在工程服务启动后，在浏览器中调用如下接口查看索引 ultraman8 的合并情况：

```
http://localhost:8080/springboot/es/indexsearch/merge/sr?indexName=ultraman8
```

请求执行后，如果在服务器中输出如下所示内容，则表明索引合并成功：

```
2019-09-07 13:59:47.318 INFO 26636 --- [nio-8080-exec-1] c.n.e.service.impl.IndexServiceImpl      : totalShards is 12;successfulShards is 6;failedShards is 0;failures size is 0
```

8.15 滚动索引

当索引较大或者数据很老旧时，可以使用 Elasticsearch 提供的滚动索引 API 将别名滚动到新的索引。

1. 构建滚动索引请求

在执行滚动索引请求前，需要构建滚动索引请求，即 RolloverRequest。RolloverRequest 的必选参数是两个字符串参数，以及一个或多个条件参数。其中，字符串参数为索引别名和新索引名称，条件参数用于确定何时回滚索引。新索引名称对应的索引会在条件参数满足时创建，并将别名指向新索引。

代码添加在 IndexServiceImpl 类中，如下所示：

```
// 构建RolloverRequest
public RolloverRequest buildRolloverRequest(String index) {
    // 指向要滚动的索引别名（第一个参数），以及执行滚动操作时的新索引名称。new index参数
    // 是可选的，可以设置为空
    RolloverRequest request = new RolloverRequest(index, index + "-2");
    // 指数年龄
    request.addMaxIndexAgeCondition(new TimeValue(7, TimeUnit.DAYS));
    // 索引中的文档数
    request.addMaxIndexDocsCondition(1000);
    // 索引的大小
    request.addMaxIndexSizeCondition(new ByteSizeValue(5, ByteSizeUnit.GB));
    return request;
}
```

在构建 RolloverRequest 时，用户还可以配置可选参数。可选参数主要有是否执行滚动、所有节点确认索引打开的超时时间、从节点连接到主节点的超时时间、请求返回前等待的活跃分片数量，以及新索引相关的设置。代码添加在 IndexServiceImpl 类中的 buildRolloverRequest 方法内，如下所示：

```
// 配置可选参数
    // 是否执行滚动（默认为true）
request.dryRun(true);
    // 所有节点确认索引打开的超时时间
request.setTimeout(TimeValue.timeValueMinutes(2));
    // 从节点连接到主节点的超时时间
request.setMasterTimeout(TimeValue.timeValueMinutes(1));
    // 请求返回前等待的活跃分片数量
request.getCreateIndexRequest().waitForActiveShards(ActiveShardCount.
```

```
    from(2));
// 请求返回前等待的活跃分片数量,重置为默认值
request.getCreateIndexRequest().waitForActiveShards(ActiveShardCount.
DEFAULT);
// 添加应用于新索引的设置,其中包括要为其创建的分片数
request.getCreateIndexRequest().settings(Settings.builder().put("index.
number_of_shards", 4));
// 添加与新索引关联的映射
String mappings = "{\"properties\":{\"field\":{\"type\":\"content\"}}}";
request.getCreateIndexRequest().mapping(mappings, XContentType.JSON);
// 添加与新索引关联的别名
request.getCreateIndexRequest().alias(new Alias(index + "-2_alias"));
```

2. 执行滚动索引请求

在 RolloverRequest 构建后,即可执行滚动索引请求。与创建索引请求类似,滚动索引请求也有同步和异步两种执行方式。

同步方式

当以同步方式执行滚动索引请求时,客户端会等待 Elasticsearch 服务器返回的查询结果 RolloverResponse。在收到 RolloverResponse 后,客户端会继续执行相关的逻辑代码。以同步方式执行的代码添加在 IndexServiceImpl 类中,如下所示:

```
// 以同步方式执行 RolloverRequest
public void executeRolloverRequest(String index) {
  RolloverRequest request = buildRolloverRequest(index);
  try {
    RolloverResponse rolloverResponse =
        restClient.indices().rollover(request, RequestOptions.DEFAULT);
    // 解析 RolloverResponse
    processRolloverResponse(rolloverResponse);
  } catch (Exception e) {
    e.printStackTrace();
  } finally {
    // 关闭 Elasticsearch 连接
    closeEs();
  }
}
```

异步方式

当以异步方式执行滚动索引请求时,高级客户端不必同步等待请求结果的返回,可以直接向接口调用方返回异步接口执行成功的结果。

为了处理异步返回的响应信息或处理在请求执行过程中引发的异常信息，用户需要指定监听器。以异步方式执行的核心代码如下所示：

```
client.indices().rolloverAsync(request, RequestOptions.DEFAULT, listener);
```

其中，listener 为监听器。

在异步请求处理后，如果请求执行成功，则调用 ActionListener 类中的 onResponse 方法进行相关逻辑的处理；如果请求执行失败，则调用 ActionListener 类中的 onFailure 方法进行相关逻辑的处理。

以异步方式执行的全部代码如下所示：

```java
// 以异步方式执行 RolloverRequest
public void executeRolloverRequestAsync(String index) {
    RolloverRequest request = buildRolloverRequest(index);
    // 构建 RolloverRequest
    ActionListener<RolloverResponse> listener = new ActionListener
        <RolloverResponse>() {
        @Override
        public void onResponse(RolloverResponse rolloverResponse) {
        }
        @Override
        public void onFailure(Exception e) {
        }
    };

    try {
        restClient.indices().rolloverAsync(request, RequestOptions.DEFAULT,
            listener);
    } catch (Exception e) {
        e.printStackTrace();
    } finally {
        // 关闭 Elastcsearch 连接
        closeEs();
    }
}
```

当然，在异步请求执行过程中可能会出现异常，异常的处理与同步方式执行情况相同。

3. 解析滚动索引请求的响应结果

不论同步方式，还是异步方式，在滚动索引请求执行后，客户端均需要对请求的响应结果 RolloverResponse 进行处理和解析。

代码共分为三层，分别是 Controller 层、Service 层和 ServiceImpl 实现层。

在 Controller 层的 IndexController 类中新增如下代码：

```
// 以同步方式执行RolloverRequest
@RequestMapping("/rollover/sr")
public String executeRolloverRequest(String indexName) {
  // 参数校验
  if (Strings.isNullOrEmpty(indexName)) {
    return "Parameters are wrong!";
  }
  indexService.executeRolloverRequest(indexName);
  return "Execute RolloverRequest success!";
}
```

在 Service 层的 IndexService 类中新增如下代码：

```
// 以同步方式执行RolloverRequest
public void executeRolloverRequest(String index);
```

在 ServiceImpl 实现层的 IndexServiceImpl 中新增如下代码：

```
// 解析RolloverResponse
private void processRolloverResponse(RolloverResponse rolloverResponse) {
  // 所有节点是否已确认请求
  boolean acknowledged = rolloverResponse.isAcknowledged();
  // 是否在超时前为索引中的每个分片启动了所需数量的分片副本
  boolean shardsAcked = rolloverResponse.isShardsAcknowledged();
  // 旧索引名称，最终被翻滚
  String oldIndex = rolloverResponse.getOldIndex();
  // 新索引名称
  String newIndex = rolloverResponse.getNewIndex();
  // 索引是否已回滚
  boolean isRolledOver = rolloverResponse.isRolledOver();
  boolean isDryRun = rolloverResponse.isDryRun();
  // 不同的条件，是否匹配
  Map<String, Boolean> conditionStatus = rolloverResponse. getConditionStatus
                                     ();
  log.info("acknowledged is " + acknowledged + ";shardsAcked is " +
      shardsAcked + ";oldIndex is "
    + oldIndex + ";newIndex is " + newIndex + ";isRolledOver is " +
      isRolledOver
    + ";isDryRun is " + isDryRun + ";conditionStatus size is "
    + (conditionStatus == null ? 0 : conditionStatus.size()));
}
```

随后编译工程，在工程根目录下输入如下命令：

```
mvn clean package
```

通过如下命令启动工程服务：

```
java -jar ./target/esdemo-0.0.1-SNAPSHOT.jar
```

在工程服务启动后，在浏览器中调用如下接口查看滚动索引的执行情况：

```
http://localhost:8080/springboot/es/indexsearch/rollover/sr?indexName=ultraman7_alias
```

请求执行后，如果在服务器中输出如下所示内容，则表明新索引已经构建成功：

```
2019-09-07 14:31:36.421  INFO 70892 --- [nio-8080-exec-1] c.n.e.service.impl.IndexServiceImpl      : acknowledged is false;shardsAcked is false;oldIndex is ultraman7;newIndex is ultraman7_alias-2;isRolledOver is false;isDryRun is true;conditionStatus size is 3
```

8.16 索引别名

在 Elasticsearch 中，有专门的接口为索引别名进行命名，即索引别名接口。当通过索引别名调用索引时，所有的接口都将自动转换为索引实际名称。

1. 构建索引别名请求

在执行索引别名请求前，需要构建索引别名请求，即 IndicatesAliasesRequest。IndicatesAliasesRequest 至少含有一个别名命名操作。一个别名命名操作必须包含当前索引名称和将要命名的别名，对应的代码添加在 IndexServiceImpl 类中，如下所示：

```
// 构建 IndicatesAliasesRequest
public IndicesAliasesRequest buildIndicatesAliasesRequest(String index,
    String indexAlias) {
  // 创建 IndicatesAliasesRequest
  IndicesAliasesRequest request = new IndicesAliasesRequest();
  // 创建别名操作，将索引的别名设为 indexAlias
  AliasActions aliasAction =new AliasActions(AliasActions.Type.ADD).index
      (index).alias(indexAlias);
  // 将别名操作添加到请求中
  request.addAliasAction(aliasAction);
  return request;
}
```

其中，AliasActions 支持的操作类型有新增别名（AliasActions.Type.ADD）、删除别名

（AliasActions.Type.REMOVE）和删除索引（AliasActions.Type. REMOVE_INDEX）。

需要指出的是，索引别名不能重复，也不能和索引名称重复；用户可以增加、删除别名，但不能修改别名。

AliasActions 还支持配置可选筛选器（filter）和可选路由（routing），对应的代码添加在 buildIndicatesAliasesRequest 方法内，如下所示：

```
// 创建别名操作，将索引的别名设为 indexAlias ADD
AliasActions aliasAction = new AliasActions(AliasActions.Type.ADD).index
    (index).alias(indexAlias).filter("{\"term\":{\"year\":2019}}").routing
    ("niudong");
```

在构建 IndicatesAliasesRequest 时，用户还可以配置可选参数，主要是所有节点确认索引操作的超时时间、从节点连接到主节点的超时时间。对应的代码添加在 IndexServiceImpl 类中的 buildIndicatesAliasesRequest 方法内，如下所示：

```
// 可选参数配置
    // 所有节点确认索引操作的超时时间
    request.timeout(TimeValue.timeValueMinutes(2));
    request.timeout("2m");
    // 从节点连接到主节点的超时时间
    request.masterNodeTimeout(TimeValue.timeValueMinutes(1));
    request.masterNodeTimeout("1m");
```

2．执行索引别名请求

在 IndicatesAliasesRequest 构建后，即可执行索引别名请求。与创建索引请求类似，索引别名请求也有同步和异步两种执行方式。

同步方式

当以同步方式执行索引别名请求时，客户端会等待 Elasticsearch 服务器返回的查询结果 IndicatesAliasesResponse。在收到 IndicatesAliasesResponse 后，客户端会继续执行相关的逻辑代码。以同步方式执行的代码添加在 IndexServiceImpl 类中，如下所示：

```
// 以同步方式执行 IndicatesAliasesRequest
  public void executeIndicatesAliasesRequest(String index, String indexAlias)
    {
    IndicesAliasesRequest request = buildIndicatesAliasesRequest(index,
        indexAlias);
    try {
      AcknowledgedResponse indicesAliasesResponse =
          restClient.indices().updateAliases(request, RequestOptions.DEFAULT);
      // 解析 AcknowledgedResponse
```

```
        processAcknowledgedResponse(indicesAliasesResponse);
    } catch (Exception e) {
        e.printStackTrace();
    } finally {
        // 关闭Elasticsearch连接
        closeEs();
    }
}
```

异步方式

当以异步方式执行索引别名请求时，高级客户端不必同步等待请求结果的返回，可以直接向接口调用方返回异步接口执行成功的结果。

为了处理异步返回的响应信息或处理在请求执行过程中引发的异常信息，用户需要指定监听器。以异步方式执行的核心代码如下所示：

```
client.indices().updateAliasesAsync(request, RequestOptions.DEFAULT, listener);
```

其中，listener 为监听器。

在异步请求处理后，如果请求执行成功，则调用 ActionListener 类中的 onResponse 方法进行相关逻辑的处理；如果请求执行失败，则调用 ActionListener 类中的 onFailure 方法进行相关逻辑的处理。

以异步方式执行的全部代码如下所示：

```
//以异步方式执行IndicatesAliasesRequest
public void executeIndicatesAliasesRequestAsync(String index, String
        indexAlias) {
    IndicesAliasesRequest request = buildIndicatesAliasesRequest(index,
        indexAlias);
    // 构建监听器
    ActionListener<AcknowledgedResponse> listener = new ActionListener
            <AcknowledgedResponse>() {
        @Override
        public void onResponse(AcknowledgedResponse indicesAliasesResponse) {
        }
        @Override
        public void onFailure(Exception e) {
        }
    };

    try {
```

```
        restClient.indices().updateAliasesAsync(request, RequestOptions.DEFAULT,
            listener);
    } catch (Exception e) {
        e.printStackTrace();
    } finally {
        // 关闭Elasticsearch连接
        closeEs();
    }
}
```

当然，在异步请求执行过程中可能会出现异常，异常的处理与同步方式执行情况相同。

3. 解析索引别名请求的响应结果

不论同步方式，还是异步方式，当索引别名请求执行后，客户端均需要对请求的响应结果 IndicatesAliaseResponse 进行处理和解析。

代码共分为三层，分别是 Controller 层、Service 层和 ServiceImpl 实现层。

在 Controller 层的 IndexController 类中新增如下代码：

```
// 以同步方式执行IndicatesAliasesRequest
@RequestMapping("/createAlias/sr")
public String executeIndicatesAliasesRequest(String indexName, String
    indexAliasName) {
    // 参数校验
    if (Strings.isNullOrEmpty(indexName) || Strings.isNullOrEmpty
        (indexAliasName)) {
        return "Parameters are wrong!";
    }
    indexService.executeIndicatesAliasesRequest(indexName, indexAliasName);
    return "Execute IndicatesAliasesRequest success!";
}
```

在 Service 层的 IndexService 类中新增如下代码：

```
// 以同步方式执行IndicatesAliasesRequest
 public void executeIndicatesAliasesRequest(String index, String indexAlias);
```

在 ServiceImpl 实现层的 IndexServiceImpl 类中不用新增代码，可以复用前文提及的 processAcknowledgedResponse 方法。

随后编译工程，在工程根目录下输入如下命令：

```
mvn clean package
```

通过如下命令启动工程服务：

```
java -jar ./target/esdemo-0.0.1-SNAPSHOT.jar
```

在工程服务启动后,在浏览器中调用如下接口查看索引 ultraman8 的别名命名情况:

```
http://localhost:8080/springboot/es/indexsearch/createAlias/sr?indexName=ultraman8&indexAliasName=ultraman8_alias
```

请求执行后,如果在服务器中输出如下所示内容,则表明索引 ultraman8 的别名 ultraman8_alias 命名成功。

```
2019-09-07 15:01:47.053  INFO 46824 --- [nio-8080-exec-1] c.n.e.service.impl.IndexServiceImpl      : acknowledged is true
```

8.17　索引别名存在校验

在 Elasticsearch 中,不仅提供了索引别名命名接口,还提供了查看索引是否有别名的接口,即索引别名存在校验接口。

1. 构建索引别名存在校验请求

在执行索引别名存在校验请求前,需要构建索引别名存在校验请求,即 GetAliasesRequest。GetAliasesRequest 的必要参数为待校验的索引别名。相应的代码添加在 IndexServiceImpl 类中,如下所示:

```java
// 构建 GetAliasesRequest
public GetAliasesRequest buildGetAliasesRequest(String indexAlias) {
    GetAliasesRequest request = new GetAliasesRequest();
    GetAliasesRequest requestWithAlias = new GetAliasesRequest(indexAlias);
    GetAliasesRequest requestWithAliases =
        new GetAliasesRequest(new String[] {indexAlias, indexAlias});
    return request;
}
```

在构建 GetAliasesRequest 时,GetAliasesRequest 的可选参数主要有带校验存在性的别名、与别名关联的一个或多个索引名称、是否本地查找和 IndicesOptions。IndicesOptions 用于控制解析不可用索引及展开通配符表达式等。相应的代码添加在 IndexServiceImpl 类的 buildGetAliasesRequest 方法内,如下所示:

```java
// 配置可选参数
    // 带校验存在性的别名
    request.aliases(indexAlias);
```

```
    // 与别名关联的一个或多个索引
    request.indices(index);
    // 设置 IndicesOptions
    request.indicesOptions(IndicesOptions.lenientExpandOpen());
    // 本地标志（默认为 false），控制是否需要在本地群集状态或所选主节点持有的群集状态中查
    // 找别名
    request.local(true);
```

2. 执行索引别名存在校验请求

在 GetAliasesRequest 构建后，即可执行索引别名存在校验请求。与索引别名请求类似，索引别名存在校验请求也有同步和异步两种执行方式。

同步方式

当以同步方式执行索引别名存在校验请求时，客户端会等待 Elasticsearch 服务器返回的布尔型查询结果。在收到查询结果后，客户端会继续执行相关逻辑代码。相应的代码添加在 IndexServiceImpl 类中，如下所示：

```
// 以同步方式执行 GetAliasesRequest
public void executeGetAliasesRequest(String indexAlias) {
  GetAliasesRequest request = buildGetAliasesRequest(indexAlias);
  try {
    boolean exists = restClient.indices().existsAlias(request,
        RequestOptions.DEFAULT);
    log.info("indexAlias exists is " + exists);
  } catch (Exception e) {
    e.printStackTrace();
  } finally {
    // 关闭 Elasticsecrch 连接
    closeEs();
  }
}
```

异步方式

当以异步方式执行索引别存在校验请求时，高级客户端不必同步等待请求结果的返回，可以直接向接口调用方返回异步接口执行成功的结果。

为了处理异步返回的响应信息或处理在请求执行过程中引发的异常信息，用户需要指定监听器。以异步方式执行的核心代码如下所示：

```
client.indices().existsAliasAsync(request, RequestOptions.DEFAULT, listener);
```

其中，listener 为监听器。

当异步请求处理后，如果请求执行成功，则调用 ActionListener 类中的 onResponse 方法进行相关逻辑的处理；如果请求执行失败，则调用 ActionListener 类中的 onFailure 方法进行相关逻辑的处理。

以异步方式执行的全部代码添加在 IndexServiceImpl 类中，如下所示：

```
// 以异步方式执行 GetAliasesRequest
public void executeGetAliasesRequestAsync(String indexAlias) {
  GetAliasesRequest request = buildGetAliasesRequest(indexAlias);
  // 构建监听器
  ActionListener<Boolean> listener = new ActionListener<Boolean>() {
    @Override
    public void onResponse(Boolean exists) {
    }
    @Override
    public void onFailure(Exception e) {
    }
  };
  try {
    restClient.indices().existsAliasAsync(request, RequestOptions.DEFAULT,
        listener);
  } catch (Exception e) {
    e.printStackTrace();
  } finally {
    // 关闭 Elasticsearch 连接
    closeEs();
  }
}
```

当然，在异步请求执行过程中可能会出现异常，异常的处理与同步方式执行情况相同。

3. 解析索引别名存在校验请求的响应结果

不论同步方式，还是异步方式，在索引别名存在校验请求执行后，客户端均需要对请求的响应结果进行处理和解析。

代码分为三层，分别是 Controller 层、Service 层和 ServiceImpl 实现层。

在 Controller 层的 IndexController 类中新增如下代码：

```
// 以同步方式执行 GetAliasesRequest
@RequestMapping("/existsAlias/sr")
public String executeGetAliasesRequest(String indexAliasName) {
  // 参数校验
  if (Strings.isNullOrEmpty(indexAliasName)) {
```

```
        return "Parameters are wrong!";
    }
    indexService.executeGetAliasesRequest(indexAliasName);
    return "Execute GetAliasesRequest success!";
}
```

在 Service 层的 IndexService 类中新增如下代码：

```
// 以同步方式执行 GetAliasesRequest
 public void executeGetAliasesRequest(String indexAlias);
```

无须在 ServiceImpl 实现层的 IndexServiceImpl 类中新增代码。

随后编译工程，在工程根目录下输入如下命令：

```
mvn clean package
```

通过如下命令启动工程服务：

```
java -jar ./target/esdemo-0.0.1-SNAPSHOT.jar
```

在工程服务启动后，在浏览器中调用如下接口查看索引别名 ultraman70_alias 和 ultraman7_alias 的存在校验情况：

```
http://localhost:8080/springboot/es/indexsearch/existsAlias/sr?indexAliasName=ultraman70_alias
 http://localhost:8080/springboot/es/indexsearch/existsAlias/sr?indexAliasName=ultraman7_alias
```

请求执行后，在服务器中分别输出如下所示内容，ultraman70_alias 索引不存在，ultraman7_alias 索引存在：

```
2019-09-07 15:23:18.452  INFO 3572 --- [nio-8080-exec-1] c.n.e.service.impl.IndexServiceImpl      : indexAlias exists is false

2019-09-07 15:33:17.791  INFO 25752 --- [nio-8080-exec-1] c.n.e.service.impl.IndexServiceImpl     : indexAlias exists is true
```

8.18 获取索引别名

在 Elasticsearch 中，不仅提供了索引别名命名接口，还提供了获取索引别名接口。

1. 构建获取索引别名请求

在执行获取索引别名请求前，需要构建获取索引别名请求，即 GetAliasesRequest。

GetAliasesRequest 的必选参数为待校验的索引别名，前面已经展示了 GetAliasesRequest 的构建方法，不再赘述。

2. 执行获取索引别名请求

在 GetAliasesRequest 构建后，即可执行获取索引别名请求。与索引别名请求类似，获取索引别名请求也有同步和异步两种执行方式。

同步方式

当以同步方式执行获取索引别名请求时，客户端会等待 Elasticsearch 服务器返回的查询结果 GetAliasesResponse。在收到 GetAliasesResponse 后，客户端会继续执行相关的逻辑代码。相应的代码添加在 IndexServiceImpl 类中，如下所示：

```java
// 以同步方式执行 GetAliasesRequest
public void executeGetAliasesRequestForAliases(String indexAlias) {
  GetAliasesRequest request = buildGetAliasesRequest(indexAlias);
  try {
    GetAliasesResponse response = restClient.indices().getAlias(request,
        RequestOptions.DEFAULT);
    // 解析 GetAliasesResponse
    processGetAliasesResponse(response);
  } catch (Exception e) {
    e.printStackTrace();
  } finally {
    // 关闭 Elasticsearch 的连接
    closeEs();
  }
}
```

异步方式

当以异步方式执行获取索引别名请求时，高级客户端不必同步等待请求结果的返回，可以直接向接口调用方返回异步接口执行成功的结果。

为了处理异步返回的响应信息或处理在请求执行过程中引发的异常信息，用户需要指定监听器。以异步方式执行的核心代码如下所示：

```java
client.indices().getAliasAsync(request, RequestOptions.DEFAULT, listener);
```

其中，listener 为监听器。

在异步请求处理后，如果请求执行成功，则调用 ActionListener 类中的 onResponse 方法进行相关逻辑的处理；如果请求执行失败，则调用 ActionListener 类中的 onFailure 方法进行相关

逻辑的处理。

以异步方式执行的全部代码如下所示：

```java
// 以异步方式执行GetAliasesRequest
public void executeGetAliasesRequestForAliasesAsync(String indexAlias) {
  GetAliasesRequest request = buildGetAliasesRequest(indexAlias);
  // 构建监听器
  ActionListener<GetAliasesResponse> listener = new ActionListener
      <GetAliasesResponse>() {
    @Override
    public void onResponse(GetAliasesResponse exists) {
    }
    @Override
    public void onFailure(Exception e) {
    }
  };

  try {
    restClient.indices().getAliasAsync(request, RequestOptions.DEFAULT,
        listener);
  } catch (Exception e) {
    e.printStackTrace();
  } finally {
    // 关闭Elasticsearch连接
    closeEs();
  }
}
```

当然，在异步请求执行过程中可能会出现异常，异常的处理与同步方式执行情况相同。

3. 解析获取索引别名请求的响应结果

不论同步方式，还是异步方式，在获取索引别名请求执行后，客户端均需要对请求的响应结果 GetAliasesResponse 进行处理和解析。

代码共分为三层，分别是 Controller 层、Service 层和 ServiceImpl 实现层。

在 Controller 层的 IndexController 类中新增如下代码：

```java
// 以同步方式执行GetAliasesRequest
@RequestMapping("/getAlias/sr")
public String executeGetAliasesRequestForAliases(String indexAliasName) {
  // 参数校验
  if (Strings.isNullOrEmpty(indexAliasName)) {
    return "Parameters are wrong!";
```

```
    indexService.executeGetAliasesRequestForAliases(indexAliasName);
    return "Execute GetAliasesRequestForAliases success!";
  }
```

在 Service 层的 IndexService 类中新增如下代码：

```
// 以同步方式执行 GetAliasesRequest
  public void executeGetAliasesRequestForAliases(String indexAlias);
```

在 ServiceImpl 实现层的 IndexServiceImpl 类中新增如下代码：

```
// 解析 GetAliasesResponse
  private void processGetAliasesResponse(GetAliasesResponse response) {
    // 检索索引及其别名的映射
    Map<String, Set<AliasMetaData>> aliases = response.getAliases();
    // 如果为空，则返回
    if (aliases == null || aliases.size() <= 0) {
      return;
    }
    // 遍历 Map
    Set<Entry<String, Set<AliasMetaData>>> set = aliases.entrySet();
    for (Entry<String, Set<AliasMetaData>> entry : set) {
      String key = entry.getKey();
      Set<AliasMetaData> metaSet = entry.getValue();
      if (metaSet == null || metaSet.size() <= 0) {
        return;
      }
      for (AliasMetaData meta : metaSet) {
        String aliaas = meta.alias();
        log.info("key is " + key + ";aliaas is " + aliaas);
      }
    }
  }
```

随后编译工程，在工程根目录下输入如下命令：

```
mvn clean package
```

通过如下命令启动工程服务：

```
java -jar ./target/esdemo-0.0.1-SNAPSHOT.jar
```

在工程服务启动后，在浏览器中调用如下接口获取索引 ultraman7_alias 的别名情况：

```
http://localhost:8080/springboot/es/indexsearch/getAlias/sr?indexAliasName=ultraman7_alias
```

在服务器中输出如下所示内容:

```
2019-09-07 15:44:09.168  INFO 11432 --- [nio-8080-exec-1] c.n.e.service.impl.IndexServiceImpl      : key is ultraman7;al
iaas is ultraman7_alias
```

8.19 索引原理解析

为了提高搜索性能,Elasticsearch 做了很多设计。下面介绍在 Elasticsearch 内部是如何索引文档的。由于第 5 章和第 6 章中曾提到部分文档的索引过程和分片过程,所以本节将从其他维度展开。

8.19.1 近实时搜索的实现

文档被索引动作和搜索该文档动作之间是有延迟的,因此,新的文档需要在几分钟后方可被搜索到,但这依然不够快,其根本原因在于磁盘。

在 Elasticsearch 中,当提交一个新的段到磁盘时需要执行 fsync 操作,以确保段被物理地写入磁盘,即使断电数据也不会丢失。不过,fsync 很消耗资源,因此它不能在每个文档被索引时都触发。

位于 Elasticsearch 和磁盘间的是文件系统缓存。内存索引缓存中的文档被写入新段的过程的资源消耗很低;之后文档会被同步到磁盘,这个操作的资源消耗很高。而一个文件一旦被缓存,它就可以被打开和读取。因此,Elasticsearch 利用了这一特性。

Lucene 允许新段在写入后被打开,以便让段中包括的文档可被搜索,而不用执行一次全量提交。这是一种比提交更轻量的过程,可以经常操作,且不会影响性能。

在 Elasticsearch 中,这种写入打开一个新段的轻量级过程,就叫作 refresh。在默认情况下,每个分片每秒自动刷新一次。这就是认为 Elasticsearch 是近实时搜索,但不是实时搜索的原因,即文档的改动不会立即被搜索,但是会在一秒内可见。

8.19.2 倒排索引的压缩

Elasticsearch 为每个文档中的字段分别建立了一个倒排索引。倒排索引示意图如图 8-20 所示。

随着文档的不断增加,倒排索引中的词条和词条对应的文档 ID 列表会不断增大,从而影响 Elasticsearch 的性能。

Elasticsearch 对词条采用了 Term Dictionary 和 Term Index 的方式来简化词条的存储和查找；同时，Elasticsearch 对词条对应的文档 ID 列表进行了必要的处理。

Elasticsearch 是如何处理这些文档 ID 列表的呢？答案很简单，即通过增量编码压缩，将大数变小数，按字节存储。

为了有效进行压缩，词条对应的文档 ID 列表是有序排列的。所谓增量编码，就是将原来的大数变成小数，仅存储增量值，示例如下。

假如原有的文档 ID 列表为：10 23 34 66 100 178。

增量编码后文档 ID 列表为：10 13 11 32 34 78。

所谓按字节存储，即查看增量编码后的数字可以用几位字节存储，而非全部用 int（4 字节）存储，从而达到压缩和节省内存的目的。当然，为了节省更多的内存，还可以对增量编码后的文档 ID 列表进行分组，再分别计算每一个存储需要的字节数。

8.20 知识点关联

在 Elasticsearch 中，有两种数据刷新操作方式，即 Refresh 和 Flush。这两种刷新操作方式有什么区别呢？

1. Refresh 方式

当我们索引文档时，文档是存储在内存中的，默认 1s 后会进入文件系统缓存。Refresh 操作本质上是对 Lucene Index Reader 调用了 ReOpen 操作，即对此时索引中的数据进行更新，使文档可以被用户搜索到。

不过，此时文档还未存储到磁盘上，如果 Elasticsearch 的服务器宕机了，那么这部分数据就会丢失。如果要对这部分数据进行持久化，则需要调用消耗较大的 Lucene Commit 操作。因此，Elasticsearch 采用可以频繁调用轻量级的 ReOpen 操作来达到近实时搜索的效果。

2. Flush 方式

虽然 Elasticsearch 采用了可以频繁调用轻量级的 ReOpen 操作来达到近实时搜索的效果，但数据终究是要持久化的。

Elasticsearch 在写入文档时，会写一份 translog 日志，基于 translog 日志可以恢复那些丢失的文档，在出现程序故障或磁盘异常时，保障数据的安全。

Flush 可高效地触发 Lucene Commit，同时清空 translog 日志，使数据在 Lucene 层面持久化。

8.21　小结

本章主要介绍了索引 API 的使用，涉及 18 个索引操作相关的接口，包括字段索引分析、索引的增删改查（创建索引、获取索引、删除索引、索引存在验证）、索引的开关（打开索引、关闭索引）、索引的容量控制（缩小索引、拆分索引、强制合并索引）、索引数据刷新（Refresh 刷新、Flush 刷新、同步 Flush 刷新、清除索引缓存、强制合并索引）和索引别名（创建索引别名、索引别名存在校验、获取索引别名）。

第三部分
Elasticsearch生态

第 9 章 Elasticsearch 插件

> 东府买舟船
> 西府买器械

在第二部分中，主要介绍了 Elasticsearch 实战，本部分主要介绍 Elasticsearch 的生态圈。本章介绍 Elasticsearch 中的插件生态。

插件是用户以自定义方式增强 Elasticsearch 功能的一种方法。Elasticsearch 插件包括添加自定义映射类型、自定义分析器、自定义脚本引擎和自定义发现等。

9.1 插件简介

Elasticsearch 插件类型包含 jar 文件、脚本和配置文件，插件必须安装在集群中的每个节点上才能使用。安装插件后，必须重新启动每个节点，才能看到插件。

在 Elasticsearch 官网上，插件被归纳为两大类，分别是核心插件和社区贡献插件。

1. 核心插件

核心插件属于 Elasticsearch 项目，插件与 Elasticsearch 安装包同时提供，插件的版本号始终与 Elasticsearch 安装包的版本号相同。这些插件是由 Elasticsearch 团队维护的。

在使用过程中，如果遇到问题，则用户可以在 GitHub 项目（打开 GiHub 官网，搜索 Elasticsearch 即可查看）的页面上进行提交。

在 Elasticsearch 项目中，核心插件列表如图 9-1 所示。

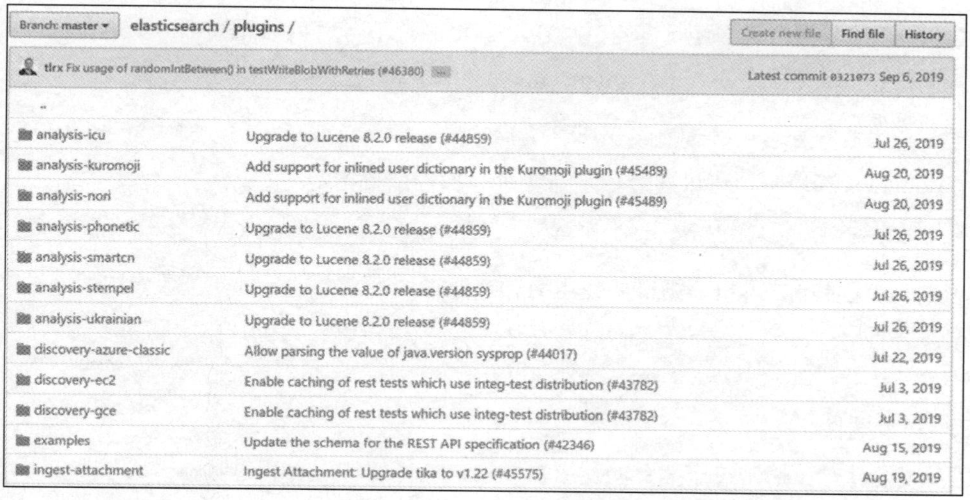

图 9-1

2．社区贡献插件

社区贡献插件属于 Elasticsearch 项目外部的插件。这些插件由单个开发人员或私人公司提供，并拥有各自的许可证及各自的版本控制系统。在使用社区贡献插件过程中，如果遇到问题，则可以在社区插件的网站上进行提交。

9.2 插件管理

Elasticsearch 提供了用于安装、查看和删除插件相关的命令，这些命令默认位于 $es_home/bin 目录中。

用户可以运行以下命令获取插件命令的使用说明：

```
sudo bin / elasticsearch - plugin - h
```

1．插件位置指定

当在根目录中运行 Elasticsearch 时，如果使用 DEB 或 RPM 包安装了 Elasticsearch，则以根目录运行 /usr/share/Elasticsearch/bin/Elasticsearch-plugin，以便 Elasticsearch 可以写入磁盘的相应文件，否则需要以拥有所有 Elasticsearch 文件的用户身份运行 bin/ Elasticsearch 插件。

当用户自定义 URL 或文件系统时，用户可以通过指定 URL 直接从自定义位置下载插件：

```
sudo bin / elasticsearch - plugin install [url]
```

其中，插件名称由其描述符确定。如在 UNIX 环境下，可以通过如下命令进行插件位置指定：

```
sudo bin/elasticsearch-plugin install file:///path/to/plugin.zip
```

在 Windows 环境下，可以通过如下命令进行插件位置指定：

```
bin\elasticsearch-plugin install file:///C:/path/to/plugin.zip
```

如果插件文件不在本机，则需要使用 HTTP 通过如下命令进行插件位置指定：

```
sudo bin/elasticsearch-plugin install http://some.domain/path/to/plugin.zip
```

2．安装插件

在安装插件时，通常每个插件的文档都包含该插件的特定安装说明。

下面以 Elasticsearch 的核心插件为例，展示插件的安装。可以通过如下命令安装 Elasticsearch 核心插件：

```
sudo bin/elasticsearch-plugin install [plugin_name]
```

例如，安装核心 ICU 插件，只需运行以下命令：

```
sudo bin/elasticsearch-plugin install analysis-icu
```

此命令将安装与用户的 Elasticsearch 版本相匹配的插件版本，并在下载时显示进度条。

3．列出当前插件

可以使用 list 命令检索当前加载插件的列表，命令如下所示：

```
sudo bin/elasticsearch-plugin list
```

4．删除插件

可以通过删除 plugins/下的相应目录或使用公共命令手动删除插件，命令如下所示：

```
sudo bin/elasticsearch-plugin remove [pluginname]
```

需要指出的是，在删除 Java 插件之后，需要重新启动节点，完成移除过程。

在默认情况下，插件配置文件（如果有）会保留在磁盘上，以防止用户在升级插件时丢失配置。如果用户希望在删除插件时清除配置文件，则使用 -p 或--purge 命令。在删除插件后，可以使用此选项删除所有延迟的配置文件。

5. 更新插件

插件是为特定版本的 Elasticsearch 构建的，因此每次更新 Elasticsearch 时都必须重新安装插件。更新命令如下所示：

```
sudo bin/elasticsearch-plugin remove [pluginname]
sudo bin/elasticsearch-plugin install [pluginname]
```

6. 其他命令行参数

插件命令还支持许多其他命令行参数，如下所示：

（1）--verbose 参数，输出更多调试信息。

（2）--silent 参数，关闭包括进度条在内的所有输出。

在命令执行过程中可能会返回以下退出代码，各代码对应的含义如下所示：

（1）0: OK。

（2）64: 未知命令或选项参数不正确。

（3）74: I/O 错误。

（4）70: 有其他错误码。

另外，某些插件的安装和运行需要更多的权限。这些插件将列出所需的权限，并在继续安装之前要求用户确认。

当基于安装自动化脚本运行插件安装脚本时，插件脚本应检测到没有从控制台调用它，并跳过确认响应，自动授予所有请求的权限。如果控制台检测失败，则用户可以通过-b 或--batch 命令强制使用批处理模式，命令如下所示：

```
sudo bin/elasticsearch-plugin install-batch [pluginname]
```

如果用户的 elasticsearch.yml 配置文件位于自定义位置，则在使用插件脚本时需要指定配置文件的路径：

```
sudo ES_PATH_CONF=/path/to/conf/dir bin/elasticsearch-plugin install <plugin name>
```

当用户通过代理安装插件时，则需要配置代码。可以用 Java 设置 http.proxyHost 和 http.proxyPort（https.proxyHost，https.proxyPort）将代理细节添到 ES_JAVA_OPTS 环境变量中，配置命令如下所示：

```
sudo ES_JAVA_OPTS="-Dhttp.proxyHost=host_name -Dhttp.proxyPort=port_number
    -Dhttps.proxyHost=host_name -Dhttps.proxyPort=https_port_number"
bin/elasticsearch-plugin install analysis-icu
```

在 Windows 环境下，命令如下所示：

```
set ES_JAVA_OPTS = "-Dhttp.proxyHost=host_name -Dhttp.proxyPort=port_number
    -Dhttps.proxyHost=host_name -Dhttps.proxyPort=https_port_number"
bin\ elasticsearch - plugin install analysis - icu
```

9.3 分析插件

9.3.1 分析插件简介

对分析器（Analyzer）而言，一般会接受一个字符串作为输入参数，分析器会将这个字符串拆分成独立的词或语汇单元（也称之为 token）。当然，在处理过程中会丢弃一些标点符号等字符，处理后会输出一个语汇单元流（也称之为 token stream）。

因此，一般分析器会包含三个部分：

（1）character filter：分词之前的预处理，过滤 HTML 标签、特殊符号转换等。

（2）tokenizer：用于分词。

（3）token filter：用于标准化输出。

Elasticsearch 为很多语言提供了专用的分析器，特殊语言所需的分析器可以由用户根据需要以插件的形式提供。Elasticsearch 内置的主要分析器有：

（1）Standard 分析器：默认的分词器。Standard 分析器会将词汇单元转换成小写形式，并且去除了停用词和标点符号，支持中文（采用的方法为单字切分）。停用词指语气助词等修饰性词语，如 the、an、的、这等。

（2）Simple 分析器：首先通过非字母字符分割文本信息，并去除数字类型的字符，然后将词汇单元统一为小写形式。

（3）Whitespace 分析器：仅去除空格，不会将字符转换成小写形式，不支持中文；不对生成的词汇单元进行其他标准化处理。

（4）Stop 分析器：与 Simple 分析器相比，增加了去除停用词的处理。

（5）Keyword 分析器：该分析器不进行分词，而是直接将输入作为一个单词输出。

（6）Pattern 分析器：该分析器通过正则表达式自定义分隔符，默认是 "\W+"，即把非字词的符号作为分隔符。

（7）Language 分析器：这是特定语言的分析器，不支持中文，支持如 English、French 和 Spanish 等语言。

通常来说，任何全文检索的字符串域都会默认使用 Standard 分析器，如果想要一个自定义分析器，则可以按照如下方式重新制作一个"标准"分析器：

```
{
    "type":      "custom",
    "tokenizer": "standard",
    "filter":    [ "lowercase", "stop" ]
}
```

在这个自定义分析器中，主要使用了 Lowercase（小写字母）和 Stop（停用词）词汇单元过滤器。

什么是 Standard 分析器呢？

一般来说，分析器会接受一个字符串作为输入。在工作时，分析器会将这个字符串拆分成独立的词或语汇单元（称之为 token），当然也会丢弃一些标点符号等字符，最终分析器输出一个语汇单元流。这就是典型的 Standard 分析器的工作模式。

分析器在识别词汇时有多种算法可供选择。最简单的是 Whitespace 分词算法，该算法按空白字符，如空格、Tab、换行符等，对语句进行简单的拆分，将连续的非空格字符组成一个语汇单元。例如，对下面的语句使用 Whitespace 分词算法分词时，会得到如下结果：

原文：You're the 1st runner home!

结果：You're、the、1st、runner、home!

而 Standard 分析器则使用 Unicode 文本分割算法寻找单词之间的界限，并输出所有界限之间的内容。Unicode 内包含的知识使其可以成功地对包含混合语言的文本进行分词。

一般来说，Standard 分析器是大多数语言分词的一个合理的起点。事实上，它构成了大多数特定语言分析器的基础，如 English 分析器、French 分析器 和 Spanish 分析器。另外，它还支持亚洲语言，只是有些缺陷，因此读者可以考虑通过 ICU 分析插件的方式使用 icu_tokenizer 进行替换。

9.3.2　Elasticsearch 中的分析插件

分析插件是一类插件，我们可通过向 Elasticsearch 中添加新的分析器、标记化器、标记过滤器或字符过滤器等扩展 Elasticsearch 的分析功能。

Elasticsearch 官方提供的核心分析插件如下。

（1）ICU 库。

读者可以使用 ICU 库扩展对 Unicode 的支持，包括更好地分析亚洲语言、Unicode 规范化、支持 Unicode 的大小写折叠、支持排序和音译。

（2）Kuromoji 插件。

读者可以使用 Kuromoji 插件对日语进行高级分析。

（3）Lucene Nori 插件。

读者可以使用 Lucene Nori 插件对韩语进行分析。

（4）Phonetic 插件。

读者可以使用 Soundex、Metaphone、Caverphone 和其他编码器/解码器将标记分析为其语音等价物。

（5）SmartCN 插件。

SmartCN 插件可用于对中文或中英文混合文本进行分析。该插件利用概率知识对简化中文文本进行最优分词。首先文本被分割成句子，然后每个句子再被分割成单词。

（6）Stempel 插件。

Stempel 插件为波兰语提供了高质量的分析工具。

（7）Ukrainian 插件。

Ukrainian 插件可用于为乌克兰语提供词干分析。

除官方的分析插件外，Elasticsearch 技术社区也贡献了不少分析插件，比较常用且著名的有：

（1）IK Analysis Plugin。

IK 分析插件将 Lucene IK Analyzer 集成到 Elasticsearch 中，支持读者自定义字典。

（2）Pinyin Analysis Plugin。

Pinyin Analysis Plugin 是一款拼音分析插件，该插件可对汉字和拼音进行相互转换。

（3）Vietnamese Analysis Plugin。

Vietnamese Analysis Plugin 是一款用于对越南语进行分析的插件。

（4）Network Addresses Analysis Plugin。

Network Addresses Analysis Plugin 可以用于分析网络地址。

（5）Dandelion Analysis Plugin。

Dandelion Analysis Plugin 可译为蒲公英分析插件，该插件提供了一个分析器（称为"蒲公英-A"），该分析器会从输入文本中提取的实体进行语义搜索。

（6）STConvert Analysis Plugin。

STConvert Analysis Plugin 可对中文简体和繁体进行相互转换。

9.3.3 ICU 分析插件

ICU 分析插件将 Lucene ICU 模块集成到 Elasticsearch 中。

1. ICU 分析插件的安装

插件必须安装在群集中的每个节点上，并且必须在安装后重新启动节点。在 Linux 环境下，可以通过如下命令安装 ICU 分析插件：

```
sudo bin/elasticsearch-plugin install analysis-icu
```

当需要删除 ICU 分析插件时，在 Linux 环境下，可以通过如下命令删除 ICU 分析插件：

```
sudo bin / elasticsearch - plugin remove analysis - icu
```

2. ICU 分析插件简介

ICU 分析插件使用 icu_normalizer char filter、icu_tokenizer 和 icu normalizer token filter 三个组件分别执行基本字符的规范化、字符标记和字符处理标准化操作。

ICU 分析插件在工作时，需要 method 和 mode 两个参数，其中，method 参数主要指的是归一化方法，接受 NFKC、NFC 或 NFKC_CF（默认）方法；mode 参数主要指的是规范化模式，接受模式的组合或分解，默认是组合模式。

3. ICU 插件的规范化字符过滤器

ICU 插件的规范化字符过滤器主要用于规范化字符，可用于所有索引，无须进一步配置。规范化的类型可以由 name 参数指定，该参数接受 NFC、NFKC 和 NFKC_CF 方法，默认为 NFKC_CF 方法。模式参数还可以设置为分解，分别将 NFC 转换为 NFD，或将 NFKC 转换为 NFKD。

在使用时，可以通过指定 unicode_set_filter 参数控制规范化的字母，该参数接受 unicodeset。下面通过一个示例，展示字符过滤器的默认用法和自定义字符过滤器：

```
PUT icu_sample
{
  "settings": {
    "index": {
      "analysis": {
        "analyzer": {
          "nfkc_cf_normalized": { // 使用默认的NFKC_CF方法
            "tokenizer": "icu_tokenizer",
            "char_filter": [
              "icu_normalizer"
```

```
                    ]
                },
                "nfd_normalized": {    //使用自定义的 nfd_normalized 令牌过滤器，该过滤器设置
                                       //为使用带分解的 NFC 方法
                    "tokenizer": "icu_tokenizer",
                    "char_filter": [
                        "nfd_normalizer"
                    ]
                }
            },
            "char_filter": {
                "nfd_normalizer": {
                    "type": "icu_normalizer",
                    "name": "nfc",
                    "mode": "decompose"
                }
            }
        }
    }
}
```

4．ICU 插件的使用方法

下面以 Windows 环境为例，展示 ICU 插件的使用方法。

不同于 Linux 环境下的插件安装命令，在 Windows 环境下的安装命令如下所示：

```
bin\elasticsearch-plugin.bat install  file:///C:\Users\牛冬\Desktop\analysis-icu-7.2.0.zip
```

其中，analysis-icu-7.2.0.zip 是下载到本地的文件，配置方式如下：

```
file:///本地路径
```

在执行安装命令后，DOS 界面会显示安装过程和安装结果，如图 9-2 所示。

```
C:\elasticsearch-7.2.0-windows-x86_642\elasticsearch-7.2.0>bin\elasticsearch-plugin.bat install  file:///C:\Users\牛冬\D
esktop\analysis-icu-7.2.0.zip
warning: ignoring JAVA_TOOL_OPTIONS=-Dfile.encoding=UTF-8
future versions of Elasticsearch will require Java 11; your Java version from [C:\Program Files\Java\jdk1.8.0_171\jre] d
oes not meet this requirement
-> Downloading file:///C:\Users\牛冬\Desktop\analysis-icu-7.2.0.zip
[=================================================] 100%??
-> Installed analysis-icu
```

图 9-2

可以通过如下命令查看插件的安装结果,即安装成功与否:

```
bin\elasticsearch-plugin.bat list
```

在命令执行后,DOS 界面会输出目前 Elasticsearch 中安装的插件,如图 9-3 所示。

```
C:\elasticsearch-7.2.0-windows-x86_642\elasticsearch-7.2.0>bin\elasticsearch-plugin.bat list
warning: ignoring JAVA_TOOL_OPTIONS=-Dfile.encoding=UTF-8
future versions of Elasticsearch will require Java 11; your Java version from [C:\Program Files\Java\jdk1.8.0_171\jre] d
oes not meet this requirement
analysis-icu
```

图 9-3

通过图 9-3 可以看出,当前安装的插件中含有 analysis-icu 插件。

在启动 Elasticsearch 时,在 Elasticsearch 的控制台也可以看到插件的加载情况,如图 9-4 所示。

```
2019-09-10T17:54:57,380][INFO ][o.e.p.PluginsService     ] [LAPTOP-1S8BALK3] loaded module [x-pack-sql]
2019-09-10T17:54:57,386][INFO ][o.e.p.PluginsService     ] [LAPTOP-1S8BALK3] loaded module [x-pack-watcher]
2019-09-10T17:54:57,397][INFO ][o.e.p.PluginsService     ] [LAPTOP-1S8BALK3] loaded plugin [analysis-icu]
```

图 9-4

下面展示 ICU 插件的使用方法。在下文提及的 Head 插件中,使用_analyze 命令,输入如下参数对字符串内容进行分析:

```
{
    "text": "我是牛冬,目前在好未来集团家长帮事业部做技术负责人/技术总监",
}
```

在使用 ICU 插件前,为了对比分析效果,我们先用 Standard 分析器对字符串文本进行分析。输入如下所示内容:

```
{
    "text": "我是牛冬,目前在好未来集团家长帮事业部做技术负责人/技术总监",
    "analyzer": "standard"
}
```

具体如图 9-5 所示。

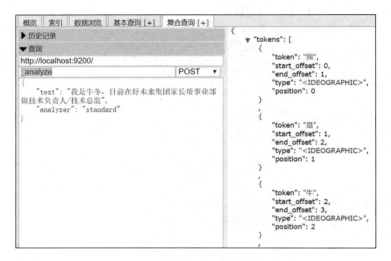

图 9-5

在请求执行后,右侧会输出分析结果,分析结果如下所示:

```
{
"tokens": [{
        "token": "我",
        "start_offset": 0,
        "end_offset": 1,
        "type": "<IDEOGRAPHIC>",
        "position": 0
    },
    {
        "token": "是",
        "start_offset": 1,
        "end_offset": 2,
        "type": "<IDEOGRAPHIC>",
        "position": 1
    },
    {
        "token": "牛",
        "start_offset": 2,
        "end_offset": 3,
        "type": "<IDEOGRAPHIC>",
        "position": 2
    },
    {
        "token": "冬",
```

```
        "start_offset": 3,
        "end_offset": 4,
        "type": "<IDEOGRAPHIC>",
        "position": 3
    },
    {
        "token": "目",
        "start_offset": 5,
        "end_offset": 6,
        "type": "<IDEOGRAPHIC>",
        "position": 4
    },
    {
        "token": "前",
        "start_offset": 6,
        "end_offset": 7,
        "type": "<IDEOGRAPHIC>",
        "position": 5
    },
    {
        "token": "在",
        "start_offset": 7,
        "end_offset": 8,
        "type": "<IDEOGRAPHIC>",
        "position": 6
    },
    {
        "token": "好",
        "start_offset": 8,
        "end_offset": 9,
        "type": "<IDEOGRAPHIC>",
        "position": 7
    },
    {
        "token": "未",
        "start_offset": 9,
        "end_offset": 10,
        "type": "<IDEOGRAPHIC>",
        "position": 8
    },
    {
        "token": "来",
        "start_offset": 10,
        "end_offset": 11,
```

```
            "type": "<IDEOGRAPHIC>",
            "position": 9
        },
        {
            "token": "集",
            "start_offset": 11,
            "end_offset": 12,
            "type": "<IDEOGRAPHIC>",
            "position": 10
        },
        {
            "token": "团",
            "start_offset": 12,
            "end_offset": 13,
            "type": "<IDEOGRAPHIC>",
            "position": 11
        },
        {
            "token": "家",
            "start_offset": 13,
            "end_offset": 14,
            "type": "<IDEOGRAPHIC>",
            "position": 12
        },
        {
            "token": "长",
            "start_offset": 14,
            "end_offset": 15,
            "type": "<IDEOGRAPHIC>",
            "position": 13
        },
        {
            "token": "帮",
            "start_offset": 15,
            "end_offset": 16,
            "type": "<IDEOGRAPHIC>",
            "position": 14
        },
        {
            "token": "事",
            "start_offset": 16,
            "end_offset": 17,
            "type": "<IDEOGRAPHIC>",
            "position": 15
```

```
    },
    {
        "token": "业",
        "start_offset": 17,
        "end_offset": 18,
        "type": "<IDEOGRAPHIC>",
        "position": 16
    },
    {
        "token": "部",
        "start_offset": 18,
        "end_offset": 19,
        "type": "<IDEOGRAPHIC>",
        "position": 17
    },
    {
        "token": "做",
        "start_offset": 19,
        "end_offset": 20,
        "type": "<IDEOGRAPHIC>",
        "position": 18
    },
    {
        "token": "技",
        "start_offset": 20,
        "end_offset": 21,
        "type": "<IDEOGRAPHIC>",
        "position": 19
    },
    {
        "token": "术",
        "start_offset": 21,
        "end_offset": 22,
        "type": "<IDEOGRAPHIC>",
        "position": 20
    },
    {
        "token": "负",
        "start_offset": 22,
        "end_offset": 23,
        "type": "<IDEOGRAPHIC>",
        "position": 21
    },
    {
```

```
            "token": "责",
            "start_offset": 23,
            "end_offset": 24,
            "type": "<IDEOGRAPHIC>",
            "position": 22
        },
        {
            "token": "人",
            "start_offset": 24,
            "end_offset": 25,
            "type": "<IDEOGRAPHIC>",
            "position": 23
        },
        {
            "token": "技",
            "start_offset": 26,
            "end_offset": 27,
            "type": "<IDEOGRAPHIC>",
            "position": 24
        },
        {
            "token": "术",
            "start_offset": 27,
            "end_offset": 28,
            "type": "<IDEOGRAPHIC>",
            "position": 25
        },
        {
            "token": "总",
            "start_offset": 28,
            "end_offset": 29,
            "type": "<IDEOGRAPHIC>",
            "position": 26
        },
        {
            "token": "监",
            "start_offset": 29,
            "end_offset": 30,
            "type": "<IDEOGRAPHIC>",
            "position": 27
        }
    ]
}
```

下面我们将 Standard 分析器换成 ICU 插件分析器，此时 analyzer 的配置参数变为 icu_analyzer，如下所示：

```
{
    "text": "我是牛冬，目前在好未来集团家长帮事业部做技术负责人/技术总监",
    "analyzer": "icu-analyzer"
}
```

Head 插件的配置情况如图 9-6 所示。

图 9-6

在请求执行后，右侧输出的分析结果如下所示：

```
{
"tokens": [{
        "token": "我是",
        "start_offset": 0,
        "end_offset": 2,
        "type": "<IDEOGRAPHIC>",
        "position": 0
    },
    {
        "token": "牛",
        "start_offset": 2,
        "end_offset": 3,
        "type": "<IDEOGRAPHIC>",
        "position": 1
    },
    {
        "token": "冬",
```

```
            "start_offset": 3,
            "end_offset": 4,
            "type": "<IDEOGRAPHIC>",
            "position": 2
        },
        {
            "token": "目前",
            "start_offset": 5,
            "end_offset": 7,
            "type": "<IDEOGRAPHIC>",
            "position": 3
        },
        {
            "token": "在",
            "start_offset": 7,
            "end_offset": 8,
            "type": "<IDEOGRAPHIC>",
            "position": 4
        },
        {
            "token": "好",
            "start_offset": 8,
            "end_offset": 9,
            "type": "<IDEOGRAPHIC>",
            "position": 5
        },
        {
            "token": "未来",
            "start_offset": 9,
            "end_offset": 11,
            "type": "<IDEOGRAPHIC>",
            "position": 6
        },
        {
            "token": "集团",
            "start_offset": 11,
            "end_offset": 13,
            "type": "<IDEOGRAPHIC>",
            "position": 7
        },
        {
            "token": "家长",
            "start_offset": 13,
            "end_offset": 15,
```

```
            "type": "<IDEOGRAPHIC>",
            "position": 8
    },
    {
            "token": "帮",
            "start_offset": 15,
            "end_offset": 16,
            "type": "<IDEOGRAPHIC>",
            "position": 9
    },
    {
            "token": "事业",
            "start_offset": 16,
            "end_offset": 18,
            "type": "<IDEOGRAPHIC>",
            "position": 10
    },
    {
            "token": "部",
            "start_offset": 18,
            "end_offset": 19,
            "type": "<IDEOGRAPHIC>",
            "position": 11
    },
    {
            "token": "做",
            "start_offset": 19,
            "end_offset": 20,
            "type": "<IDEOGRAPHIC>",
            "position": 12
    },
    {
            "token": "技术",
            "start_offset": 20,
            "end_offset": 22,
            "type": "<IDEOGRAPHIC>",
            "position": 13
    },
    {
            "token": "负责",
            "start_offset": 22,
            "end_offset": 24,
            "type": "<IDEOGRAPHIC>",
            "position": 14
```

```
        },
        {
            "token": "人",
            "start_offset": 24,
            "end_offset": 25,
            "type": "<IDEOGRAPHIC>",
            "position": 15
        },
        {
            "token": "技术",
            "start_offset": 26,
            "end_offset": 28,
            "type": "<IDEOGRAPHIC>",
            "position": 16
        },
        {
            "token": "总监",
            "start_offset": 28,
            "end_offset": 30,
            "type": "<IDEOGRAPHIC>",
            "position": 17
        }
    ]
}
```

通过上述两个分析器的对比，可以明显看出 ICU 插件的分析效果更好，更符合中文的语境。

9.3.4 智能中文分析插件

智能中文分析插件（SmartCN），本质上是将 Lucene 的智能中文分析模块集成到 Elasticsearch 中。

SmartCN 提供了一个中文文本或中英文混合文本的分析器。该分析器利用概率知识对简化中文文本进行最优分词。首先文本被分成句子，然后每个句子再被分割成单词。

1. SmartCN 插件的安装

我们可以使用插件管理器安装 SmartCN 插件，安装命令如下所示：

```
sudo bin/elasticsearch-plugin install analysis-smartcn
```

需要指出的是，插件必须安装在集群的每个节点上，并且必须在安装后重新启动节点。

我们还可以通过其他方式安装（如 Windows 方式，在后面介绍）。从本地安装时，需要先下载插件到本地。

当需要删除该插件时，可以使用如下命令删除：

```
sudo bin/elasticsearch-plugin remove analysis-smartcn
```

需要指出的是，在删除插件之前必须停止节点。

2. SmartCN 插件的简介

SmartCN 插件提供了不可配置的 SmartCN 分析器、SmartCN Tokenizer 和 SmartCN Stop Token 过滤器。

SmartCN 分析器是 Lucene 4.6 版本之后自带的，中文分词效果不错，但英文分词有问题。与 IKAnalyzer 分析器相比，SmartCN 分析器在分词时会带来较多碎片，且目前不支持自定义词库。

SmartCN 分析器支持重新实现和扩展分析器。在使用时，可以将 SmartCN 分析器重新实现为自定义分析器，扩展和配置代码如下所示：

```
{
  "settings": {
    "analysis": {
      "analyzer": {
        "rebuilt_smartcn": {
          "tokenizer": "smartcn_tokenizer",
          "filter": [
            "porter_stem",
            "smartcn_stop"
          ]
        }
      }
    }
  }
}
```

此外，SmartCN 分析器还支持用户指定的自定义通用词，配置方式如下所示：

```
{
  "settings": {
    "index": {
      "analysis": {
        "analyzer": {
          "smartcn_with_stop": {
```

```
            "tokenizer": "smartcn_tokenizer",
            "filter": [
              "porter_stem",
              "my_smartcn_stop"
            ]
          }
        },
        "filter": {
          "my_smartcn_stop": {
            "type": "smartcn_stop",
            "stopwords": [
              "_smartcn_",
              "stack",
              "的"
            ]
          }
        }
      }
    }
}
```

3. SmartCN 分析器实战

前面介绍了 SmartCN 分析器在 Linux 环境下的安装和卸载，下面以 Windows 环境为例，展示该插件的使用。

Windows 环境下的安装命令如下所示：

```
bin\elasticsearch-plugin.bat install  file:///C:\Users\牛冬\Desktop\analysis-smartcn-7.2.0.zip
```

其中，analysis-smartcn-7.2.0.zip 是下载到本地的文件，配置方式如下：

```
file:///本地路径
```

执行安装命令后，DOS 界面会显示安装过程和安装结果，如图 9-7 所示。

```
C:\elasticsearch-7.2.0-windows-x86_642\elasticsearch-7.2.0>bin\elasticsearch-plugin.bat install  file:///C:\Users\牛冬\Desktop\analysis-smartcn-7.2.0.zip
warning: ignoring JAVA_TOOL_OPTIONS=-Dfile.encoding=UTF-8
future versions of Elasticsearch will require Java 11; your Java version from [C:\Program Files\Java\jdk1.8.0_171\jre] does not meet this requirement
-> Downloading file:///C:\Users\牛冬\Desktop\analysis-smartcn-7.2.0.zip
[=================================================] 100%??
-> Installed analysis-smartcn
```

图 9-7

可以通过如下命令查看插件的安装结果，即安装成功与否：

```
bin\elasticsearch-plugin.bat list
```

命令执行后，DOS 界面会输出目前 Elasticsearch 中安装的插件，内容如图 9-8 所示。

```
C:\elasticsearch-7.2.0-windows-x86_642\elasticsearch-7.2.0>bin\elasticsearch-plugin.bat list
warning: ignoring JAVA_TOOL_OPTIONS=-Dfile.encoding=UTF-8
future versions of Elasticsearch will require Java 11; your Java version from [C:\Program Files\Java\jdk1.8.0_171\jre] d
oes not meet this requirement
analysis-icu
analysis-smartcn
```

图 9-8

通过图 9-8 可以看到，在当前 Elasticsearch 安装的插件中含有 analysis-smartcn 插件。

在启动 Elasticsearch 时，在 Elasticsearch 的控制台可以看到插件的加载情况，如图 9-9 所示。

```
[2019-09-10T18:04:41,869][INFO ][o.e.p.PluginsService     ] [LAPTOP-1S8BALK3] loaded plugin [analysis-icu]
[2019-09-10T18:04:41,869][INFO ][o.e.p.PluginsService     ] [LAPTOP-1S8BALK3] loaded plugin [analysis-smartcn]
```

图 9-9

下面介绍 SmartCN 插件的使用。在 Head 插件中，使用 _analyze 命令输入如下参数，对字符串内容进行分析：

```
{
    "text": "我是牛冬，目前在好未来集团家长帮事业部做技术负责人/技术总监",
}
```

此时"analyzer"参数配置为"smartcn"，即使用 SmartCN 插件对字符串进行分析。如图 9-10 所示。

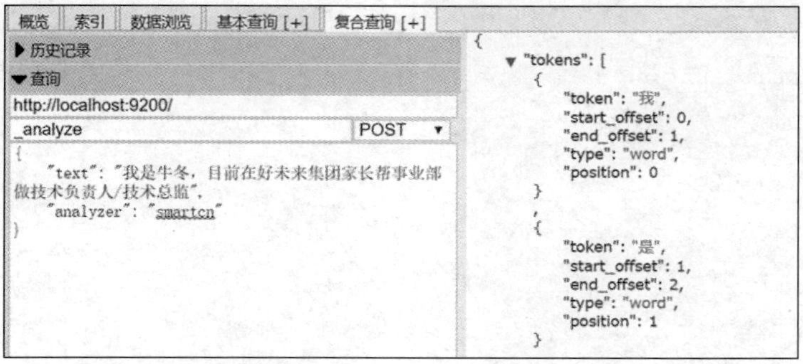

图 9-10

在请求执行后，右侧会显示使用 SmartCN 插件对字符串进行分析的结果，如下所示：

```
{
```

```
"tokens": [{
        "token": "我",
        "start_offset": 0,
        "end_offset": 1,
        "type": "word",
        "position": 0
    },
    {
        "token": "是",
        "start_offset": 1,
        "end_offset": 2,
        "type": "word",
        "position": 1
    },
    {
        "token": "牛",
        "start_offset": 2,
        "end_offset": 3,
        "type": "word",
        "position": 2
    },
    {
        "token": "冬",
        "start_offset": 3,
        "end_offset": 4,
        "type": "word",
        "position": 3
    },
    {
        "token": "目前",
        "start_offset": 5,
        "end_offset": 7,
        "type": "word",
        "position": 5
    },
    {
        "token": "在",
        "start_offset": 7,
        "end_offset": 8,
        "type": "word",
        "position": 6
    },
    {
        "token": "好",
```

```
            "start_offset": 8,
            "end_offset": 9,
            "type": "word",
            "position": 7
        },
        {
            "token": "未来",
            "start_offset": 9,
            "end_offset": 11,
            "type": "word",
            "position": 8
        },
        {
            "token": "集团",
            "start_offset": 11,
            "end_offset": 13,
            "type": "word",
            "position": 9
        },
        {
            "token": "家长",
            "start_offset": 13,
            "end_offset": 15,
            "type": "word",
            "position": 10
        },
        {
            "token": "帮",
            "start_offset": 15,
            "end_offset": 16,
            "type": "word",
            "position": 11
        },
        {
            "token": "事业",
            "start_offset": 16,
            "end_offset": 18,
            "type": "word",
            "position": 12
        },
        {
            "token": "部",
            "start_offset": 18,
            "end_offset": 19,
```

```
            "type": "word",
            "position": 13
    },
    {
            "token": "做",
            "start_offset": 19,
            "end_offset": 20,
            "type": "word",
            "position": 14
    },
    {
            "token": "技术",
            "start_offset": 20,
            "end_offset": 22,
            "type": "word",
            "position": 15
    },
    {
            "token": "负责人",
            "start_offset": 22,
            "end_offset": 25,
            "type": "word",
            "position": 16
    },
    {
            "token": "技术",
            "start_offset": 26,
            "end_offset": 28,
            "type": "word",
            "position": 18
    },
    {
            "token": "总监",
            "start_offset": 28,
            "end_offset": 30,
            "type": "word",
            "position": 19
    }
]
```

与前面两个分析器（Standard 分析器和 ICU 分析器）相比，可以看出 SmartCN 的分析效果显然更好，更符合中文的语境。

9.4 API 扩展插件

如果 Elasticsearch 内置的接口不够用，则可以使用 API 扩展插件。

API 扩展插件通过添加新的、与搜索有关的 API 或功能，实现对 Elasticsearch 新功能的添加。Elasticsearch 社区人员陆陆续续贡献了不少 API 扩展插件编辑器，汇总如下。

（1）Carrot2 Plugin。

该插件用于结果聚类，读者可访问 GitHub 官网，搜索 elasticsearch-carrot2，查看配套代码。

（2）Elasticsearch Trigram Accelerated Regular Expression Filter。

该插件包括查询、过滤器、原生脚本、评分函数，以及用户最终创建的任意其他内容。通过该插件可以让搜索变得更好。读者可访问 GitHub 官网，搜索 search-extra 获取插件。

（3）Elasticsearch Experimental Highlighter。

该插件是用 Java 编写的，用于文本高亮显示。读者可访问 GitHub 官网，搜索 search-highlighter 获取插件。

（4）Entity Resolution Plugin。

该插件使用 Duke（Duke 是一个用 Java 编写的、快速灵活的、删除重复数据的引擎）进行重复检测。读者可访问 GitHub 官网，搜索 elasticsearch-entity-resolution 获取插件。

（5）Entity Resolution Plugin(zentity)。

该插件用于实时解析 Elasticsearch 中存储的实体信息。读者可访问 GitHub 官网，搜索 zentity 获取插件。

（6）PQL language Plugin。

该插件允许用户使用简单的管道查询语法对 Elasticsearch 进行查询。读者可访问 GitHub 官网，搜索 elasticsearch-pql 获取插件。

（7）Elasticsearch Taste Plugin。

该插件基于 Mahout Taste 的协同过滤算法实现。读者可访问 GitHub 官网，搜索 elasticsearch-taste 获取插件。

（8）WebSocket Change Feed Plugin。

该插件允许客户端创建到 Elasticsearch 节点的 WebSocket 连接，并从数据库接收更改的提要。读者可访问 GitHub 官网，搜索 es-change-feed-plugin 获取插件。

9.5 监控插件

在使用 Elasticsearch 的过程中，CPU 的使用率可能会意外增加，这会导致应用服务的响应时间变长；也可能会出现 HTTP 返回码（如 503 的错误）的数量迅速上升，此时，Elasticsearch 的索引速度会直线下降。

幸好 Elasticsearch 有监控和报警功能，使得用户可以实时了解 Elasticsearch 的状态。

监控插件允许用户监视 Elasticsearch 索引，并在违反阈值时触发警报。那么警报是如何触达用户的呢？

Elasticsearch 内置了电子邮件、PagerDuty、Slack 和 HipChat 的相关功能，用户可以在多项告警选项中自由选择。此外，Elasticsearch 还有强大的 WebHook 输出功能，可以与用户现有的监控基础设施或任意第三方系统集成。

另外，用户可以对告警功能进行配置，将搜索中的相关信息包含在通知内，告警功能还支持简易模板。

同时，用户还可以通过 Kibana 联通 Elasticsearch 的监控和报警功能。

Elasticsearch 中内置了监控和报警插件，即 X-Pack。X-Pack 允许用户根据数据中的更改需求采取操作，其设计原则是，如果用户在 Elasticsearch 中查询某些内容，就对其发出警报。因此，用户只需定义一个查询、条件、计划和将要采取的操作，X-Pack 就可以完成剩下的工作。

此外，Elasticsearch 社区还提供了一些知名的插件，如 Head 插件和 Cerebro 插件，它们都能对 Elasticsearch 进行监控。这两个插件将在后面进行介绍。

9.6 数据提取插件

当我们不想或不能通过 API 接口向 Elasticsearch 中存储数据时，可以使用数据提取插件向 Elasticsearch 中写入数据。

Elasticsearch 中内置的核心数据提取插件主要如下。

（1）附件提取插件。

附件提取插件允许用户使用 Apache 文本提取库 Tika 以通用格式（如 PPT、XLS 和 PDF 等）提取文件中的数据。

（2）Geoip 数据提取插件。

Geoip 数据提取插件默认在 Geoip 字段中。根据来自 MaxMind 数据库的数据添加有关 IP

地址、地理位置等信息。当前，该插件已经内置在 Elasticsearch 中。

（3）提取用户代理 user_agent 插件。

该插件用于从用户代理头值中提取详细信息。当前，该插件已经内置在 Elasticsearch 中。

此外，Elasticsearch 技术社区也贡献了一些数据提取插件，如提取 CSV 格式文件数据的插件等。

由于附件提取插件尚未成为 Elasticsearch 内置的插件，因此在使用时，用户需要自行安装。在 Linux 环境下的安装命令如下所示：

```
sudo bin/elasticsearch-plugin install ingest-attachment
```

需要指出的是，插件必须安装在群集的每个节点上，并且必须在安装后重新启动节点。

如果不是实时下载安装，而是离线下载后安装，则可以从 Elastcsearch 官网下载插件。

当插件不再使用时，可以使用以下命令删除插件：

```
sudo bin/elasticsearch-plugin remove ingest-attachment
```

需要指出的是，在删除插件之前必须停止节点。

9.7　常用插件实战

本节介绍 Head 插件和 Cerebro 插件的使用。这两个插件都能监控 Elasticsearch 集群中每个节点的情况，也都提供了 API 搜索相关的功能。下面一一介绍。

9.7.1　Head 插件

Head 插件，全称为 elasticsearch-head，是一个界面化的集群操作和管理工具，可以对集群进行"傻瓜式"操作。既可以把 Head 插件集成到 Elasticsearch 中，也可以把 Head 插件当成一个独立服务。

Head 插件主要有三方面的功能：

（1）显示 Elasticsearch 集群的拓扑结构，能够执行索引和节点级别的操作。

（2）在搜索接口能够查询 Elasticsearch 集群中原始 JSON 或表格格式的数据。

（3）能够快速访问并显示 Elasticsearch 集群的状态。

1. Head 插件的安装

Head 插件的安装方式有两种，方式一是通过 Elasticsearch 自带的 plugin 命令进行安装，如下所示：

```
elasticsearch/bin/elasticsearch-plugin -install mobz/elasticsearch-head
```

通过命令安装时会实时下载安装包进行安装。

方式二是离线安装，即用户将 Head 插件下载到本地后再进行安装。读者可访问 GitHub 官网，搜索 elasticsearch-head，获取 Head 插件。

在 Head 插件页面，可以看到如图 9-11 所示内容。

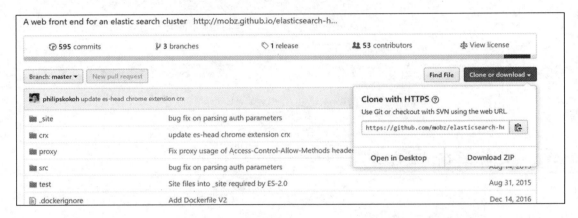

图 9-11

可以使用 git 命令下载插件：

```
git clone https://github.com/mobz/elasticsearch-head.git
```

也可以单击图 9-11 中所示的"Download ZIP"按钮，进行下载。

将 Head 插件下载到本地后，需将其解压缩到某一目录下。解压缩后的目录如图 9-12 所示。

名称	修改日期	类型	大小
_site	2019/8/14 19:27	文件夹	
crx	2019/8/14 19:27	文件夹	
proxy	2019/8/14 19:27	文件夹	
src	2019/8/14 19:27	文件夹	
test	2019/8/14 19:27	文件夹	
.dockerignore	2019/8/14 19:27	DOCKERIGNORE 文...	1 KB
.gitignore	2019/8/14 19:27	文本文档	1 KB
.jshintrc	2019/8/14 19:27	JSHINTRC 文件	1 KB
Dockerfile	2019/8/14 19:27	文件	1 KB
Dockerfile-alpine	2019/8/14 19:27	文件	1 KB
elasticsearch-head.sublime-project	2019/8/14 19:27	SUBLIME-PROJECT ...	1 KB
grunt_fileSets.js	2019/8/14 19:27	JavaScript 文件	4 KB
Gruntfile.js	2019/8/14 19:27	JavaScript 文件	3 KB
index.html	2019/8/14 19:27	搜狗高速浏览器HTM...	2 KB
LICENCE	2019/8/14 19:27	文件	1 KB
package.json	2019/8/14 19:27	JSON 文件	1 KB
plugin-descriptor.properties	2019/8/14 19:27	PROPERTIES 文件	1 KB
README.textile	2019/8/14 19:27	TEXTILE 文件	7 KB

图 9-12

随后，开启 Elasticsearch 服务，在浏览器中输入如下 URL：

```
http://localhost:9200/_plugin/head/
```

上述安装方式在 Linux 系统环境下和在 Windows 系统环境下均相同，不同的是，在 Windows 环境下，可以直接单击图 9-12 中的 index.html 打开对应的页面，如图 9-13 所示。

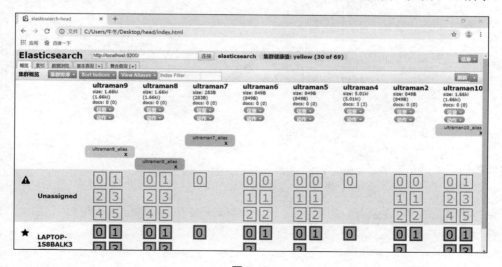

图 9-13

在默认情况下，Head 插件将立即尝试连接位于 http://localhost:9200/的集群节点。如果用户需要更改连接节点的信息，则可以在"连接"框中输入正确的 Elasticsearch 其他节点地址，并单击"连接"按钮进行连接。

在 Windows 环境下，如果直接打开 Head 插件首页，则在浏览器的开发者模式下可以看到跨域的报错，因此需要对 Elasticsearch 跨域请求访问进行配置。配置是通过修改 elasticsearch/config/elasticsearch.yml 文件实现的，内容如下所示：

```
http.cors.enabled: true
http.cors.allow-origin: "*"
```

配置后，即可正常访问 Head 插件首页。在首页中，可以看到如图 9-14 所示的几部分内容。

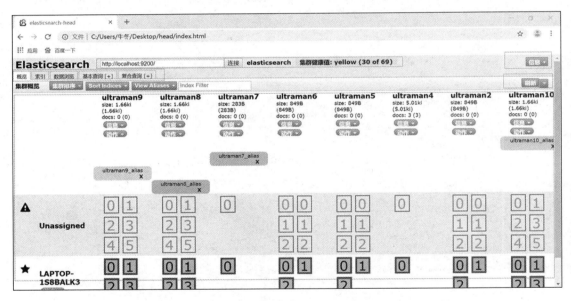

图 9-14

从图 9-14 可以看出，Head 插件首页由 4 部分组成：节点地址输入区域、信息刷新区域、导航条和概览中的集群信息汇总。

第一部分是节点地址输入区域。

第二部分是信息刷新区域，可以查看 Elasticsearch 相关的信息和刷新插件的信息。其中，插件提供的数据刷新方式如图 9-15 所示。

图 9-15

用户可以选择手动刷新、快速刷新、每 5 秒刷新或每 1 分钟刷新。

此外，在"信息"按钮部分，可以查看 Elasticsearch 相关的信息，如图 9-16 所示。

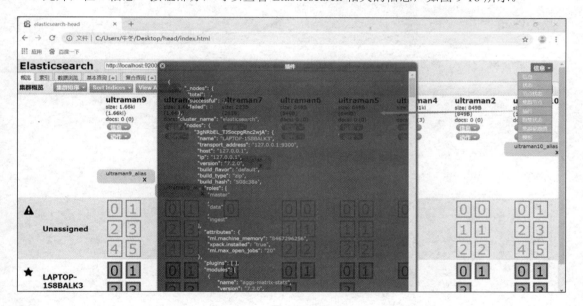

图 9-16

从图 9-16 可以看到 Elasticsearch 相关的信息，包括集群节点信息、节点状态、集群状态、集群信息、集群健康值等内容。单击对应的按钮，即可查看对应的信息。

第三部分是导航条，从图 9-16 中可以看到概览、索引、数据浏览、基本查询和复合查询五个 Tab 导航，默认为概览。

第四部分是概览中的集群信息汇总。我们可以看到 Elasticsearch 已经创建的索引，这些索引信息包含了索引的名称、索引的大小和索引的数据量，并且通过"信息"和"动作"两个按钮可以查看索引信息，或者给索引创建别名。

以 ultraman9 为例，单击索引 ultraman9 下的"信息"→"索引状态"选项，可以看到如图 9-17 所示内容。

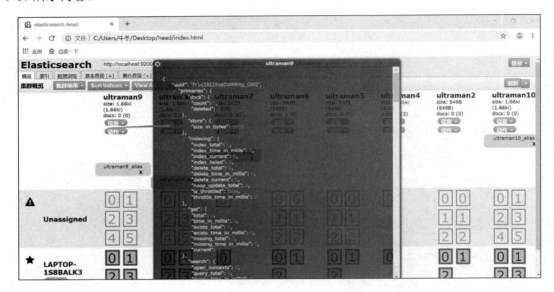

图 9-17

单击索引 ultraman9 下的"信息"→"索引信息"选项，可以看到如图 9-18 所示内容。

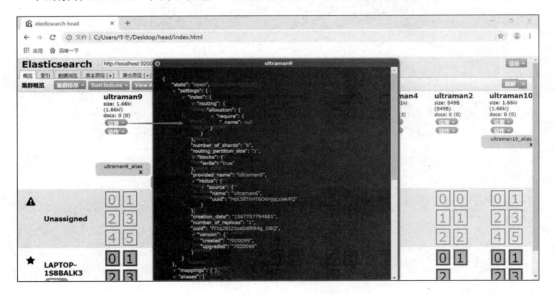

图 9-18

而"动作"选项下的信息有创建别名、快照、关闭等功能，如图9-19所示。

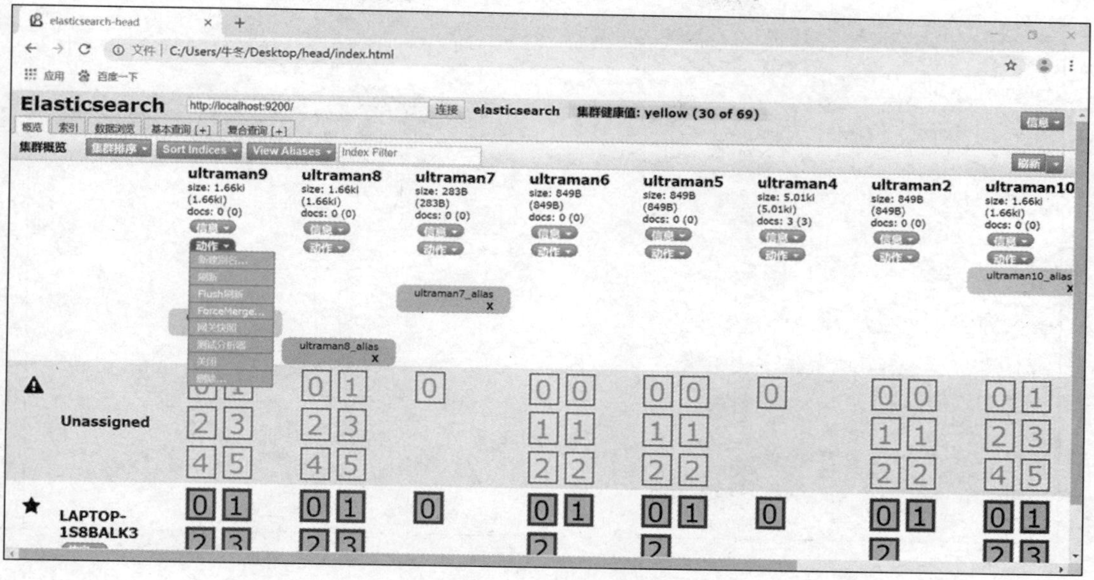

图9-19

图9-17中带有感叹号的Unassigned表示未分配的节点，带有星星的表示主节点，其节点名称叫作LAPTOP-1S8BALK3。

切换到"索引"标签页，可以查看当前Elasticsearch集群中的索引情况，如图9-20所示。

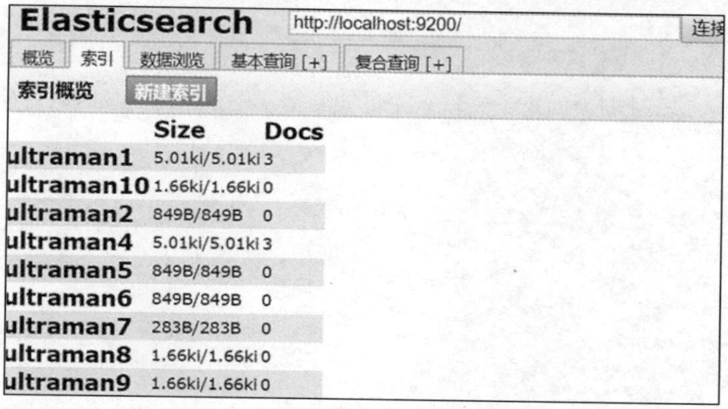

图9-20

切换到"数据浏览"标签页，可以查看特定索引下的存储数据，如图 9-21 所示。

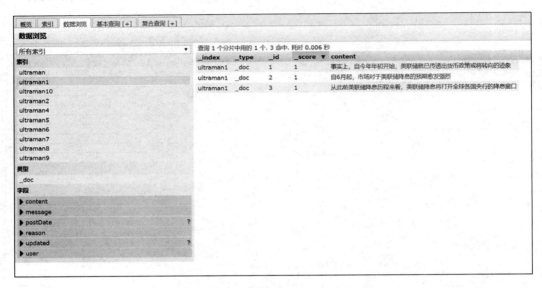

图 9-21

在图 9-21 中，我们既可以查看索引 ultraman1 中存储的数据，也可以基于字段进行数据筛选，如图 9-22 所示。

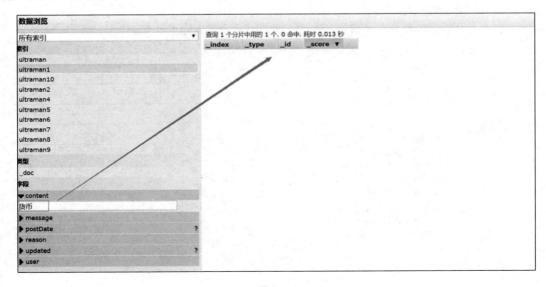

图 9-22

切换到"基本查询"标签页,用自由拼接条件进行简单的数据查询,如图 9-23 所示。

图 9-23

当用多个查询条件进行搜索或查询时,需要注意多个查询条件间的匹配方式。匹配方式主要有 3 种,即 must、should 和 mus_tnot。三种匹配方式的说明如下所示。

(1) must 子句:文档必须匹配 must 查询条件,相当于"="。

(2) should 子句:文档应该匹配 should 子句查询的一个或多个条件。

(3) must_not 子句:文档不能匹配该查询条件,相当于"!="。

在图 9-23 中选择 match 方式,常用的匹配方式还有 term、text 和 range 等。

term 表示的是精确匹配,wildcard 表示的是通配符匹配,prefix 表示的是前缀匹配,range 表示的是区间查询。

在图 9-23 中的"+""-"按钮用于增加查询条件或减少查询条件。

在查询结果展示区域中,用户可以设置数据的呈现形式,如 table、JSON、CVS 表格等,还可以勾选"显示查询语句"选项,呈现通过表单内容拼接的搜索语句。

当切换到"复合查询"标签页时,可以自由拼接条件,进行复杂的数据查询。

"复合查询"标签页为用户提供了编写 RESTful 接口风格的请求,用户可以使用 JSON 进行复杂的查询，比如发送 PUT 请求新增及更新索引，使用 delete 请求删除索引等。总之，在"复合查询"页面，用户可对 Elasticsearch 中的数据或者索引进行各种增删改查等操作请求。如图 9-24 所示。

图 9-24

从图 9-24 可以看到，页面中有一个输入窗口，允许用户任意调用 Elasticsearch 的 RESTful API。

在 Elasticsearch 中，RESTFul API 的基本格式如下所示：

```
http://ip:port/索引/类型/文档 ID
```

在 Elasticsearch 中，以 POST 方法自动生成 ID，而 PUT 方法需要指明 ID。

配置接口包含以下四个选项：

（1）请求方法，与 HTTP 的请求方法相同，如 GET、PUT、POST、DELETE 等。还可以配置查询 JSON 请求数据、请求对应的 Elasticsearch 节点和请求路径。

（2）支持配置 JSON 验证器对用户输入的 JSON 请求数据进行 JSON 格式校验。

（3）支持重复请求计时器配置重复请求的频率和时间。

（4）在结果转换器中支持使用 JavaScript 表达式变换结果。

下面展示一个拼接 JSON 查询语句的示例，如图 9-25 所示。

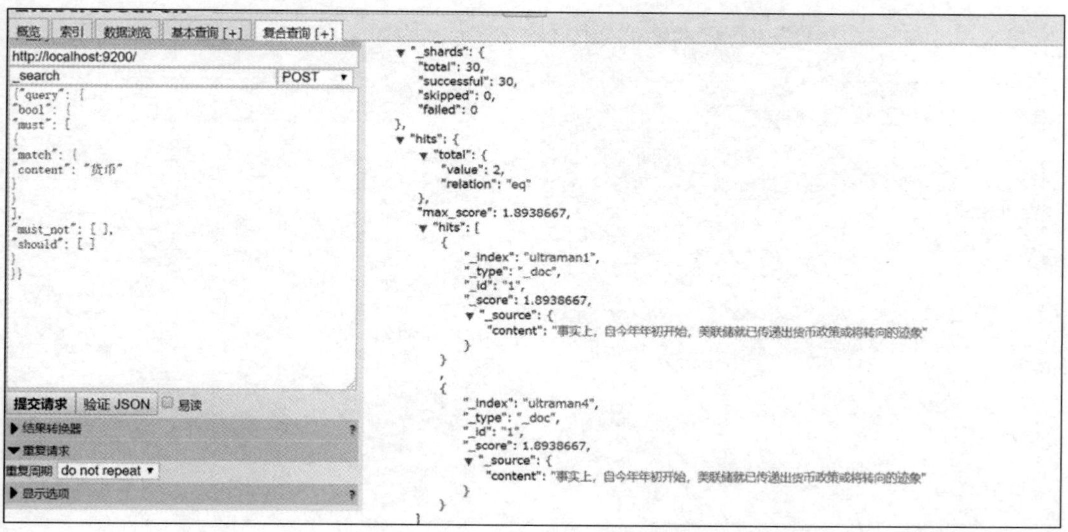

图 9-25

下面以索引 ultraman1 为例，展示对相关索引的操作。如查看索引 ultraman1 中编号为 1 的文档，页面配置如图 9-26 所示。

图 9-26

从图 9-26 可见，文档 1 已经被搜索并展示在右侧窗口。

在索引 ultraman1 中提交数据的示例如图 9-27 所示，增加文档编号为 10 的内容。

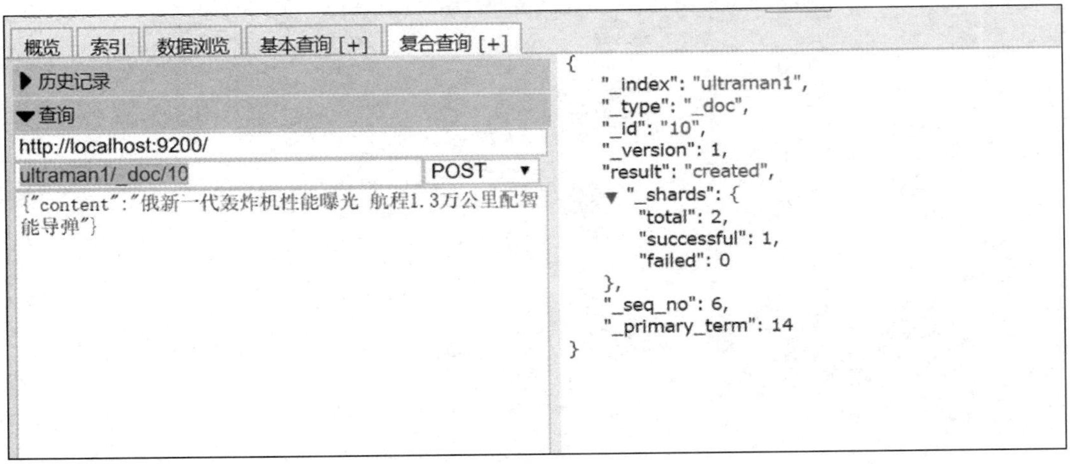

图 9-27

从图 9-27 可见，新增文档编号为 10 的请求已经执行成功，文档编号为 10，当前文档版本编号为 1。

下面通过 GET 请求查看索引 ultraman1 中文档编号为 10 的数据，如图 9-28 所示。

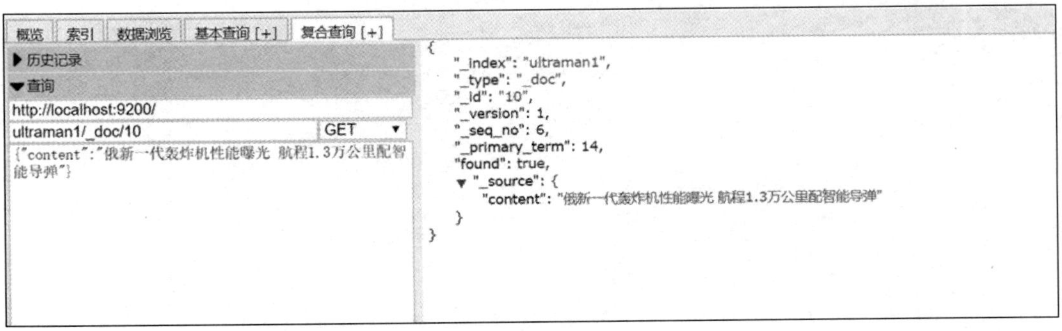

图 9-28

从图 9-28 可见，文档编号为 10 的数据确实已经被索引到 ultraman1 中了。

在上面的示例中我们新增文档时，指明了文档编号。前文曾提及，插入数据使用 POST 或者 PUT 方法，Elasticsearch 会用 POST 方法自动生成 ID，而 PUT 方法则需要指明 ID。下面展示不指明文档编号就新增文档的场景，如图 9-29 所示。

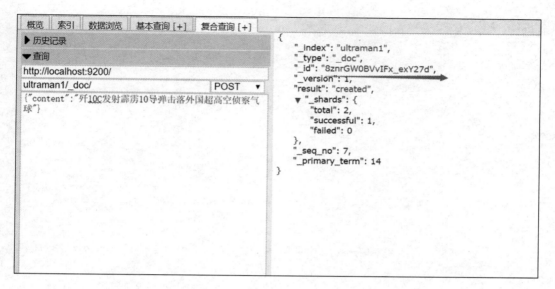

图 9-29

从图 9-29 可见，文档的编号是个随机值。那么使用 PUT 方法时的场景呢？我们依然不指明文档编号，执行请求后，结果如图 9-30 所示。

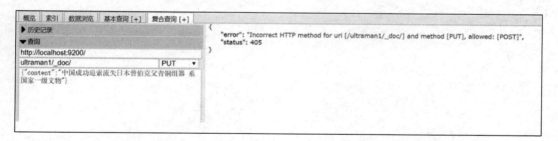

图 9-30

从图 9-30 可见，在索引 ultraman1 中使用 PUT 方法且不指明文档编号为新增文档时会失败。

下面展示在索引 ultraman1 中使用 PUT 方法且指明文档编号的新增文档的场景，如图 9-31 所示。

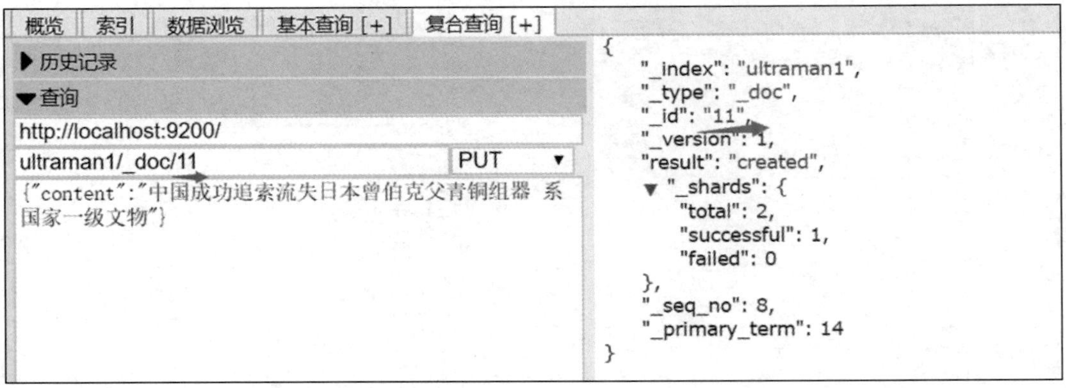

图 9-31

从图 9-31 可见，在索引 ultraman1 中使用 PUT 方法，且指明文档编号的新增文档执行成功。

下面展示删除文档的方法，如图 9-32 所示。

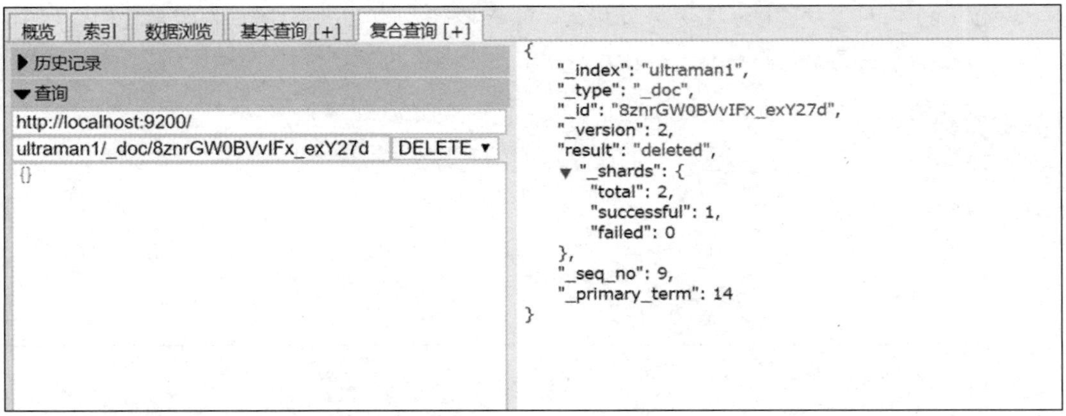

图 9-32

从图 9-32 可见，文档删除请求执行成功。

需要指出的是，由于 Head 插件可以对数据进行增删改查，因此在实际开发时尽量不要使用，如果一定要用，则至少要限制 IP 地址。

删除文档之后还能查到吗？我们尝试一下，查询结果如图 9-33 所示。

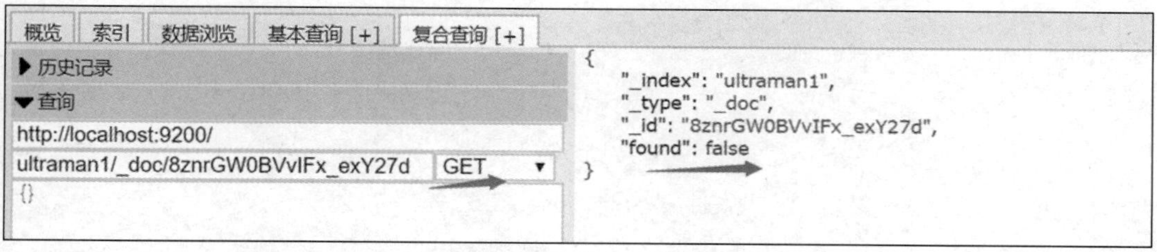

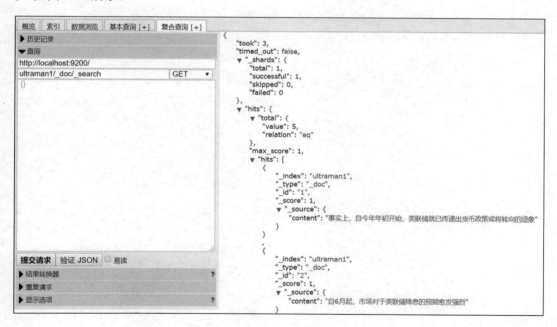

图 9-33

从图 9-33 可见，在索引 ultraman1 中查询已删除文档的请求执行失败。也就是说，已删除的文档不能再被查询。

除可以删除对单个索引单个文档的请求外，还可以对单个索引单个类型的所有数据进行查询，如图 9-34 所示。

图 9-34

在简单查询页面可以进行多个查询条件的匹配；同理，在复杂查询页面，也能进行相关查询，如使用布尔查询。布尔查询的基本格式如下所示：

```
{
    "bool": {
        "must": [],
```

```
        "should": [],
        "must_not": [],
        "filter": {}
    }
}
```

（1）must。文档必须匹配这些条件才能被搜索出来。

（2）must_not。文档必须不匹配这些条件才能被搜索出来。

（3）should。如果满足这些语句中的任意语句，则将增加搜索排名结果_score；否则，对查询结果无任何影响。其主要作用是修正每个文档的相关性得分。

（4）filter。表示必须匹配，但它是以不评分的过滤模式进行的。这些语句对评分没有贡献，只是根据过滤标准排除或包含文档。

需要指出的是，如果没有 must 语句，那么需要至少匹配其中的一条 should 语句。但如果存在至少一条 must 语句，则对 should 语句的匹配没有要求。

布尔查询示例如图 9-35 所示。

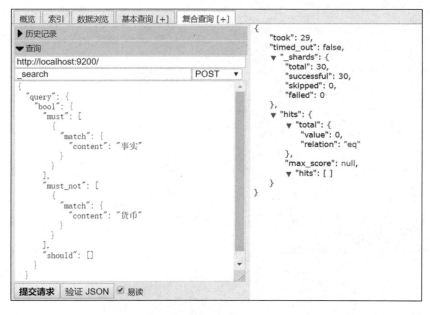

图 9-35

由图 9-35 可知，查询匹配"事实"、且不匹配"货币"的文档为空，这与存储在所有索引中的数据相匹配。

9.7.2 Cerebro 插件

Cerebro 插件是插件工具 kopf 的升级版本。Cerebro 插件中包含了 kopf 的功能，如监控工具，并包含了 Head 插件的部分功能，可以图形化地进行新建索引等操作。

被替换的 kopf 早已不再更新；而 Cerebro 插件则在不断维护和升级中。

读者可以选择对应操作系统环境的版本进行下载。如在 Windows 环境下，可以下载 cerebro-0.8.4.tgz。下载到本地后，将文件解压缩至根目录，根目录内容如图 9-36 所示。

图 9-36

在使用 Cerebro 插件之前，可以进行必要的配置。当然不进行配置，采用默认配置也是可以的。配置文件为 cerebro-0.8.4\cerebro-0.8.4\conf\application.conf。application.conf 的配置内容如下所示：

```
# Secret will be used to sign session cookies, CSRF tokens and for other encryption utilities.
# It is highly recommended to change this value before running cerebro in production.
secret = "ki:s:[[@=Ag?QI`W2jMwkY:eqvrJ]JqoJyi2axj3ZvOv^/KavOT4ViJSv?6YY4[N"
# Application base path
basePath = "/"
# Defaults to RUNNING_PID at the root directory of the app.
# To avoid creating a PID file set this value to /dev/null
#pidfile.path = "/var/run/cerebro.pid"
pidfile.path=/dev/null
# Rest request history max size per user
rest.history.size = 50 // defaults to 50 if not specified
# Path of local database file
#data.path: "/var/lib/cerebro/cerebro.db"
data.path = "./cerebro.db"
es = {
```

```
    gzip = true
  }
  # Authentication
  auth = {
    # either basic or ldap
    type: ${?AUTH_TYPE}
    settings {
      # LDAP
      url = ${?LDAP_URL}
      # OpenLDAP might be something like "ou=People,dc=domain,dc=com"
      base-dn = ${?LDAP_BASE_DN}
      # Usually method should be "simple" otherwise, set it to the SASL
      # mechanisms to try
      method = ${?LDAP_METHOD}
      # user-template executes a string.format() operation where
      # username is passed in first, followed by base-dn. Some examples
      #  - %s => leave user untouched
      #  - %s@domain.com => append "@domain.com" to username
      #  - uid=%s,%s => usual case of OpenLDAP
      user-template = ${?LDAP_USER_TEMPLATE}
      // User identifier that can perform searches
      bind-dn = ${?LDAP_BIND_DN}
      bind-pw = ${?LDAP_BIND_PWD}
      group-search {
        // If left unset parent's base-dn will be used
        base-dn = ${?LDAP_GROUP_BASE_DN}
        // Attribute that represent the user, for example uid or mail
        user-attr = ${?LDAP_USER_ATTR}
        // Define a separate template for user-attr
        // If left unset parent's user-template will be used
        user-attr-template = ${?LDAP_USER_ATTR_TEMPLATE}
        // Filter that tests membership of the group. If this property is empty
        // then there is no group membership check
        // AD example => memberOf=CN=mygroup,ou=ouofthegroup,DC=domain,DC=com
        // OpenLDAP example => CN=mygroup
        group = ${?LDAP_GROUP}
      }
      # Basic auth
      username = ${?BASIC_AUTH_USER}
      password = ${?BASIC_AUTH_PWD}
    }
  }
  # A list of known hosts
  hosts = [
```

```
#{
#   host = "http://localhost:9200"
#   name = "Localhost cluster"
#   headers-whitelist = [ "x-proxy-user", "x-proxy-roles", "X-Forwarded-
#   For" ]
#}
# Example of host with authentication
#{
#   host = "http://some-authenticated-host:9200"
#   name = "Secured Cluster"
#   auth = {
#     username = "username"
#     password = "secret-password"
#   }
#}
]
```

主要参数的配置说明如下。

（1）pidfile.path：服务运行的 pid 的存放位置。如果要避免产生 pid，则可以使用/dev/null。

（2）data.path：Cerebro 存储数据的位置，默认为 Cerebro 安装目录。

（3）auth -> settings -> username：Cerebro Web 服务的账号。

（4）auth -> settings -> password：Cerebro Web 服务的密码。

（5）hosts -> host：Elasticsearch 集群的 host 地址。

（6）hosts -> name：Elasticsearch 集群的名称。

配置 hosts 的代码如下所示：

```
hosts = [
  {
    host = "http:// localhost:9200"
    name = "Test Elasticsearch Cluster for niudong"
  }
```

配置后，切换到 cerebro-0.8.4 目录下并启动 Cerebro 服务，切换命令和启动命令如图 9-37 所示。

```
PS C:\Users\牛冬> cd C:\Users\牛冬\Desktop\cerebro-0.8.4\cerebro-0.8.4
PS C:\Users\牛冬\Desktop\cerebro-0.8.4\cerebro-0.8.4> bin/cerebro
Picked up JAVA_TOOL_OPTIONS: -Dfile.encoding=UTF-8
[info] play.api.Play - Application started (Prod) (no global state)
[info] p.c.s.AkkaHttpServer - Listening for HTTP on /0:0:0:0:0:0:0:0:9000
```

图 9-37

当 Cerebro 服务启动后,即可在浏览器中访问如下 URL,打开 Cerebro 图形界面:

```
http://localhost:9000/
```

从图 9-37 可以看出,Cerebro 会默认连接 http://localhost:9200/。用户可以根据自己环境的情况连接对应的 Elasticsearch 节点。配置后,单击"Connect"按钮进行连接,如图 9-38 所示。

图 9-38

连接 Elasticsearch 节点成功后,会进入 Cerebro 的首页,如图 9-39 所示。

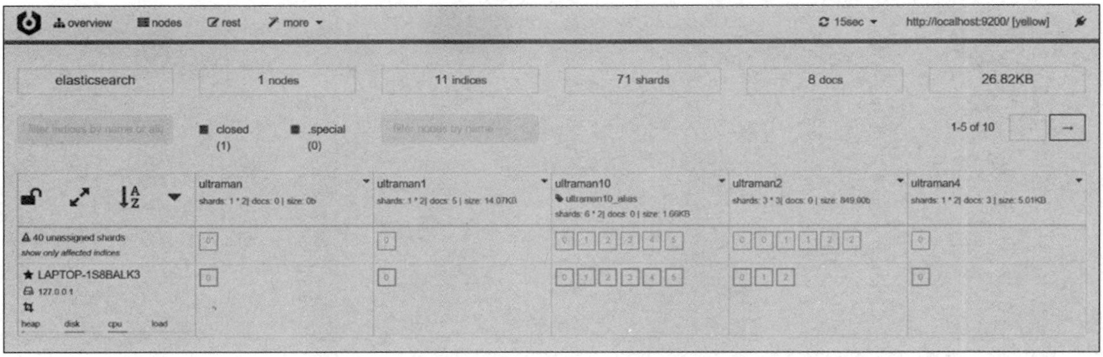

图 9-39

从图 9-39 可见,Cerebro 首页中有"Overview"、"nodes"、"rest"和"more"四个标签页。默认当前是"Overview"页面。

在"Overview"页面中，可以看到 Elasticsearch 集群的各个 Node 节点的详细信息。该标签页分为三部分，顶端的线条、Elasticsearch 集群内各种信息的统计、各个节点的信息。其中：

（1）顶端的线条：颜色释义与 Head 插件中的颜色释义相同，共绿色、红色、黄色三种，绿色代表集群工作正常。

（2）Elasticsearch 集群内各种信息的统计：包括集群名称、节点数量、索引数量、分片数量、文档数量和索引所占存储空间的大小等信息。

（3）各个节点的信息：在最下方表格中，每行代表一个节点，每列代表一个索引。

单击某个索引，即可查看该索引下的信息，如图 9-40 所示。

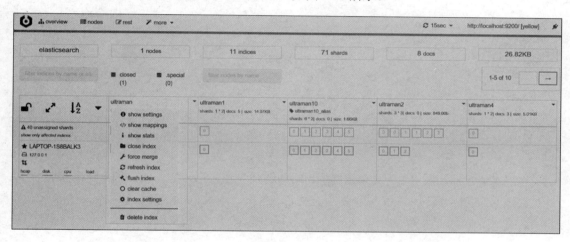

图 9-40

切换到"Nodes"标签页，可以看到各节点的资源使用情况，如图 9-41 所示。

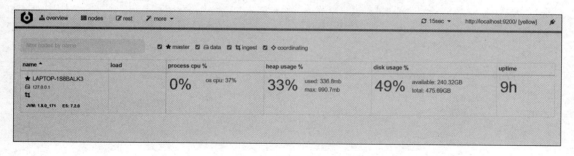

图 9-41

从图 9-41 可见，各节点的资源使用情况主要有 CPU、堆、磁盘使用、数据更新时间等。

切换到"Rest"标签页，用户可以向 Elasticsearch 集群发出 RESTful 格式的 API 请求，如图 9-42 所示。

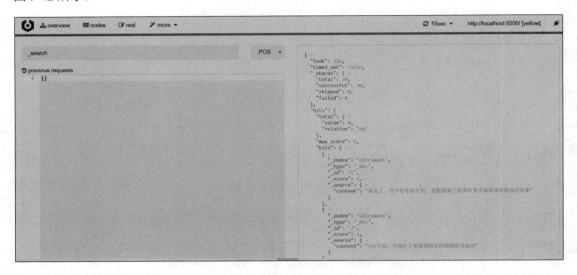

图 9-42

在提交请求前，既可以配置请求的刷新时间和频率（这些配置与 Head 插件类似），也可以通过如图 9-43 所示按钮，分别进行复制 URL（cURL）、请求 JSON 格式化（format）和发出请求（send）操作。

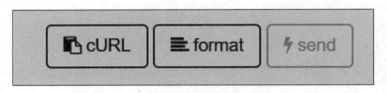

图 9-43

切换到"More"标签页，可以做更多的操作，如图 9-44 所示。

图 9-44

单击"create index"选项,即可创建索引,域名如图 9-45 所示。

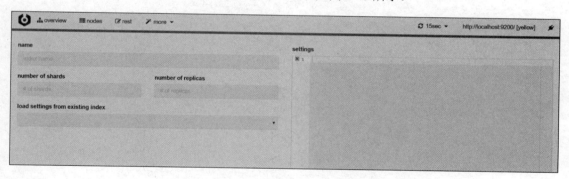

图 9-45

填写相关字段即可创建对应的索引。

单击"cluster settings"选项,用户可以对 Elasticsearch 集群进行设置,设置参数如图 9-46 所示。

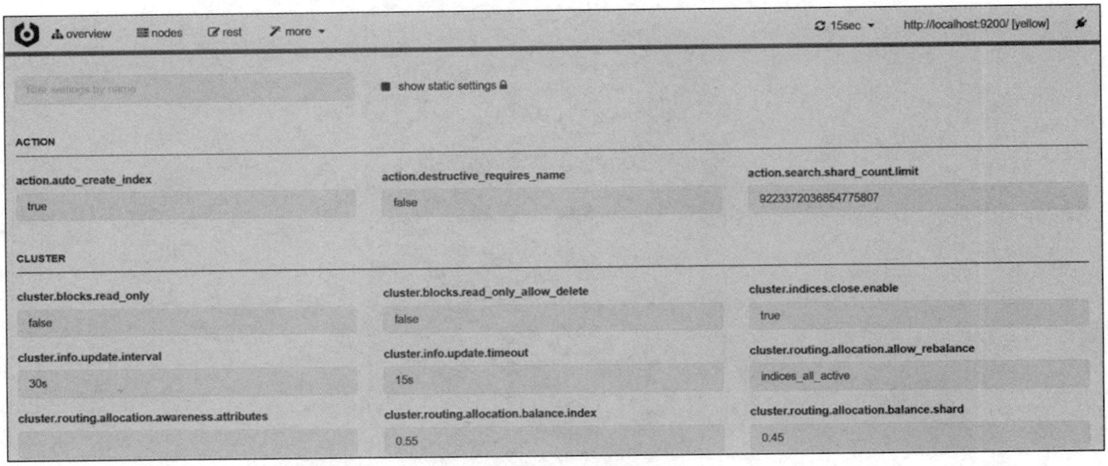

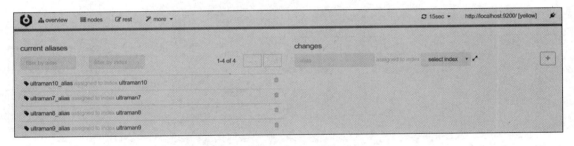

图 9-46

单击"aliases"选项,可以为索引维护其别名信息,如图 9-47 所示。

图 9-47

单击"analysis"选项,可以进行字符串分析操作,如图 9-48 所示。

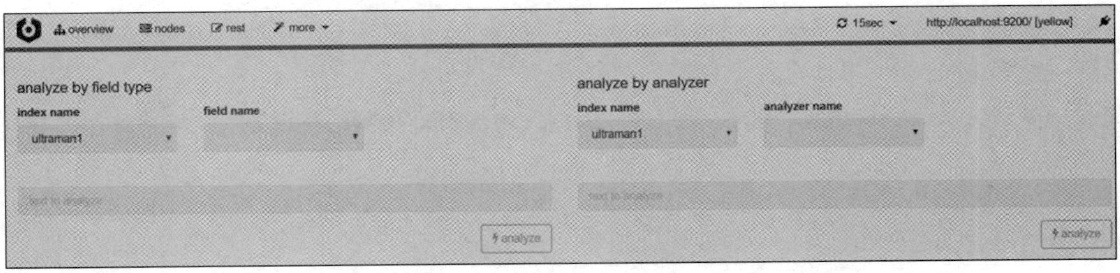

图 9-48

9.8 知识点关联

插件，即 Plug-in，又称为 addin、add-in、addon 或 add-on，在游戏或手机等场景中被称为"外挂"，是一个多称谓的词汇。

一般来说，插件是一种程序代码的统称。插件代码的编写会遵循一定规范，特别是载体的应用程序接口。因此，插件只能运行在程序规定的系统平台下，像 Java 一样"一次编译、到处运行"的插件几乎是不存在的。

此外，插件一般不能脱离指定的平台单独运行，毕竟插件需要由载体系统提供必要的本地函数库和数据。与之相反的是，应用程序无须依赖插件即可运行，因而当插件加载到应用程序上时，动态更新不会对应用程序造成影响。

最早的插件出现在 20 世纪 70 年代中期的 EDT 文本编辑器中。EDT 文本编辑器在 Univac 90/60 系列大型机上运行 Unisys VS/9 操作系统时，可以提供运行插件代码的功能，该功能允许插件程序进入编辑器的缓冲，允许插件程序染指内存中正在编辑的任务。插件程序使得编辑器可以在缓冲区上进行文本编辑，而这个缓冲区是编辑器和插件共同享用的。

在个人电脑时代，第一个带有插件的应用软件是 1987 发行的、安装在苹果电脑上的 HyperCard 和 QuarkXPress。

现在，我们常用的浏览器中就有各种各样的插件程序，如广告过滤、智能填写表单，甚至是火车票抢票插件等。

插件的工作原理是什么呢？

一般来说，已公开的应用程序接口提供一个标准的接口或函数库，允许其他人编写插件与应用程序互动，并提供一个稳定的应用程序接口，允许其他插件正常运行。

使用插件的好处很多，特别是在产品功能拓展、生态构建过程中。

其一，一般插件的载体程序会提供类似总线的架构，接口清晰，易于开发者理解；多个插件安装后也是独立运行、互不干扰。

其二，插件和载体程序是可插拔关系，组合很灵活，因而插件和载体程序都容易修改，方便升级和维护。

其三，插件程序的编写一般都是基于单一职责原则，因此插件的可移植性较强、重用机会多。也正是因为基于单一职责原则，所以多个插件之间是松耦合关系。

需要指出的是，与插件概念类相似的一个概念是组件。组件和插件有明显的区别，插件属于程序接口；组件是一种控件、对象，复用程度更高。

9.9 小结

本章主要介绍了 Elasticsearch 的插件生态，包括官方维护的插件和技术社区维护的插件；介绍了插件的统一管理方法，如安装和卸载等。

本章先后介绍了分析插件、API 扩展插件、监控插件和数据提取插件，重点介绍了分析插件中的 ICU 分析插件和智能中文分析插件、监控插件中的 Head 插件和 Cerebro 插件。

第 10 章
Elasticsearch 生态圈

天际浮云入思深
物情生态看销沉

第 9 章介绍了 Elasticsearch 的插件生态，插件生态是依托于 Elasticsearch 内部的，属于一种相对狭义、微观的生态；本章主要介绍 Elasticsearch 的宏观生态。

10.1 ELK

提到 Elasticsearch 生态，很多人第一反应就是 ELK Stack。什么是 ELK Stack 呢？很简单，ELK Stack 指的就是 Elastic Stack。

10.1.1 Elastic Stack

"ELK"是三个开源项目的首字母缩写，这三个项目分别是：Elasticsearch、Logstash 和 Kibana，如图 10-1 所示。当然，这并非是 Elastic Stack 的全部，读者可以根据需要在生态中添加 Redis、Kafka、Filebeat 等软件。

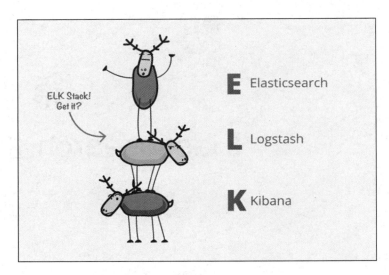

图 10-1

当前的 Elastic Stack 其实是 ELK Stack 的更新换代产品。2015 年，ELK Stack 中加入了一系列轻量级的单一功能数据采集器，并把它们叫作 Beats。

Beats 加入 ELK 家族后，再叫 ELK 显然不太合适，那么新的家族叫什么呢？BELK？BLEK？ELKB？

其实，Elastic 官方当时也的确想继续沿用首字母缩写的方式，但 Elastic 又相继收购了 APM 公司 Opbeat、机器学习公司 Prelert、SaaS 服务公司 Found、搜索服务公司 Swiftype、终端安全公司 Endgame 等来扩大自己的商业版图。

对于 Elastic 扩展速度如此之快的生态而言，一直采用首字母缩写的确不是长久之计。于是，Elastic Stack 这个"一劳永逸"般的名字就诞生了。

10.1.2 Elastic Stack 版本的由来

Elasticsearch 的版本号从 2 直接升到 5 是怎么回事呢？

在最初的 Elastic Stack 生态中，Elasticsearch、Logstash、Kibana 和 Beats 有各自的版本号，如当 Elasticsearch 和 Logstash 的版本号为 V2.3.4 时，Kibana 的版本号是 V4.5.3，而 Beats 的版本号是 V1.2.3。

因此，Elastic Stack 官方将产品版本号也进行了统一，从 V5.0 开始。因为当时的 Kibana 版本号已经是 4.x 了，其下个版本只能是 5.0，所以其他产品的版本号也随之"跳级"，于是 V5.0 版本的 Elastic Stack 在 2016 年就面世了。

10.1.3　ELK 实战的背景

在实际使用过程中，什么场景适合使用 ELK 呢？

在实战中，我们既可以用 ELK 管理和分析日志，也可以用 ELK 分析索引中的数据。

在当前的软件开发过程中，业务发展节奏越来越快，服务器梳理越来越多，随之而来的就是各种访问日志、应用日志和错误日志。随着时间的流逝，日志的累积也越来越多。

此时，会出现这样的问题：运维人员无法很好地管理日志；开发人员排查业务问题时需要到服务器上查询大量日志；当运营人员需要一些业务数据时，需要到服务器上分析日志。

在上述场景中，通常意义上的"awk"和"grep"命令已经力不从心，而且效率很低。这时 ELK 就可以"隆重登场"啦！

ELK 的三个组件是如何分工协作的呢？

首先，我们使用 Logstash 进行日志的搜集、分析和过滤。一般工作方式为 C/S 架构，Client 端会被安装在需要收集日志的主机上，Server 端则负责收集的各节点的日志数据，并进行过滤、修改和分析等操作，预处理过的数据会一并发到 Elasticsearch 上。

随后将 Kibana 接入 Elasticsearch，并为 Logstash 和 Elasticsearch 提供日志分析友好的 Web 界面，帮助用户汇总、分析和搜索重要数据的日志。

10.1.4　ELK 的部署架构变迁

ELK 架构为数据分布式存储、可视化查询和日志解析创建了一个功能强大的管理链。ELK 架构为用户建立了集中式日志收集系统，将所有节点上的日志统一收集、管理和访问。三者相互配合，取长补短，共同完成分布式大数据处理工作。

当前官方推荐的 ELK 部署架构并非一步到位，而是经过迭代演进发展而来的。下面简单介绍 ELK 架构的发展历程。

最简单的一种 ELK 部署架构方式如图 10-2 所示。

首先由分布于各个服务节点上的 Logstash 搜集相关日志和数据，经过 Logstash 的分析和过滤后发送给远端服务器上的 Elasticsearch 进行存储。Elasticsearch 将数据以分片的形式压缩存储，并提供多种 API 供用户进行查询操作。用户还可以通过配置 Kibana Web Portal 对日志进行查询，并根据数据生成报表。

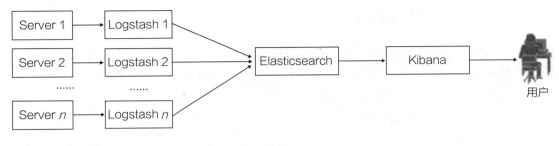

图 10-2

该架构最显著的优点是搭建简单,易于上手。但缺点同样很突出,因为 Logstash 消耗资源较大,所以在运行时会占用很多的 CPU 和内存。并且系统中没有消息队列缓存等持久化手段,因而存在数据丢失隐患。因此,一般这种部署架构通常用于学习和小规模集群。

基于第一种 ELK 部署架构的优缺点,第二种架构引入了消息队列机制,如图 10-3 所示。

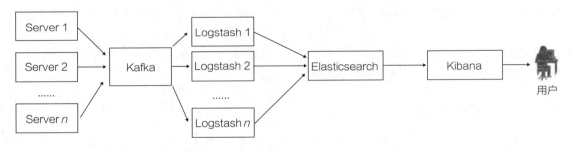

图 10-3

位于各个节点上的 Logstash 客户端先将数据和日志等内容传递给 Kafka,当然,也可以用其他消息机制,如各类 MQ(Message Queue)和 Redis 等。

Kafka 会将队列中的消息和数据传递给 Logstash,经过 Logstash 的过滤和分析等处理后,传递给 Elasticsearch 进行存储。最后由 Kibana 将日志和数据呈现给用户。

在该部署架构中,Kafka 的引入使得即使远端 Logstash 因故障而停止运行,数据也会被存储下来,从而避免数据丢失。

第二种部署架构解决了数据的可靠性问题,但 Logstash 的资源消耗依然较多,因而引出第三种架构。第三种架构引入了 Logstash-forwarder,如图 10-4 所示。

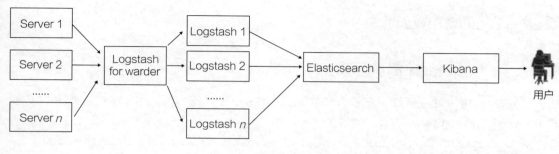

图 10-4

Logstash-forwarder 将日志数据搜集并统一后发送给主节点上的 Logstash，Logstash 在分析和过滤日志数据后，把日志数据发送至 Elasticsearch 进行存储，最后由 Kibana 将数据呈现给用户。

这种架构解决了 Logstash 在各计算机点上占用系统资源较多的问题。与 Logstash 相比，Logstash-forwarder 所占系统的 CPU 和内存几乎可以忽略不计。

而且，Logstash-forwarder 的数据安全性更好。Logstash-forwarder 和 Logstash 之间的通信是通过 SSL 加密传输的，因此安全有保障。

随着 Beats 组件引入 ELK Stack，第四种部署架构应运而生，如图 10-5 所示。

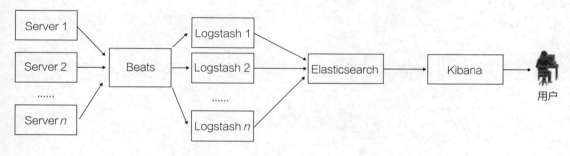

图 10-5

在实际使用中，Beats 平台在满负荷状态时所耗系统资源和 Logstash-forwarder 相当，但其扩展性和灵活性更好。Beats 平台目前包含 Packagebeat、Topbeat 和 Filebeat 三个产品，均为 Apache 2.0 License。同时用户可以根据需要进行二次开发。

与前面三个部署架构相比，显然第四种架构更灵活，可扩展性更强。

用户可以根据自己的需求搭建自己的 ELK。

10.2 Logstash

10.2.1 Logstash 简介

Logstash 由三部分组成，即输入模块（INPUTS）、过滤器模块（FILTERS）和输出模块（OUTPUTS），如图 10-6 所示。

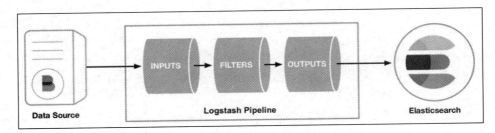

图 10-6

Logstash 能够动态地采集、转换和传输数据，不受格式或复杂度的影响。利用 Grok 从非结构化数据中派生出结构，从 IP 地址解码出地理坐标，匿名化或排除敏感字段，并简化整体处理过程。

从官网下载 Logstash 安装包。下载完成后，在本地解压缩。解压缩后的根目录内容如 10-7 所示。

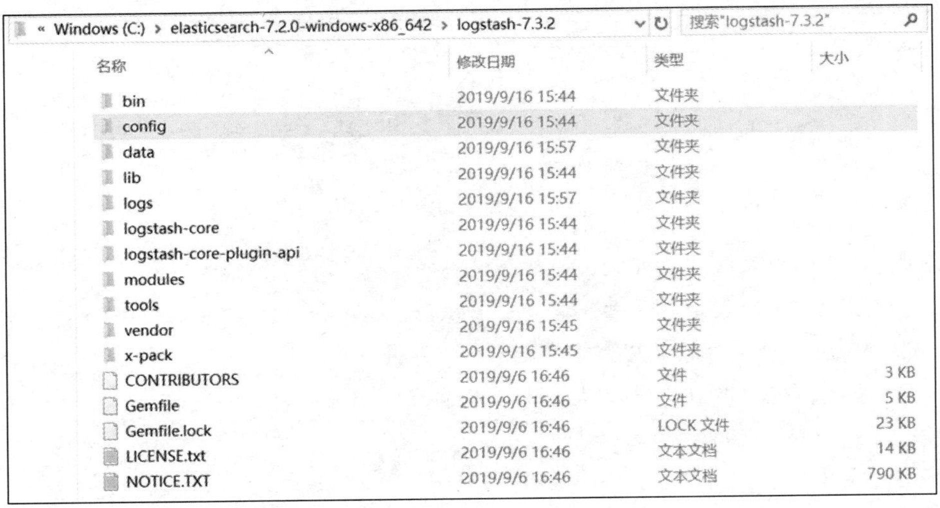

图 10-7

根目录下有 bin、config、data、lib、logstash core 和 tools 等内容。

在 Logstash 启动后，会自动创建 logs 目录。随后配置 config 目录下的 logstash.conf 文件。首次配置时可参考同目录的 logstash-simple.conf 示例进行配置。配置后，执行 bin/logstash -f logstash.conf 即可启动 Logstash 服务，如下所示：

```
PS C:\elasticsearch-7.2.0-windows-x86_642\logstash-7.3.2> bin/logstash -f logstash.conf
warning: ignoring JAVA_TOOL_OPTIONS=$JAVA_TOOL_OPTIONS
Thread.exclusive is deprecated, use Thread::Mutex
Sending Logstash logs to C:/elasticsearch-7.2.0-windows-x86_642/logstash-7.3.2/logs which is now configured via log4j2.properties
[2019-09-16T16:00:27,478][WARN ][logstash.config.source.multilocal] Ignoring the 'pipelines.yml' file because modules or command line options are specified
[2019-09-16T16:00:27,532][INFO ][logstash.runner          ] Starting Logstash {"logstash.version"=>"7.3.2"}
[2019-09-16T16:00:29,268][INFO ][logstash.config.source.local.configpathloader] No config files found in path {:path=>"C:/elasticsearch-7.2.0-windows-x86_642/logstash-7.3.2/logstash.conf"}
[2019-09-16T16:00:29,287][ERROR][logstash.config.sourceloader] No configuration found in the configured sources.
[2019-09-16T16:00:29,911][INFO ][logstash.agent           ] Successfully started Logstash API endpoint {:port=>9600}
```

此时，在浏览器中输入 http://localhost:9600/，浏览器的页面中即可输出如下内容：

```
{
    "host": "LAPTOP-1S8BALK3",
    "version": "7.3.2",
    "http_address": "127.0.0.1:9600",
    "id": "e0903a8b-e533-4836-98bb-79d905720920",
    "name": "LAPTOP-1S8BALK3",
    "ephemeral_id": "a96193e8-2ce9-4eea-8a79-90972c92085c",
    "status": "green",
    "snapshot": false,
    "pipeline": {
        "workers": 4,
        "batch_size": 125,
        "batch_delay": 50
    },
    "build_date": "2019-09-06T16:42:57+00:00",
    "build_sha": "1071165b526bcd43b475caed0b4289e3d32dc52f",
    "build_snapshot": false
}
```

需要指出的是，Logstash 文件夹存放的路径中不能有中文命名的文件夹，否则会给出如下错误提示：

```
Logstash - java.lang.IllegalStateException: Logstash stopped processing because of an error: (ArgumentError) invalid byte sequence in US-ASCII
```

10.2.2 Logstash 的输入模块

Logstash 的输入模块用于采集各种样式、大小和来源的数据。一般来说，数据往往以各种各样的形式，或分散或集中地存储于很多系统中。Logstash 支持各种输入选择，可以在同一时间从众多常用来源捕捉事件，能够以流式传输方式，轻松地从用户的日志、指标、Web 应用、数据存储及各种 AWS 服务中采集数据。

为了支持各种数据输入，Logstash 提供了很多输入插件，汇总如下。

（1）azure_event_hubs：该插件从微软 Azure 事件中心接收数据。读者可访问 GitHub 官网，搜索 logstash-input-azure_event_hubs 获取插件。

（2）beats：该插件从 Elastic Beats 框架接收数据。读者可访问 GitHub 官网，搜索 logstash-input-beats 获取插件。

（3）cloudwatch：该插件从 Amazon Web Services CloudWatch API 中提取数据。读者可访问 GitHub 官网，搜索 logstash-input-cloudwatch 获取插件。

（4）couchdb_changes：该插件从 CouchDB 更改 URI 的流式处理事件中获取数据。读者可访问 GitHub 官网，搜索 logstash-input-couchdb_changes 获取插件。

（5）dead_letter_queue：该插件从 logstash 的 dead letter 队列中读取数据。读者可访问 GitHub 官网，搜索 logstash-input-dead_letter_queue 获取插件。

（6）elasticsearch：该插件从 ElasticSearch 群集中读取查询结果。读者可访问 GitHub 官网，搜索 logstash-input-elasticsearch 获取插件。

（7）exec：该插件将 shell 命令的输出捕获为事件，并获取数据。读者可访问 GitHub 官网，搜索 logstash-input-exec 获取插件。

（8）file：该插件从文件流式处理中获取数据。读者可访问 GitHub 官网，搜索 logstash-input-file 获取插件。

（9）ganglia：该插件通过 UDP 数据包读取 ganglia 中的数据包来获取数据。读者可访问 GitHub 官网，搜索 logstash-input-ganglia 获取插件。

（10）gelf：该插件从 graylog2 中读取 gelf 格式的消息获取数据。读者可访问 GitHub 官网，搜索 logstash-input-gelf 获取插件。

（11）http：该插件通过 HTTP 或 HTTPS 接收事件获取数据。读者可访问 GitHub 官网，搜索 logstash-input-http 获取插件。

（12）jdbc：该插件通过 JDBC 接口从数据库中获取数据。读者可访问 GitHub 官网，搜索 logstash-input-jdbc 获取插件。

（13）kafka：该插件从 Kafka 主题中读取事件，从而获取数据。读者可访问 GitHub 官网，搜索 logstash-input-kafka 获取插件。

（14）log4j：该插件通过 TCP 套接字从 Log4J SocketAppender 对象中读取数据。读者可访问 GitHub 官网，搜索 logstash-input-log4j 获取插件。

（15）rabbitmq：该插件从 RabbitMQ 数据交换中提取数据。读者可访问 GitHub 官网，搜索 logstash-input-rabbitmq 获取插件。

https://github.com/logstash-plugins/logstash-input-rabbitmq

（16）redis：该插件从 redis 实例中读取数据。读者可访问 GitHub 官网，搜索 logstash-input-redis 获取插件。

10.2.3　Logstash 过滤器

Logstash 过滤器用于实时解析和转换数据。

在数据从源传输到存储库的过程中，Logstash 过滤器能够解析各个数据事件，识别已命名的字段，构建对应的数据结构，并将它们转换成通用格式，以便更轻松、更快速地进行分析，实现商业价值。

Logstash 过滤器有以下特点：

（1）利用 Grok 从非结构化数据中派生出结构。

（2）从 IP 地址破译出地理坐标。

（3）将 PII 数据匿名化，完全排除敏感字段。

（4）简化整体处理，不受数据源、格式或架构的影响。

为了处理各种各样的数据源，Logstash 提供了丰富多样的过滤器库，常用的过滤器插件汇总如下。

（1）aggregate：该插件用于从一个任务的多个事件中聚合信息。读者可访问 GitHub 官网，搜索 logstash-filter-aggregate 获取插件。

（2）alter：该插件对 mutate 过滤器不处理的字段执行常规处理。读者可访问 GitHub 官网，搜索 logstash-filter-alter 获取插件。

（3）bytes：该插件将以计算机存储单位表示的字符串形式，如"123MB"或"5.6GB"，解析为以字节为单位的数值。读者可访问 GitHub 官网，搜索 logstash-filter-bytes 获取插件。

（4）cidr：该插件根据网络块列表检查 IP 地址。读者可访问 GitHub 官网，搜索 logstash-filter-cidr 获取插件。

（5）cipher：该插件用于对事件应用增加或移除密钥。读者可访问 GitHub 官网，搜索

logstash-filter-cipher 获取插件。

（6）clone：该插件用于复制事件。读者可访问 GitHub 官网，搜索 logstash-filter-clone 获取插件。

（7）csv：该插件用于将逗号分隔的值数据解析为单个字段。读者可访问 GitHub 官网，搜索 logstash-filter-csv 获取插件。

（8）date：该插件用于分析字段中的日期，多用于事件日志中存储的时间戳。读者可访问 GitHub 官网，搜索 logstash-filter-date 获取插件。

（9）dns：该插件用于执行正向或反向 DNS 查找。读者可访问 GitHub 官网，搜索 logstash-filter-dns 获取插件。

（10）elasticsearch：该插件用于将 Elasticsearch 日志事件中的字段复制到当前事件中。读者可访问 GitHub 官网，搜索 logstash-filter- elasticsearch 获取插件。

（11）geoip 该插件用于添加有关 IP 地址的地理信息。读者可访问 GitHub 官网，搜索 logstash-filter- geoip 获取插件。

（12）json：该插件用于解析 JSON 事件。读者可访问 GitHub 官网，搜索 logstash-filter-json 获取插件。

（13）kv：该插件用于分析键值对。读者可访问 GitHub 官网，搜索 logstash-filter-kv 获取插件。

（14）memcached：该插件用于提供与 memcached 中数据的集成。读者可访问 GitHub 官网，搜索 logstash-filter- memcached 获取插件。

（15）split：该插件用于将多行消息拆分为不同的事件。读者可访问 GitHub 官网，搜索 logstash-filter- split 获取插件。

10.2.4 Logstash 的输出模块

Logstash 的输出模块用于将目标数据导出到用户选择的存储库。

在 Logstash 中，尽管 Elasticsearch 是 Logstash 官方首选的，但它并非唯一选择。

Logstash 提供众多输出选择，用户可以将数据发送到指定的地方，并且能够灵活地解锁众多下游用例。

（1）csv：该插件以 CVS 格式将结果数据写入磁盘。读者可访问 GitHub 官网，搜索 logstash-output-csv 获取插件。

（2）mongodb：该插件将结果数据写入 MongoDB。读者可访问 GitHub 官网，搜索 logstash-output-mongodb 获取插件。

（3）elasticsearch：该插件将结果数据写入 Elasticsearch。读者可访问 GitHub 官网，搜索 logstash-output-elasticsearch 获取插件。

（4）email：该插件将结果数据发送到指定的电子邮件。读者可访问 GitHub 官网，搜索 logstash-output- email 获取插件。

（5）kafka：该插件将结果数据写入 Kafka 的 Topic 主题。读者可访问 GitHub 官网，搜索 logstash-output- kafka 获取插件。

（6）file：该插件将结果数据写入磁盘上的文件。读者可访问 GitHub 官网，搜索 logstash-output- file 获取插件。

（7）redis：该插件使用 redis 中的 rpush 命令将结果数据发送到 redis 队列。读者可访问 GitHub 官网，搜索 logstash-output- redis 获取插件。

10.3 Kibana

Kibana 是一个基于 Web 的图形界面，可以让用户在 Elasticsearch 中使用图形和图表对数据进行可视化。

在实际使用过程中，Kibana 一般用于搜索、分析和可视化存储在 Elasticsearch 指标中的日志数据。Kibana 利用 Elasticsearch 的 REST 接口检索数据，不仅允许用户创建自己的数据定制仪表板视图，还允许他们以特殊的方式查询和过滤数据。可以说从跟踪、查询、负载到理解请求如何流经整个应用，Kibana 都能轻松完成。

10.3.1 Kibana 简介

Kibana 提供了基本内容服务、位置分析服务、时间序列服务、机器学习服务，以及图表和网络服务。

（1）基本内容服务：指的是 Kibana 核心产品中搭载的一批经典功能，如基于筛选数据绘制柱形图、折线图、饼图、旭日图等。

（2）位置分析服务：主要借助 Elastic Maps 探索位置数据。另外，还可以获得创意，对定制图层和矢量形状进行可视化。

（3）时间序列服务：借助 Kibana 团队精选的时序数据 UI，对用户所用 Elasticsearch 中的数据执行高级时间序列分析。因而，用户可以利用功能强大、简单易学的表达式来描述查询、转换和可视化。

（4）机器学习服务：主要是借助非监督型机器学习功能检测隐藏在用户所用 Elasticsearch 数据中的异常情况，并探索那些对用户有显著影响的属性。

（5）图表和网络服务：凭借搜索引擎的相关性功能，结合 Graph 关联分析，揭示用户所用 Elasticsearch 数据中极其常见的关系。

此外，Kibana 还支持用户把 Kibana 可视化内容分享给他人，如团队成员、老板、客户、合规经理或承包商等，进而让每个人都感受到 Kibana 的便利。

除分享链接外，Kibana 还有其他内容输出形式，如嵌入仪表板，导出为 PDF、PNG 或 CSV 等格式文件，以便把这些文件作为附件发送给他人。

我们可从官网下载 Kibana，下载完成后，即可在本地进行解压缩。解压缩后的 Kibana 的根目录如图 10-8 所示。

图 10-8

10.3.2 连接 Elasticsearch

由于 Kibana 服务需要连接到 Elasticsearch，因此在启动 Kibana 前，需要先启动 Elasticsearch，否则，Kibana 在启动过程中会显示如下所示错误：

```
[warning][admin][elasticsearch]     Unable     to     revive     connection:
http://localhost:9200/
```

在启动 Elasticsearch 后，需要对 config/kibana.yml 文件进行配置。主要是配置 elasticsearch.hosts 属性，该属性在无配置的情况下默认连接到 http://localhost:9200。

在配置好 elasticsearch.hosts 属性后，即可通过如下命令启动 Kibana：

```
bin/kibana
```

如果是在 Windows 环境下启动，则使用如下命令启动 Kibana：

```
bin\kibana.bat
```

当 Kibana 启动成功后，输出内容如下所示：

```
Ready
  log   [09:38:23.046] [info][status][plugin:snapshot_restore@7.2.0] Status changed from uninitialized to yellow - Waiti
ng for Elasticsearch
  log   [09:38:23.061] [info][status][plugin:data@7.2.0] Status changed from uninitialized to green - Ready
  log   [09:38:23.442] [info][status][plugin:timelion@7.2.0] Status changed from uninitialized to green - Ready
  log   [09:38:23.452] [info][status][plugin:ui_metric@7.2.0] Status changed from uninitialized to green - Ready
  log   [09:38:24.983] [info][status][plugin:elasticsearch@7.2.0] Status changed from yellow to green - Ready
  log   [09:38:25.049] [info][license][xpack] Imported license information from Elasticsearch for the [data] cluster: mo
de: basic | status: active
  log   [09:38:25.073] [info][status][plugin:xpack_main@7.2.0] Status changed from yellow to green - Ready
  log   [09:38:25.074] [info][status][plugin:graph@7.2.0] Status changed from yellow to green - Ready
  log   [09:38:25.075] [info][status][plugin:searchprofiler@7.2.0] Status changed from yellow to green - Ready
  log   [09:38:25.079] [info][status][plugin:ml@7.2.0] Status changed from yellow to green - Ready
  log   [09:38:25.079] [info][status][plugin:tilemap@7.2.0] Status changed from yellow to green - Ready
  log   [09:38:25.083] [info][status][plugin:watcher@7.2.0] Status changed from yellow to green - Ready
  log   [09:38:25.083] [info][status][plugin:grokdebugger@7.2.0] Status changed from yellow to green - Ready
  log   [09:38:25.084] [info][status][plugin:logstash@7.2.0] Status changed from yellow to green - Ready
  log   [09:38:25.085] [info][status][plugin:beats_management@7.2.0] Status changed from yellow to green - Ready
  log   [09:38:25.086] [info][status][plugin:index_management@7.2.0] Status changed from yellow to green - Ready
  log   [09:38:25.087] [info][status][plugin:index_lifecycle_management@7.2.0] Status changed from yellow to green - Rea
dy
  log   [09:38:25.117] [info][status][plugin:rollup@7.2.0] Status changed from yellow to green - Ready
  log   [09:38:25.118] [info][status][plugin:remote_clusters@7.2.0] Status changed from yellow to green - Ready
  log   [09:38:25.119] [info][status][plugin:cross_cluster_replication@7.2.0] Status changed from yellow to green - Read
y
  log   [09:38:25.120] [info][status][plugin:snapshot_restore@7.2.0] Status changed from yellow to green - Ready
  log   [09:38:25.122] [info][kibana-monitoring][monitoring] Starting monitoring stats collection
  log   [09:38:25.140] [info][status][plugin:maps@7.2.0] Status changed from yellow to green - Ready
  log   [09:38:30.029] [    ][reporting] Generating a random key for xpack.reporting.encryptionKey. To prevent pendin
g reports from failing on restart, please set xpack.reporting.encryptionKey in kibana.yml
  log   [09:38:30.043] [info][status][plugin:reporting@7.2.0] Status changed from uninitialized to green - Ready
  log   [09:38:30.143] [    ][task_manager] This Kibana instance defines an older template version (7020099) than is
currently in Elasticsearch (7030299). Because of the potential for non-backwards compatible changes, this Kibana instanc
e will only be able to claim scheduled tasks with "kibana.apiVersion" <= 1 in the task metadata.
  log   [09:38:30.196] [    ][task_manager] This Kibana instance defines an older template version (7020099) than is
currently in Elasticsearch (7030299). Because of the potential for non-backwards compatible changes, this Kibana instanc
e will only be able to claim scheduled tasks with "kibana.apiVersion" <= 1 in the task metadata.
  log   [09:38:30.379] [    ][task_manager] This Kibana instance defines an older template version (7020099) than is
currently in Elasticsearch (7030299). Because of the potential for non-backwards compatible changes, this Kibana instanc
e will only be able to claim scheduled tasks with "kibana.apiVersion" <= 1 in the task metadata.
  log   [09:38:30.638] [info][migrations] Creating index .kibana_1.
  log   [09:38:31.049] [info][migrations] Pointing alias .kibana to .kibana_1.
  log   [09:38:31.136] [info][migrations] Finished in 508ms.
  log   [09:38:31.139] [info][listening] Server running at http://localhost:5601
  log   [09:38:31.863] [info][status][plugin:spaces@7.2.0] Status changed from yellow to green - Ready
```

我们可以在浏览器的地址栏中输入 http://localhost:5601，打开 Kibana 页面。此时，打开的页面是一个欢迎页面。

我们既可以单击"Try our sample data"按钮体验 Kibana 的功能，也可以单击"Explore on my own"按钮使用自己的数据体验 Kibana 的功能。

下面以单击"Try our sample data"按钮为例,体验 Kibana 的功能。

单击"Try our sample data"按钮,进入如图 10-9 所示页面。

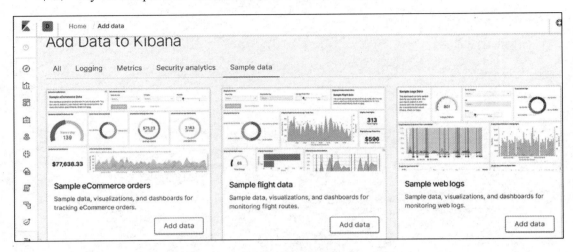

图 10-9

我们选择第一种展现类型作为示例,单击"Add Data"按钮后,Kibana 开始对数据进行加载,这个过程会持续数十秒。当数据加载后,"Add Data"按钮会变成如图 10-10 所示的"View Data"按钮。

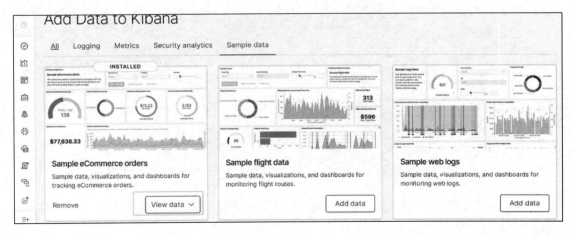

图 10-10

单击"View Data"按钮，进入如图 10-11 所示页面。

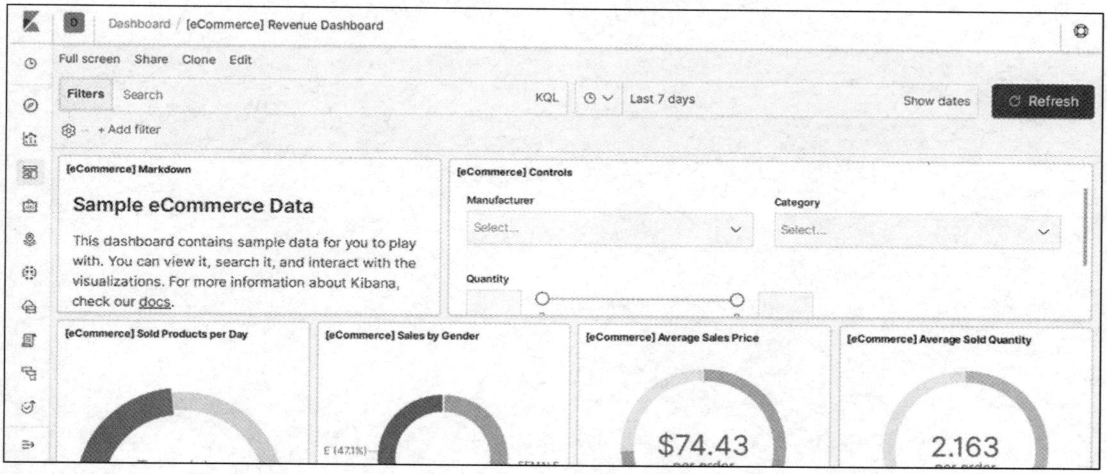

图 10-11

在图 10-13 所示页面中，可以通过筛选数据，查看细粒度的呈现。主要有关键词和时间两个筛选维度。单击"Discover"按钮，切换到如图 10-12 所示页面，查看数据的时间轴信息和数据详情信息。

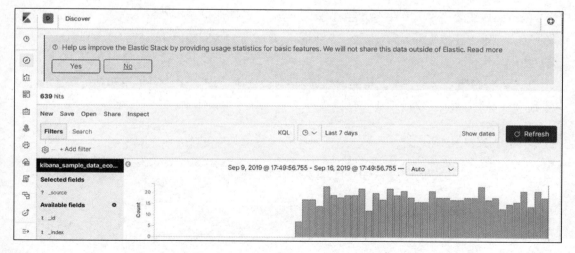

图 10-12

在实际使用中，需要配置必要的 Elasticsearch 索引信息，以便与 Kibana 进行数据联通。

配置索引的页面如图 10-13 所示。

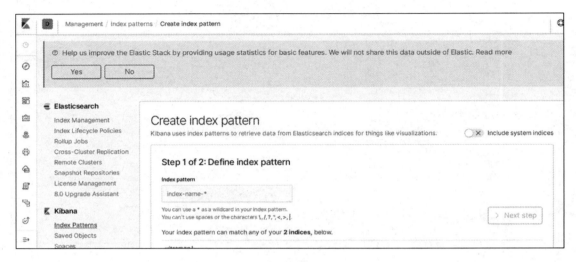

图 10-13

10.4　Beats

Logstash 在数据收集上并不出色，而作为代理，其性能也并不达标。于是，Elastic 官方发布了 Beats 系列轻量级采集组件。至此，Elastic 形成了一个完整的生态链和技术栈，成为大数据市场的佼佼者。

Beats 平台集合了多种单一用途的数据采集器。它们从成百上千台机器和系统向 Logstash 或 Elasticsearch 发送数据。

10.4.1　Beats 简介

Beats 是一组轻量级采集程序的统称，如图 10-14 所示。

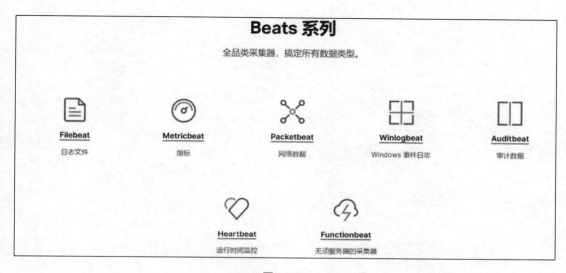

图 10-14

Beats 中包括但不限于以下组件：

（1）Filebeat：该组件会进行文件和目录的采集，主要用于收集日志数据。

（2）Metricbeat：该组件会进行指标采集。这里说的指标可以是系统的，也可以是众多中间件产品的。主要用于监控系统和软件的性能。

（3）Packetbeat：该组件通过网络抓包和协议分析，对一些请求响应式的系统通信进行监控和数据收集，可以收集到很多常规方式无法收集到的信息。

（4）Winlogbeat：该组件专门针对 Windows 的 event log 进行数据采集。

（5）Audibeat：该组件用于审计数据场景，收集审计日志。

（6）Heartbeat：该组件用于系统间连通性检测，如 ICMP、TCP、HTTP 等的连通性监控。

（7）Functionbeat：该组件用于无须服务器的采集器。

以上是 Elastic 官方支持的 7 种组件，事实上，借助于开源的力量，互联网上早已创造出大大小小几十甚至上百种组件，只有我们没想到的，没有 Beats 做不到的。官方不负责维护的 Beats，社区统一称之为 Community Beats。

官方支持的 7 种组件与 ELK 的数据流转关系如图 10-15 所示。

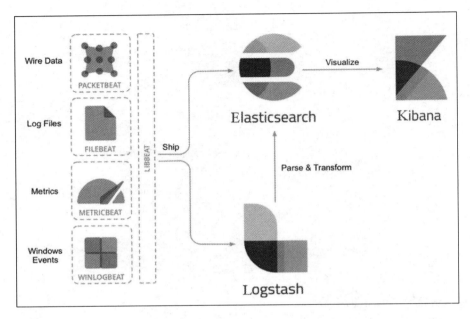

图 10-15

更多组件可以从 Elasticsearch 官网下载。

10.4.2 Beats 轻量级设计的实现

前面提过，Beats 是一组轻量级采集程序的统称，那么 Beats 是如何做到轻量级的呢？

（1）数据处理简单。在数据收集层面，Beats 并不进行过于复杂的数据处理，只是将数据简单的组织并上报给上游系统。

（2）并发性好、便于部署。Beats 采用 Go 语言开发而成。众所周知，Go 语言是一种系统编程语言，能够在不依赖虚拟机的情况下运行，包通常比较小。在跨平台方面，Beats 与 Go 语言保持一致，支持多种操作系统，如 Linux、Windows、FreeBSD 和 macOS。

因此，Beats 的性能显著好于 Logstash。

10.4.3 Beats 的架构

Beats 之所以有上乘的性能及良好的可扩展性，能获得如此强大的开源支持，其根本原因在于它有一套设计良好的代码框架。

Beats 的架构设计如图 10-16 所示。

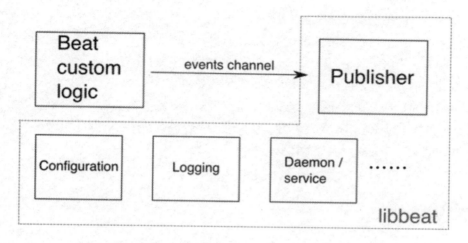

图 10-16

libbeat 是 Beats 的核心包。

在 Beats 架构中，有输出模块（Publisher）、数据收集模块（Logging）、配置文件模块（Configuration）、日志处理模块和守护进程模块（Daemon/ service）。

其中，输出模块负责将收集到的数据发送给 Logstash 或者 Elasticsearch。

因为 Go 语言天然就有 channel，所以收集数据的逻辑代码与输出模块都是通过 channel 通信的。也就是说，两个模块的耦合度最低。因此，当开发一个收集器时，完全不需要知道输出模块的存在，当程序运行时，自然就"神奇"地把数据发往服务端了。

除此之外，配置文件模块、日志处理模块、守护进程模块等功能为开发者扩展 Beats 的功能提供了极大的空间。

10.5 知识点关联

本章主要介绍了 Elasticsearch 的生态圈，不难看出其生态圈十分繁荣。不仅官方在维护 Elasticsearch 的生态，社区技术人员也在积极地贡献自己的力量，正所谓"众人拾柴火焰高"！

而这背后正是生态思维。

在技术圈中，Java 技术栈中的 Spring 生态发展得也是花团锦簇。从 Spring Boot 到 Eureka、Hystrix、Zuul、Archaius、Consul、Sleuth、Spring Cloud ZooKeeper、Feign、Ribbon……终于，以 Spring Boot 为核心的 Spring Cloud 微服务生态建立起来，并持续发展。

安卓系统、微信等也都在积极建立属于自己的生态体系。安卓系统生态中数以万计的开发者贡献了众多的 App。微信在拓展和完善功能的同时，通过小程序的生态打造，连接了众多的线下场景，孕育了无限想象空间。

技术如此，商业如此，同理，技术人员的发展也是如此。

求职之路漫漫，因此在求职或者跳槽时不能光盯着眼前的利益，更要看重未来的个人"生态"：找到适合自己的公司，找到有能力、有意愿、有方法培养自己的领导，找到一群"技术派"的同事，找到一个技术氛围好的团队，等等。

10.6 小结

本章主要介绍了 Elasticsearch 的生态圈，即 ELK Stack。先后介绍了 ELK Stack 的背景、ELK 的实战部署架构设计，以及 Logstash、Kibana 和 Beats。